Methods in Molecular Biology

Series Editor
John M. Walker
School of Life and Medical Sciences
University of Hertfordshire
Hatfield, Hertfordshire, UK

For further volumes:
http://www.springer.com/series/7651

For over 35 years, biological scientists have come to rely on the research protocols and methodologies in the critically acclaimed *Methods in Molecular Biology* series. The series was the first to introduce the step-by-step protocols approach that has become the standard in all biomedical protocol publishing. Each protocol is provided in readily-reproducible step-by-step fashion, opening with an introductory overview, a list of the materials and reagents needed to complete the experiment, and followed by a detailed procedure that is supported with a helpful notes section offering tips and tricks of the trade as well as troubleshooting advice. These hallmark features were introduced by series editor Dr. John Walker and constitute the key ingredient in each and every volume of the *Methods in Molecular Biology* series. Tested and trusted, comprehensive and reliable, all protocols from the series are indexed in PubMed.

Mucins

Methods and Protocols

Edited by

Akihiko Kameyama

Cellular and Molecular Biotechnology Research Institute, National Institute of Advanced Industrial Science and Technology (AIST), Tsukuba, Ibaraki, Japan

Humana Press

Editor
Akihiko Kameyama
Cellular and Molecular Biotechnology
Research Institute
National Institute of Advanced Industrial
Science and Technology (AIST)
Tsukuba, Ibaraki, Japan

ISSN 1064-3745 ISSN 1940-6029 (electronic)
Methods in Molecular Biology
ISBN 978-1-0716-3669-5 ISBN 978-1-0716-3670-1 (eBook)
https://doi.org/10.1007/978-1-0716-3670-1

This Humana imprint is published by the registered company Springer Science+Business Media, LLC, part of Springer Nature.
The registered company address is: 1 New York Plaza, New York, NY 10004, U.S.A.

Paper in this product is recyclable.

Preface

Mucin, a major component of mucus and mucous membranes, exhibits a slimy property that is one of the characteristics of living organisms and plays an important role in epithelial tissues lubrication. It is also thought to play an important role in the border region between the outside and inside of the body, contributing to infection and mucosal defense, and in relation to the maintenance and changes in the intestinal microflora. Mucins are classified into gel-forming mucins and membrane-bound mucins. The latter is associated with tumor metastasis and prognosis. In particular, tumor-associated MUC 1 is a potential diagnostic and prognostic marker and an antitumor vaccine target.

In the early 2000s, with the advent of the post-genomic era, we were under the illusion that proteomic analysis could identify every protein expressed in the body. However, many proteins undergo posttranslational modifications, which are regulated by multiple genes and complex regulatory mechanisms; therefore, the true nature of the modifications cannot be revealed via proteomic analysis. Mucins, in particular, have complex structural features, such as gene repeats, a lack of specific protease-sensitive sequences, and variable glycosylated tandem repeat domains, that hinder proteomic analysis. Owing to their complex structures, high molecular weight, and physicochemical properties, elucidating the structure and functions of mucins still remains challenging.

This edition of *Mucins: Methods and Protocols* is divided into eight parts and covers a wide range of topics, including mucin extraction, isolation, physicochemical property analysis, and experimental methods. Glycosylation plays an essential role in mucin structure and function. One of the features of this edition is that many chapters are directly or indirectly related to glycosylation. Origin of mucins addressed in the edition includes jellyfish (Chap. 1), feces (Chap. 2), saliva and salivary glands (Chaps. 3 and 34), bronchi (Chap. 4), stomach (Chaps. 5 and 25), intestines (Chaps. 6 and 35), and cervical tract (Chap. 23). Quantification was performed using the chromogenic (Chap. 2), dye (Chap. 10), and MS methods using stable isotopes (Chap. 11). For glycan analysis, several practical methods for glycan release, derivatization, and analysis have been included to allow for the selection of the most appropriate protocol, specific to the research subjects and available resources (Chaps. 12–19). Organic synthesis of peptides glycosylated at specific site (Chap. 16) and algorithmic tool for estimating the glycosylation sites (Chap. 20) would be useful to clarify the importance of glycosylation sites in mucin functions. Many antibodies used to stain mucins in tissue sections are also dependent on glycosylation (Chap. 8). The molecular biology part includes analysis of mucin gene expression and methylation-specific electrophoresis of genes (Chaps. 21–23), as well as the production of glycoengineered mucins by genome editing (Chap. 24). Mucin glycans interact with endogenous lectins and lectin-like molecules present in pathogens and are a source of nutrients for intestinal bacteria. Chapters 25 and 27–30 cover such topics. Novel methods of investigation include NMR analysis of the interactions between MUC-1 glycopeptides and antibodies (Chap. 26), imaging of mucin networks by AFM (Chap. 31), structural analysis of mucin glycopeptides by MD simulation (Chap. 32), creation of nanoparticles using mucins (Chap. 33), rheological analysis of saliva (Chap. 34), and rheological analysis of mucins using AFM (Chap. 35). Experimental methods using supported molecular matrix electrophoresis, a simple method of mucin analysis performed in our laboratory, are also discussed (Chaps. 7 and 9).

All chapters follow a consistent format and are rich in "Notes" based on the authors' experiences, enhancing practical utility of described methods in mucin research, which remains challenging. Although it is useful to read chapters specific to intended research, we believe that reading the entire volume will provide insights on how to conduct mucin research.

Tsukuba, Japan *Akihiko Kameyama*

Contents

Contributors

LAURA ABRIL-PARREÑO • *Physiology of Reproduction Group, Department of Physiology, Faculty of Veterinary Medicine, International Excellence Campus for Higher Education and Research (Campus Mare Nostrum), University of Murcia, Murcia, Spain*

YOICHIRO ARATA • *Faculty of Pharma-Science, Teikyo University, Tokyo, Japan*

HISASHI ASHIDA • *Faculty of Biology-Oriented Science and Technology, Kindai University, Kinokawa, Japan*

PATCHARAPORN BOOTTANUN • *Cellular and Molecular Biotechnology Research Institute, National Institute of Advanced Industrial Science and Technology (AIST), Tsukuba, Japan*

JEROME CARPENTER • *Department of Pathology and Laboratory Medicine, Marsico Lung Institute, University of North Carolina at Chapel Hill, Chapel Hill, NC, USA*

MARSHALL J. COLVILLE • *Robert Frederick Smith School of Chemical and Biomolecular Engineering, Cornell University, Ithaca, NY, USA; Department of Clinical Sciences, College of Veterinary Medicine, Cornell University, Ithaca, NY, USA*

WEIJIE DONG • *College of Basic Medical Sciences, Dalian Medical University, Dalian, China*

SEAN FAIR • *Department of Biological Sciences, Biomaterials Research Cluster, Faculty of Science and Engineering, Bernal Institute, University of Limerick, Limerick, Ireland*

YUKINOBU GOSO • *Department of Applied Bioscience, Kanagawa Institute of Technology, Atsugi, Japan*

MICHIYO HIGASHI • *Department of Pathology, Field of Oncology, Graduate School of Medical and Dental Sciences, Kagoshima University, Kagoshima, Japan*

MOMOKA HORIKIRI • *Department of Applied Physics, Graduate School of Engineering, Osaka University, Osaka, Japan*

LING-TING HUANG • *Robert Frederick Smith School of Chemical and Biomolecular Engineering, Cornell University, Ithaca, NY, USA*

TAKAFUMI ICHIKAWA • *Department of Biochemistry, Kitasato University School of Allied Health Science, Sagamihara, Japan*

KATSUNORI IIJIMA • *Department of Gastroenterology, Akita University Graduate School of Medicine, Akita, Japan*

HIROSHI INUI • *Department of Chemistry, School of Science, Kitasato University, Sagamihara, Japan*

SHIORI KAISE • *Department of Chemistry, School of Science, Kitasato University, Sagamihara, Japan*

AKIHIKO KAMEYAMA • *Cellular and Molecular Biotechnology Research Institute, National Institute of Advanced Industrial Science and Technology (AIST), Tsukuba, Japan*

TAKUMA KANEKO • *Department of Chemistry, School of Science, Kitasato University, Sagamihara, Japan*

TAKANE KATAYAMA • *Graduate School of Biostudies, Kyoto University, Kyoto, Japan*

TOSHIHIKO KATOH • *Graduate School of Biostudies, Kyoto University, Kyoto, Japan*

REI KAWASHIMA • *Department of Biochemistry, Kitasato University School of Allied Health Science, Sagamihara, Japan*

MEHMET KESIMER • *Department of Pathology and Laboratory Medicine, Marsico Lung Institute, University of North Carolina at Chapel Hill, Chapel Hill, NC, USA*

CEREN KIMNA • *School of Engineering and Design, Department of Materials Engineering, Technical University of Munich, Garching, Germany; Center for Protein Assemblies (CPA) and Munich Institute of Biomedical Engineering, Technical University of Munich, Garching, Germany*
SACHIKO KOIZUMI • *Cellular and Molecular Biotechnology Research Institute, National Institute of Advanced Industrial Science and Technology (AIST), Tsukuba, Japan*
RYOKA KOKUBU • *Division of Structural Glycobiology, Institute of Molecular Biomembrane and Glycobiology, Tohoku Medical and Pharmaceutical University, Sendai, Japan*
ATSUSHI KUNO • *Cellular and Molecular Biotechnology Research Institute, National Institute of Advanced Industrial Science and Technology (AIST), Tsukuba, Japan*
MAKOTO KURIHARA • *Department of Applied Bioscience, Kanagawa Institute of Technology, Atsugi, Japan*
OLIVER LIELEG • *School of Engineering and Design, Department of Materials Engineering, Technical University of Munich, Garching, Germany; Center for Protein Assemblies (CPA) and Munich Institute of Biomedical Engineering, Technical University of Munich, Garching, Germany*
THERESA M. LUTZ • *School of Engineering and Design, Department of Materials Engineering, Technical University of Munich, Garching, Germany; Center for Protein Assemblies (CPA) and Munich Institute of Biomedical Engineering, Technical University of Munich, Garching, Germany*
NORIYOSHI MANABE • *Division of Structural Glycobiology, Institute of Molecular Biomembrane and Glycobiology, Tohoku Medical and Pharmaceutical University, Sendai, Japan*
ATSUSHI MATSUDA • *Sysmex Corporation, Reagent Engineering, Protein Technology Group, Kobe, Japan*
KEI MATSUO • *Department of Pathology, Kagoshima University Graduate School of Medical and Dental Sciences, Kagoshima, Japan*
TAKAHISA MATSUZAKI • *Department of Applied Physics, Graduate School of Engineering, Osaka University, Osaka, Japan; Center for Future Innovation, Graduate School of Engineering, Osaka University, Osaka, Japan*
LUISA GRACIA MAZUCA • *Bioinformatics Program, The University of Texas at El Paso, El Paso, TX, USA*
ANRI MOCHIZUKI • *Department of Chemistry, School of Science, Kitasato University, Sagamihara, Japan*
JONATHON E. MOHL • *Bioinformatics Program, The University of Texas at El Paso, El Paso, TX, USA*
TATSUYA MORITA • *College of Agriculture, Academic Institute, Shizuoka University, Shizuoka, Japan*
TAKAO MUKAI • *Department of Animal Science, School of Veterinary Medicine, Kitasato University, Towada, Aomori, Japan*
TARO MUKAIBO • *Division of Oral Reconstruction and Rehabilitation, Kyushu Dental University, Kitakyushu, Japan*
MISUGI NAGAI • *Cellular and Molecular Biotechnology Research Institute, National Institute of Advanced Industrial Science and Technology (AIST), Tsukuba, Japan*
IZURU NAGASHIMA • *Cellular and Molecular Biotechnology Research Institute, National Institute of Advanced Industrial Science and Technology (AIST), Tsukuba, Japan*

NAOMICHI NISHIMURA • *College of Agriculture, Academic Institute, Shizuoka University, Shizuoka, Japan*

KEITA NISHIYAMA • *Laboratory of Animal Food Function, Graduate School of Agricultural Science, Tohoku University, Sendai, Miyagi, Japan; Livestock Immunology Unit, International Education and Research Center for Food Agricultural Immunology (CFAI), Graduate School of Agricultural Science, Tohoku University, Sendai, Miyagi, Japan*

SHIHO OHNO • *Division of Structural Glycobiology, Institute of Molecular Biomembrane and Glycobiology, Tohoku Medical and Pharmaceutical University, Sendai, Japan*

SHOTA OKAMOTO • *Department of Microbiology and Immunology, Graduate School of Medicine, Osaka University, Osaka, Japan*

RYU OKUMURA • *Department of Microbiology and Immunology, Graduate School of Medicine, Osaka University, Osaka, Japan*

MATTHEW PASZEK • *Robert Frederick Smith School of Chemical and Biomolecular Engineering, Cornell University, Ithaca, NY, USA; Kavli Institute at Cornell for Nanoscale Science, Cornell University, Ithaca, NY, USA*

GIORGIA RADICIONI • *Department of Pathology and Laboratory Medicine, Marsico Lung Institute, University of North Carolina, Chapel Hill, NC, USA*

HIROKI SHIMIZU • *Cellular and Molecular Biotechnology Research Institute, National Institute of Advanced Industrial Science and Technology (AIST), Tsukuba, Japan*

JUN SHOJI • *Division of Ophthalmology, Department of Visual Sciences, Nihon University School of Medicine, Tokyo, Japan*

TAKANORI SUGIURA • *Division of Oral and Maxillofacial Surgery, Ushiku Aiwa General Hospital, Ushiku, Japan; Department of Oral Oncology, Oral and Maxillofacial Surgery, Ichikawa General Hospital, Tokyo Dental College, Ichikawa, Japan*

MINAMI SUGIYAMA • *Department of Chemistry, School of Science, Kitasato University, Sagamihara, Japan*

HIROMI TAKADA • *Graduate School of Biostudies, Kyoto University, Kyoto, Japan*

MAYUMI TAMURA • *Faculty of Pharma-Science, Teikyo University, Tokyo, Japan*

HIROKI TANABE • *Department of Nutritional Sciences, Faculty of Health and Welfare Science, Nayoro City University, Nayoro, Japan*

SHINRA TANAKA • *Department of Chemistry, School of Science, Kitasato University, Sagamihara, Japan*

MUGEN TANIGUCHI • *Department of Microbiology and Immunology, Graduate School of Medicine, Osaka University, Osaka, Japan*

AKIHIDE TANIMOTO • *Department of Pathology, Kagoshima University Graduate School of Medical and Dental Sciences, Kagoshima, Japan*

DAIGO TSUBOKAWA • *Department of Parasitology and Tropical Medicine, Kitasato University School of Medicine, Sagamihara, Japan*

KIMINORI USHIDA • *Department of Chemistry, School of Science, Kitasato University, Sagamihara, Japan*

KEITA YAMADA • *The Laboratory of Toxicology, Faculty of Pharmacy, Osaka Ohtani University, Osaka, Japan*

MIKIO YAMADA • *Division of Oral Reconstruction and Rehabilitation, Kyushu Dental University, Kitakyushu, Japan*

SATORU YAMAGAMI • *Division of Ophthalmology, Department of Visual Sciences, Nihon University School of Medicine, Tokyo, Japan*

YOSHIKI YAMAGUCHI • *Division of Structural Glycobiology, Institute of Molecular Biomembrane and Glycobiology, Tohoku Medical and Pharmaceutical University, Sendai, Japan*
SEIYA YOKOYAMA • *Department of Pathology, Kagoshima University Graduate School of Medical and Dental Sciences, Kagoshima, Japan*
HIROSHI Y. YOSHIKAWA • *Department of Applied Physics, Graduate School of Engineering, Osaka University, Osaka, Japan*

Part I

Extraction and Separation

Chapter 1

Preparation of Jellyfish Mucin

Kiminori Ushida, Hiroshi Inui, Takuma Kaneko, Shinra Tanaka, Anri Mochizuki, Shiori Kaise, and Minami Sugiyama

Abstract

A mucin-type glycoprotein extracted from various species of jellyfish (JF) is named qniumucin (Q-mucin). Compared with general mucins, most of which are from mammals including humans, Q-mucin can be collected on a relatively large scale with high yield. Owing to its simple structure with low heterogeneity, Q-mucin has a potential to be developed into material mucins which opens various applications valuable to humans. On the basis of our present knowledge, here, we describe our protocol for the extraction of Q-mucin, which can be extracted from any JF species worldwide. Experimental protocols to identify the structure of Q-mucin are also introduced.

Key words Jellyfish mucin, Mucin extraction, Mucin purification, Glycoform analysis, Mass spectrometry, Anion exchange chromatography, Amino acid composition analysis, Mucin, Extracellular matrix

1 Introduction

Mucins optimized for material science and engineering (which we here call *material mucins*) [1], if exist, are attractive macromolecules showing various specific properties that cannot be realized in any other substances. We expect their various valuable applications in various fields, including medical, biological, industrial, and environmental fields. For example, one possible use is as artificial mucus, which helps defense mechanisms on various mucosal surfaces in humans.

However, since we still have technical difficulty in generating artificial mucins by either chemical synthesis or biotechnological production, extraction of natural mucins from some organisms is the most effective strategy available today. On the other hand, as the removal of contaminants after extraction is a high technical barrier, this methodology brings serious problems in the purity and homogeneity of obtained mucins for use as material mucins.

Akihiko Kameyama (ed.), *Mucins: Methods and Protocols*, Methods in Molecular Biology, vol. 2763,
https://doi.org/10.1007/978-1-0716-3670-1_1,

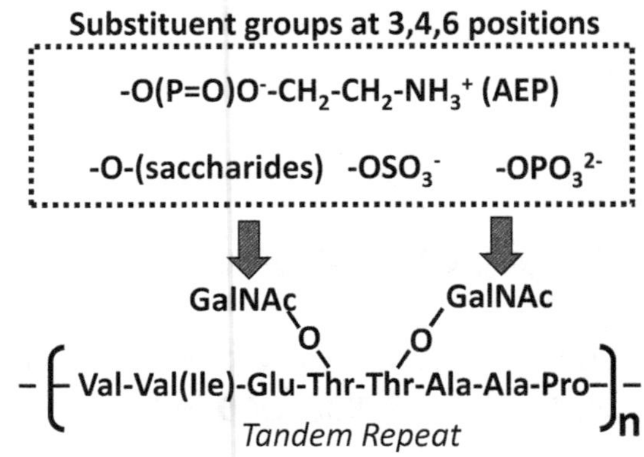

Fig. 1 General structure of JF mucin (Qniumucin: Q-mucin). The main part is the sequence of tandem repeats (TRs) of eight amino acid residues. Basically, the glycan chains consist of *N*-acetylgalactosamine (GalNAc), which is randomly modified with various substituent groups: 2-aminoethylphosphonates (AEPs), some saccharides, sulfates, and phosphates

Mucins are inherently heterogeneous substances, which makes effective purification based on their physicochemical properties difficult.

In 2007 [2], a novel mucin-type glycoprotein (MTGP) (*see* **Note 1**) was extracted from several species of jellyfish (JF) for the first time and was named "qniumucin" (Q-mucin) [1, 2], whose structure is shown in Fig. 1 [1–5].

Compared with other conventional mucins obtained from domestic animals, the structure of Q-mucin is simple and less heterogeneous. About at least 90% of the main peptide chain seems to be composed of two types of tandem repeat (TR) (Val-Val-Glu-Thr-Thr-Ala-Ala-Pro or Val-Ile-Glu-Thr-Thr-Ala-Ala-Pro) [2, 3], and most of the glycan chains connected at two Thr positions are as short as a monosaccharide and composed of *N*-acetyl galactosamine (GalNAc) [4] and its derivatives. As mentioned later, this sequence is conserved in almost all JFs. The normal yield of Q-mucin is relatively large (e.g., 100 mg from 1 kg of wet JF on average) and a substantial amount of a homogeneous mucin can be obtained from a single extraction. Since Q-mucin is not involved in liquid mucus but in solid (jellied) extracellular matrix (ECM), possibility of contamination is supposed to be lower than those extracted from mucus fluid. This merit is similar to that of various submaxillary gland mucins, which are directly extracted from the solid organ. These characteristics are advantageous for obtaining a well-defined material mucin. Therefore Q-mucin is a candidate material mucin producible on a large industrial scale.

Furthermore, glycan chains of Q-mucin are modified with three types of charged moiety: sulfates, phosphates, and 2-aminoethylphosphonates (AEPs) without any sialic acids [1–

5]. Together with glutamic acid on the main peptide chain, they provide both positive and negative charges on Q-mucin. These groups have different pK_a (or pK_b) values (*see* supporting information of ref. [5]) and the density of charges on Q-mucin is affected by the pH in the bulk solution (*see* **Note 2**).

Although Q-mucin has a very simple structure with low heterogeneity and is not an MTGP from mucus, its framework is formed with the most basic elements present in so-called PTS (abundant with proline, threonine, and serine residues) regions of mucin, i.e., a simple main peptide chain composed of simple TRs, dense *O*-glycan chains, and charged groups on glycan chains. Accordingly, Q-mucin also has peculiar features in common with general mucins as follows.

1. Q-mucin has no fixed tertiary structure but a rather ribbon-like flexible structure similar to PTS regions of general mucins. (Q-mucin is a model compound that simulates PTS regions.)
2. Most of the peptide chain of Q-mucin is composed of only two kinds of simple TR (VVETTAAP or VIETTAAP) with eight amino acid residues.
3. For 1 and 2, the peptide portions of Q-mucin itself and their decomposed fragments are less likely to disturb biological signaling molecules, which mainly sense the sequence and tertiary structure of peptides, such as enzymes, antigens, and antibodies.
4. On the other hand, *O*-glycan chains showing glycoforms easily interact with surrounding glyco-sensing molecules, such as lectins.
5. Since glycan chains surrounding the main peptide chain block other molecules such as proteinases, Q-mucin is rather stable against proteinases and other hydrolytic enzymes.
6. Q-mucin is a polymeric surfactant and increases the wettability of its aqueous solutions, including natural mucus.
7. Groups with negative charges such as sulfates, phosphates, and phosphonates on glycan chains of Q-mucin interact with and sometimes capture various mineral ions existing in the bulk solution. They can control the pH of the aqueous solution, balancing the concentrations of counter ions such as sodium and potassium ions.
8. Q-mucin is essentially biodegradable and sustainable with less impact on the environment.

Q-mucin originally exists as part of the ECM of the entire JF body (mesoglea) [5]. Since it resembles the proteoglycan (PG) in the mammalian cartilage, we speculate that Q-mucin is an ancestor molecule of PG. In the present protocol described here, we only collect free Q-mucin dissociated from ECM by some spontaneous elimination process. Since large amounts of partial Q-mucin components remain in the residual ECM after extraction, the yield of Q-mucin may significantly increase in the future.

Similar to PG contacting with blood system, Q-mucin also seems to capture calcium ions existing in seawater and living JFs effectively collect mineral cations in surrounding seawater through their swimming motion which also stirs the seawater effectively. Because of this activity, extracted Q-mucin may contain various mineral cations initially involved in seawater. Some cations whose binding constants to negatively charged moieties are small can be removed by chelate reagents such as ethylenediamine-*N, N, N′, N′*-tetraacetic acid (EDTA). However, some cations such as Ca^{2+} bind to Q-mucin tightly and cannot be removed completely by the protocols described here. The tightly binding cations erase the negative charges on the glycan chains and are expected to change various physicochemical properties of Q-mucin. In the present protocol, we cannot control the compositions of mineral cations involved in final Q-mucin. We recommend the readers to perform microelement analysis using ICP.

Q-mucin has been found in various scyphozoan [6] and cubozoan JFs. We confirmed the existence of some glycoproteins resembling Q-mucin in only a few species of hydrozoan JF (*see* **Note 3**). Surprisingly, the sequences of two types of TR are conserved in all scyphozoan and cubozoan species after more than 0.5 M years of development. Glycan chains and their modification were found to be dependent on species and their ecological activities. As a result, the positions of Q-mucin peaks appearing in anion exchange (AEX) chromatograms shifted depending on the source of Q-mucin. The combination of the peptide chain with the constant sequence and the glycan chains with the flexible composition and connectivity is the most important aspect of Q-mucin, and also, of other general mucins.

In this chapter, we describe the protocol to isolate Q-mucin from some JF species and the analysis of the glycan chains. We describe the methods of catching, storage, extraction of semipurified (SP) Q-mucin, and further purification. Since Q-mucin is extracted on a larger scale than the general mucins, its purification and standard analyses depend on various high performance liquid chromatography (HPLC) techniques without using gradient centrifugation. We also exclude electrophoresis techniques (SDS-PAGE (Sodium Dodecyl Sulfate-Polyacrylamide Gel Electrophoresis) or SMME (Supported Molecular Matrix Electrophoresis) [7]) because Q-mucin is poorly stained whereas other impurities existing in JFs are strongly stained.

Normally, each glycan chain of Q-mucin is a monosaccharide (GalNAc) and only very small portions of oligosaccharides exist. The exact reason why such results are obtained remains to be clarified. On the other hand, a large portion of glycans are modified with charged groups. Therefore, the method of characterizing glycans in Q-mucin is different from those developed for other conventional mucins.

The entire protocol proceeds as follows: (i) catch JFs and extract SP Q-mucin, (ii) find Q-mucin by anion exchange chromatography (AEXC) with amino acid composition analysis (AACA), purify the SP Q-mucin by fractionation by AEXC, and (iv) analyze glycan chains systematically using a combination of several methodologies. Although we indicate the instruments together with their provider's name that we use in our own laboratory, we do not think each of them as the only one choice to perform the operations successfully. We also indicate some conditions of HPLC and HPLC-ESI-MS (Electrospray Ionization-Mass Spectrometry) for the reference of readers. We hope readers to design their own methods that may be better than ours, considering the spec of the instruments and the accompanying parameters indicated in this chapter.

2 Materials

2.1 Catching JF

To date, we have confirmed the existence of Q-mucin in almost all scyphozoan and cubozoan species (Table 1) that we have ever caught around Japanese islands. We believe that Q-mucin can be

Table 1
Summary of species in which the existence of Q-mucin is confirmed

Class	Species	Maximum individual size (diameter) (mm)	Maximum individual weight (g)
Scyphozoa	*Aurelia coerulea*	~350	~1000
	Aurelia limbata	~350	~1000
	Chrysaora pacifica	~350	~1000
	Chrysaora sp.	~500	~2000
	Chrysaora fuscescens	~400	~1000
	Chrysaora melanaster	~350	~1000
	Cyanea capillata	~800	~5000
	Mastigias albipunctata	~300	~1000
	Rhopilema asamushi	~1000	~10,000
	Nemopilema nomurai	~2000	~200,000
Cubozoa	*Carybdea brevipedalia*	~30	~10
	Chironex yamaguchii	~50	~200
	Morbakka virulenta	~400	~3000

extracted from other related species distributed worldwide using our protocol. In all JF species, the amino acid sequences of the TR part were identical, i.e., VVETTAAP or VIETTAAP. In some species, however, depending on the composition of glycan chains, the peak corresponding to Q-mucin in AEXC shifts from sample to sample. If easy Q-mucin extraction is desired, species belonging to the genus *Aurelia* (e. g, *Aurelia aurita*) are most convenient because they are widely abundant all over the world and provide Q-mucin of uniform quality (*see* **Note 4**). In this chapter, we describe the method of extraction of Q-mucin in the species of the genus *Aurelia* as the standard method. The bulk amount of this JF caught reaches more than 100 kg in 1 day at a single location. Although its species mainly appear in early summer, we can find some individuals of *Aurelia coerulea* throughout the year, for example, everywhere in Tokyo Bay.

1. Soft brail net or dipper.
2. Plastic sieve.

2.2 Storage of Raw JF

JFs appear occasionally and irregularly. Sometimes more than 10^4 kg of JFs invade a single spot in 1 day. We cannot determine the location, the species, the amount, and the schedule for the collection of raw JFs beforehand (*see* **Note 5**). Therefore, we must store raw JFs by freezing for sustainable production. Even at freezing temperatures (≤ -20 °C) where the activities of microorganisms are minimal, collagens of JF bodies are gradually decomposed by the activities of endogenous proteinases (*see* **Note 6**). Eighty percent of the solid (jelly) portion is lost after 1–2 months at −20 °C and turns to liquid after thawing [5]. Since degraded fragments may interfere with the extraction of Q-mucin as water-soluble impurities, the addition of an inhibitor (chelate reagents) sometimes provides better results. At the same time, however, chelate reagents also remove some of the counter cations from acidic groups on glycan chains of Q-mucin, thereby lowering the pH of the liquid after crushing. The composition of counter cations in raw JFs markedly depends on the environment where they inhabit and the binding constant of each cation against various acid groups on Q-mucin (sulfates, phosphates, phosphonates, and carboxylates).

A low pH may cause partial hydrolysis of peptide chains, glycan chains, and moieties on saccharides, and pH adjustment to 7 at each step of the extraction may be desired. However, since Q-mucins from some species contain many sulfate groups on their glycan chains, local acidity near glycan chains cannot be removed completely.

Modification of the protocol concerning the addition of an inhibitor and pH control may be necessary depending on the species used as raw materials.

1. 100 mM EDTA solution: 372 g of EDTA·2Na·2H_2O was dissolved in 1 L of distilled water. Since this concentration almost saturates at room temperature, the addition of NaOH is recommended to dissolve all reagents instantly. pH may shift toward 8.
2. Antiseptic solution: A commercial antiseptic solution with NaClO and NaOH as the main constituents is preferred. We use KAO Haiter Bleach Regular®. The pH of the commercial solution is adjusted to the optimum value for long-period storage. Products from other brands are acceptable, but we must use a simple antiseptic solution without unwanted additives such as surfactants.
3. 0.1 M NaOH solution: 4 g of NaOH pellets is dissolved in 1 L of water.
4. Zippered bag: We use Ziploc® freezer bags (large/gallon) from S.C. Johnson & Son Inc. About 2 kg of JFs can be packed in one bag of Ziploc® freezer bags of this size. We preferred the double-zipper bags of sufficient thickness not easily broken by sharp edges of iced samples. Equivalent products are acceptable.
5. Freezer: The temperature setting must be lower than −20 °C. Lower temperatures delay the degradation of raw JFs.

2.3 Extraction of SP Q-mucin

Since the decomposition of raw JFs is suppressed with the pretreatment using EDTA, all the processes can be performed at room temperature except for ultracentrifugation and dropwise addition of ethanol (EtOH) during which the temperature may increase. The following instruments and materials are used for 2 kg of raw JF to obtain 10–200 mg of SP Q-mucin.

1. Blender for crushing JFs: An approximately 1 L commercial food blender for family use may be used.
2. 0.1 M NaOH solution: The same as described in Subheading 2.2.
3. pH meter: A handy type is convenient.
4. Plastic bottle with spigot: Used as a substitute for dropping funnel.
5. 5 L glass bottle with GL45 screw cap: A glass bottle of this size is easy to handle.
6. EtOH for precipitation: Isopropyl alcohol (IPA, 2-propanol) contaminating EtOH is used to reduce the cost.
7. Ultracentrifuge for large volume preparation: In our laboratory, a 4 × 1 L size rotor was used.
8. Ultracentrifuge for small volume preparation: In our laboratory, a 4 × 50 mL size rotor was used.

9. Showcase refrigerator: It is convenient for the dropwise addition of EtOH.
10. Stirrer and stirring magnet.
11. Dialysis membrane: We use Spectra/Por® 5, with molecular weight cutoff (MWCO) = 12–14 kDa.
12. Digital ion (salt) meter: High-resolution (0.01%) type based on conductivity measurement. We use an ES-421 and a PAL-SALT from Atago Co., Ltd.
13. Large-volume freeze dryer: We use FDL-200 from Tokyo Rikakikai Co. Ltd., (3 L with a −80 °C trap) for eight eggplant-shaped flasks, together with a prefreezer, PFR-1000 + PFM-1000.
14. Eggplant-shaped flask.

2.4 Anion Exchange Chromatography (AEXC)

AEXC performed on an HPLC system is used in the following.

1. Identification of the candidate peak(s) for Q-mucin in chromatograms using a continuous gradient program.
2. Isolation of the candidate peak(s) in chromatograms with stepwise gradient program.
3. Fractionation of the candidate peak(s).

Common instruments and materials are shared in these operations.

The results of AEXC of Q-mucin from various JF resources depend on not only the species but also the environment where they live (*see* **Note 7**). The retention time of Q-mucin peaks varies significantly depending on the density of charges on the glycan chains. Some species show more than one peak of Q-mucin in the chromatograms.

Since Q-mucin contains almost no aromatic amino acids, the candidate peaks show strong UV absorption at 215 nm with very low intensities at 275 nm. The UV absorption at 215 nm originates from peptide bonds on the main peptide chain and amide bonds on GalNAc. If we assume that the extinction coefficients of these bonds are approximately the same, 215 nm absorption roughly indicates the molar number of amino acids and amino saccharide monomers. We choose one or several candidate peaks in AEXC and final confirmation will be made by AACA described in the next section. Strong and sharp peaks corresponding to polyglutamic acid (PGA) sometimes appear with train of peaks because the hard parts of JFs are believed to be composed of PGA.

The interaction between Q-mucin and AEX resin is essentially through the strong adsorption of Q-mucin onto the resin via the anionic charged groups on glycan chains. Since many interaction points exist on a single Q-mucin molecule, a long equilibrium period is required for absorption and desorption. Therefore, a continuous increase in the ion strength of the mobile phase is occasionally inadequate for the isolation of Q-mucin peaks. The most effective method is to apply a stepwise increase in the ion strength optimized for isolation and fractionation of the Q-mucin candidate peak.

1. HPLC system: Any HPLC systems that can monitor dual UV absorptions can be used. Two single UV absorption detectors or one multichannel detector can be used. In our laboratory, the GL-7400 series from GL Sciences Inc. is used.
2. Column: Any column of AEX resin with weak interactions via amino groups can be used. In our laboratory, TSK-gel DEAE-5PW from Tosoh Corp. is used with a corresponding guard column. Both columns for analysis (7.5 × 75 mm) and fractionation (21.5 × 150 mm) are used.
3. Eluent A: 10 mM phosphate buffer whose pH is adjusted to 7. Dissolve 600 mg of NaH_2PO_4 (or 780 mg of $NaH_2PO_4 \cdot 2H_2O$) and 1790 mg of $Na_2HPO_4 \cdot 12H_2O$ in distilled water up to a final volume of 1 L. Normally, the pH is smaller than 7 and add 0.1 M NaOHaq to adjust the pH to 7. The eluent is degassed with a vacuum pump before its use.
4. Eluent B: 0.5 M NaCl in 10 mM phosphate buffer whose pH is adjusted to 7. Dissolve 600 mg of NaH_2PO_4 (or 780 mg of $NaH_2PO_4 \cdot 2H_2O$), 1790 mg of $Na_2HPO_4 \cdot 12H_2O$, and 29.3 g of NaCl in distilled water up to a final volume of 1 L. Normally, the pH is smaller than 7 and add 0.1 M NaOHaq to adjust the pH to 7. The eluent is degassed with a vacuum pump before its use.

2.5 Batch Purification by Adsorption and Desorption on AEX Resin

Once the ion strength for the desorption of Q-mucin from the AEX resin is determined, a batch process is applicable to further purification of SP Q-mucin. By this method, a large amount of Q-mucin (>50 mg) can be processed in a single batch. If the ion strengths of two stripping buffers are determined from the result of stepwise fractionation in AEXC, a sustainable production of Q-mucin from a single species is possible.

1. AEX resin: The resin is of the same type as that used in the column for AEXC. We use Toyopearl® DEAE-650M from Tosoh corp.
2. Basic buffer: 10 mM phosphate buffer. The same as Eluent A (*see* Subheading 2.4, **item 3**).
3. Washing buffer: 1 M of NaCl is added to the basic buffer.
4. Stripping buffer #1: The ion strength of the basic buffer (IS_1) is adjusted with NaCl to be slightly lower than that for the desorption of Q-mucin. Mixing the basic buffer with the washing buffer is simple and quick.

5. Stripping buffer #2: The ion strength of the basic buffer (IS_2) is adjusted with NaCl to be slightly higher than that for the desorption of Q-mucin. Mixing the basic buffer with the washing buffer is simple and quick.
6. For determination of IS_1 and IS_2 values, *see* Subheading 3.3.3.

2.6 Amino Acids Composition Analysis (AACA)

To confirm that the candidate peak indeed corresponds to Q-mucin, AACA is performed. The total molar percentage of the six amino acids (Val, Ile, Glu, Thr, Ala, and Pro) that are supposed to be contained in a TR is examined, and we use this value to evaluate the quality of the purified sample. The value sometimes exceeds 90% after fractionation by AEXC. The molar ratio of these six amino acids is close to that of the TR: Val + Ile:Glu:Thr:Ala: Pro = 2:1:2:2:1. The existence of the TR can also be confirmed by other methodologies including sequencing with Edman degradation [2] (*see* **Note 8**), NMR spectroscopy [3] (*see* **Note 9**), and limited degradation by proteinases combined with MS as described in Subheading 2.11.

Q-mucin is hydrolyzed basically by the classical Miller's method [8]. However, since the peptide chain of Q-mucin is protected by surrounding glycan chains, special care must be taken to complete the hydrolysis of Q-mucin without any side reactions. Moreover, glycan chains of mucins contain relatively large amounts of oxygen atoms, and unwanted oxidation or dehydration may cause large errors. Therefore, the complete removal of oxygen from a sample is essential. We use freeze-pump-thaw cycling method for this purpose.

Amount of AEP is also estimated on an amino acid analyzer or our homemade HPLC system if the calibration was made beforehand, although the yield of AEPs from the hydrolysis of AEP esters in saccharides is unknown.

1. Vacuum line: A branched glass line equipped with a greaseless cock and a male greaseless joint on each branch is used. A rotary pump is connected to the line. A cold trap at liquid N_2 temperature is recommended to prevent HCl leakage.
2. Hand burner with oxygen line: This burner is used for sealing glass ampules.
3. Dewar pot: A small steel Dewar pot with a tea-cup shape is convenient.
4. 6 M HCl solution including 1% phenol: The use of stock solutions is not recommended. Instead, we used ampules of 6 M HCl solution commercially available (1 mL ampules of 6 N hydrochloric acid of sequanal grade from Thermo Fisher Scientific).
5. Glass tube: We connect a greaseless female joint to a test tube of 8 ϕ whose total length is about 30 cm (*see* Fig. 2b #1).

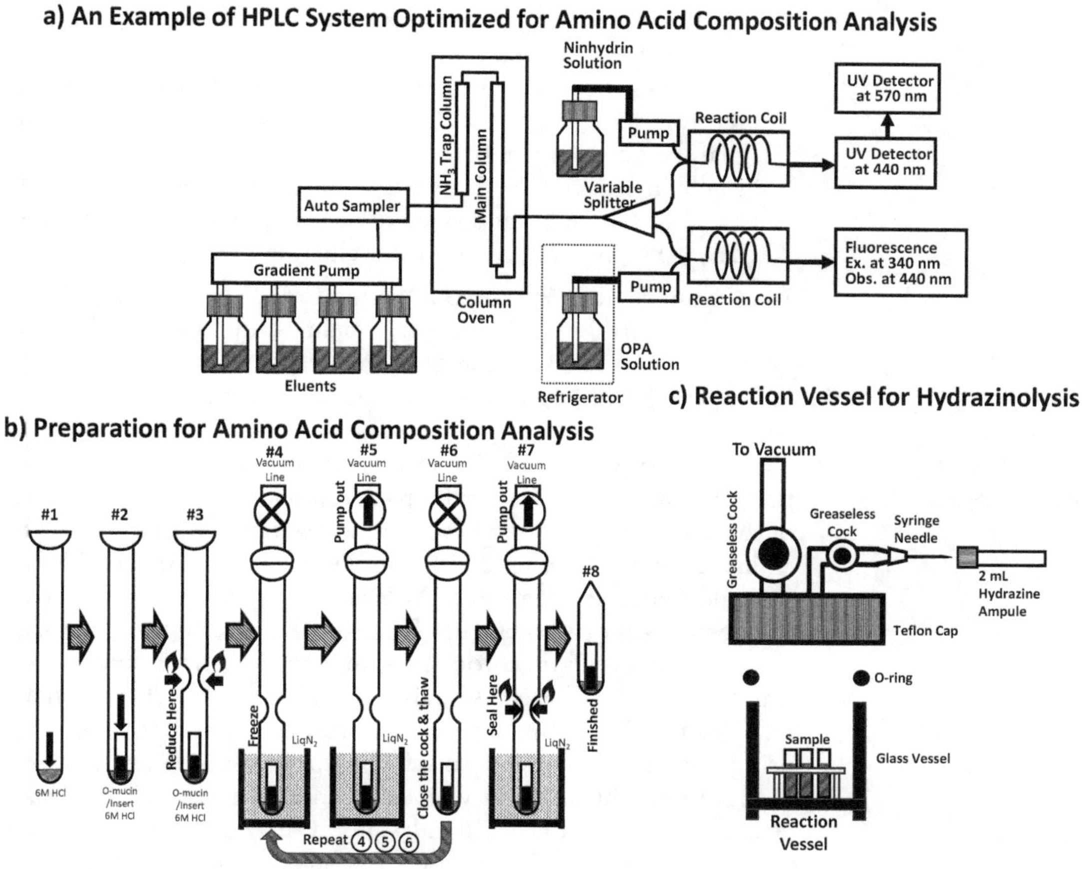

Fig. 2 Instruments optimized for the analysis of Q-mucin. (**a**) High performance liquid chromatography (HPLC) system for amino acid composition analysis. (**b**) Detailed protocol for freeze-pump-thaw cycling method to remove oxygen by Miller's method. (**c**) Reaction vessel for hydrazinolysis in gas phase

6. Glass inserts: In our laboratory, we use a 300 μL Crimp top SCI-VI insert from Thermo Fisher Scientific Inc. (*see* Fig. 2b #2).
7. Heat block incubator: It is used for 12 ϕ ampules.
8. Aluminum foil.
9. 1.5 mL plastic microtube.
10. Centrifuge concentrator.
11. 0.02 M HCl.
12. Amino acid analyzer: Any commercial automatic analyzer based on the postcolumn ninhydrin (NYN) method can be applied for this purpose. For example, LA8080 Amino-SAAYA® from Hitachi High Tech. Sci. Corp. is applicable. The NYN method is indispensable for the quantification of Pro (*see* **Note 10**).

13. HPLC system: In our laboratory, a homemade special amino acid analyzer is used (*see* Fig. 2a). The flow is divided into two flows after the separation column toward two postcolumn reactors using NYN and *O*-phthalaldehyde (OPA). It is useful that we can compare two chromatograms synchronously for a single injection. OPA is useful to obtain five TR components, Val, Ilu, Glu, Thr, and Ala, with background-free intensities. The NYN line is necessary for evaluating the amount of Pro.
14. Columns for 9: In our laboratory, we use #2619PH (4.0 × 150 mm) for the main column and #2619L (4.0 × 150 mm) for the ammonia-trapping column. Both are from Hitachi High-Technologies. Sci. Co. Corp.

2.7 Monosaccharide Composition Analysis

Monosaccharide composition analysis is performed with a conventional method using 4 M trifluoro acetic acid followed by labeling of the anomeric position of each monosaccharide with an aromatic amine (2-aminopyrizine (2AP), 2-aminobenzene (2AB), or ABEE: *p*-amino benzene ethyl ester (ABEE)) to detect the adducts with UV absorption or fluorescence [9, 10]. The determination of saccharide compounds on chromatograms depends on the commercial standard sample kit used as reference. Since monosaccharides have many isomers with the same exact mass, monosaccharide analysis using standard samples is also valuable for HPLC-MS analysis.

As for Q-mucin from various JF species, most of the glycan chains are single monosaccharides, i.e., GalNAc monomers which are modified with AEPs, phosphates, sulfates, and some saccharides. Among these, phosphates and sulfates are easily removed during hydrolysis with 4 M TFA. As a result, two prominent peaks attributed to GalNAc and AEP-GalNAc appear in chromatograms.

Moreover, small portions of other hexoses (Gal and Glc) and pentoses (Ara) are found in chromatograms. We cannot exactly conclude that they indeed originate from Q-mucin itself not from impurities or byproducts formed after hydrolysis.

1. HPLC system: An HPLC system equipped with a UV absorption detector, or an emission detector can be used. The emission detector is more sensitive than the UV absorption detector. The emission detector is more sensitive and the amount of samples necessary for analysis is smaller. In our laboratory, the GL-7400 series from GL Sciences Inc. is used.
2. Centrifugal evaporator/concentrator: This device is equipped with a cold trap and a vacuum pump.
3. Screwed-capped 1.5 mL plastic tubes.
4. A heat block incubator that can be used for **item 3**.

5. 8 M TFA solution: 6 mL of TFA is diluted with distilled water up to a final volume of 50 mL.
6. 2-propanol: used as received.
7. 10 mM Ammonium acetate buffer: 14 μL of acetic acid and 19 mg of ammonium acetate are diluted with distilled water up to a final volume of 50 mL.
8. 2,4,6-methylpyridine: used as received.
9. Acetic anhydride: used as received.
10. Standard monosaccharide mixture: In our laboratory, monosaccharide mixture-11 unlabeled from MGC Corp. (0.5 mL with 200 μM each of saccharide) is used. Similar standard mixtures from other sources can be used.
11. ABEE reaction reagent.
    (a) 250 mg of 4-aminobenzoic acid ethyl ester (ABEE) is added to a mixture of 0.5 mL of methanol and 170 μL of acetic acid and dissolved completely.
    (b) 145 μL of the borane-pyridine complex is added and dissolved completely.
12. Chloroform: used as received.
13. HPLC column for reverse phase chromatography: In our laboratory, we use Honenpack C18 (4.6 × 75 mm) from MGC Corp.
14. Eluent C [0.2 M Potassium tetraborate buffer (pH 8.9):acetonitrile = 93:7(v/v)]
    (a) 61.1 g of potassium tetraborate tetrahydrate is dissolved in distilled water of almost 1 L.
    (b) The pH is adjusted to 8.9 by adding 0.1 M NaOH solution.
    (c) The total solution is diluted with distilled water up to a final volume of 1 L.
    (d) 70 mL of acetonitrile (HPLC grade is preferred) is mixed with 930 mL of (c).
    (e) The eluent is degassed with a vacuum pump before its use.
15. Eluent D [0.02% trifluoroacetic acid:acetonitrile = 50:50 (v/v)].
    (a) 0.2 mL of trifluoroacetic acid is diluted with distilled water up to a final volume of 1 L.
    (b) Equal volumes of (a) and acetonitrile (HPLC grade is preferred) are mixed.
    (c) The eluent is degassed with a vacuum pump before its use.

2.8 Glycan Chain (Oligosaccharides) Composition Analysis After Hydrazinolysis

Gas phase hydrazinolysis is performed using a commercial system optimized for this purpose. The reaction vessel was previously provided by Taitec Corp., but is now unavailable. Instead, we suggest another design of the reaction vessel as shown in Fig. 2c which can be easily reproduced. The reaction glass vessel can accommodate several microtubes and can be operated under vacuum condition. Therefore, the reaction vessel must be unbreakable and airtight in vacuum operation. Hydrazine anhydrous in ampules of 2 mL is provided from Tokyo Chemical Industry Co. Ltd. We can introduce it directly to the evacuated vessel through a syringe needle without touching any dangerous reagent. We also need to prepare a heat block incubator applicable to the reaction vessel.

We used two types of HPLC analysis.

Since molecular masses of the isomers of saccharides are the same, we cannot distinguish them by HPLC-MS. The HPLC based on UV absorption and emission used in the monosaccharide analysis is appropriate for identifying each isomer at different retention times using standard samples. Therefore, the comparison of HPLC-MS results with HPLC-UV absorption (or emission) results on the same HPLC instrument with an identical column is required. Therefore, ABEE labeling of anomeric positions with aromatic amines is also performed after hydrazinolysis and the obtained samples are conveyed to both HPLC-UV absorption (or emission) and HPLC-MS systems.

Since sulfates, AEPs, and phosphates are partly preserved during hydrazinolysis, HPLC-MS analysis is valuable for the identification of these esters. Sulfate and phosphate esters, whose integer masses are the same, are distinguishable by high-resolution MS. We used both the UV absorption and a high-resolution MS system.

1. HPLC system: We use an HPLC system equipped with a high-resolution ESI-MS (LC-ESI-MS) system, a UV absorption detector, or an emission detector. We use the Orbitrap system from Thermo Fisher Sci. Corp. in our laboratory including LC: Ultimate3000 (RS pump, RS autosampler, RS column compartment, diode array detector, DIONEX), ESI-MS: Exactive Plus (Thermo Scientific), and the software Xcalibur.
2. Centrifugal evaporator/ concentrator: Equipped with a cold trap and a vacuum pump.
3. Reaction vessel designed for hydrazinolysis (*see* Fig. 2c).
4. Screwed-capped 1.5 mL plastic tubes.
5. Hydrazine anhydrous: 2 mL ampules from Tokyo Chemical Industry Co., Ltd. are useful.
6. Heat block incubator: It can be used for both the reaction vessel and the plastic tubes.

7. 10 mM Ammonium acetate buffer: *See* Subheading 2.7, **item 10**.
8. 2,4,6-Methylpyridine: used as received.
9. Acetic anhydride: used as received.
10. Acetic acid.
11. ABEE reaction reagent: *See* Subheading 2.7, **item 11**.
12. HPLC column: Any ODS columns can be used, but HPLC conditions must be tuned. We use ULTRON® VX-ODS 5 mm (4.6 × 150 mm) from Shinwa Chemical Industries Ltd.
13. Eluent E [0.1 M acetic acid]: 6 mL of glacial acetic acid is diluted to a final volume of 1 L. The eluent is degassed with a vacuum pump before its use.
14. Eluent F [0.1 M acetic acid: acetonitrile = 50:50 (v/v)]: 1:1 (v/v) mixture of eluent E and acetonitrile (HPLC grad is preferred). The eluent is degassed with a vacuum pump before its use.
15. Eluent G [0.1 M acetic acid: acetonitrile = 91:9 (v/v)]: 91:9 (v/v) mixture of eluent E and acetonitrile (HPLC grad is preferred). The eluent is degassed with a vacuum pump before its use.

2.9 Glycan Composition Analysis by β-elimination

β-elimination is also carried out for secondary glycan analysis. Since products generated after β-elimination are fragile, separation by HPLC is impossible. Therefore, the analysis of the total mixture by MS is performed. Normally, the number of detectable glycans generated after β-elimination is smaller than that after hydrazinolysis.

1. High-resolution ESI-MS system: The same system as that used in Subheading 2.8 (the HPLC system is not used).
2. Centrifugal evaporator/ concentrator: Equipped with a cold trap and a vacuum pump.
3. Disposable 1.5 mL plastic tubes.
4. 1 M $NaBH_4$ + 0.5 M NaOH solution: 2 g of NaOH and 3.8 g of $NaBH_4$ are dissolved in distilled water up to a final volume of 100 mL.
5. Heat block incubator: It can be used for 1.5 mL plastic tube.
6. 4 M aqueous solution of acetic acid.
7. 1% (w/w) acetic acid/methanol solution. This solution must be prepared immediately prior to use.
8. Cation exchange cartridge with additional cation exchange resin: Strong acidic cation exchange resin 50Wx2 200–400 mesh (H form) from Fujifilm Wako Chemicals. Corp. is added to an Oasis MCX 3 cc (60 mg) from Waters Corp. to

make the total cation exchange capacity larger than the amount of sodium ion in the sample.

9. Eluent H [0.1% formic acid solution]: We use 0.1% formic acid in water, Optima® obtained from Thermo Fisher Sci. Corp. The eluent is degassed with a vacuum pump before its use.

2.10 Identification of Each Single TR with Limited Degradation

Since one glutamic acid is present in a single TR (VVETTAAP or VIETTAAP) of Q-mucin, endoproteinase Glu-C can be used to cut off a single TR together with two glycan chains on either of two threonine residues. However, mucin is inherently resistant to proteinases because of the glycan chains protecting the main peptide chain. The yield is relatively low and self-degradation of enzymes may be observed. Normally, the TR part occupies a large mass-portion in an adduct; it can be detected in both positive and negative modes. ESI-MS results of total solution indicate the partial detachment of glycans from threonine. It is unclear whether this detachment occurs during the enzymic reaction or ESI. Minor adducts with several varieties of glycan chains can be separated by HPLC-ESI-MS, although the separations between their corresponding peaks are small in resultant chromatograms.

1. Endoproteinase Glu-C stock solution: 0.5 mL of iced ultrapure water is added to 2 mg (the whole amount of a single package) of endoproteinase Glu-C (V8 proteinase) of *Staphylococcus aureus* V8 from Roche and stirred gently with a pipette so as not to produce foam until the enzyme is dissolved completely. The stock solution is preserved at −40 °C.
2. 10 mM phosphate buffer. The same as Eluent A (*see* Subheading 2.4, **item 3**).
3. 10 mM Ammonium acetate buffer: *See* Subheading 2.7, **item 10**.
4. Heat block incubator.
5. Centrifugal filter device with regenerated cellulose membrane, 3000 NMWL: In our laboratory, we use Amicon® Ultra-2 mL Centrifugal Filters Ultracel® -3 K from Merck Milipore.
6. LC-ESI-MS system: The same system as that used in Subheadings 2.8 and 2.9.
7. HPLC column: In our laboratory, inertsil® ODS-3 (4.6 × 75 mm) from GL Sciences Inc. is used.
8. Eluent I [0.1% (w/v) formic acid solution]: The same as Eluent H. The eluent is degassed with a vacuum pump before its use.
9. Eluent J [Acetonitrile]: An HPLC grade solvent is used as received. The eluent is degassed with a vacuum pump before its use.

2.11 Quantitation of Sulfate Group

Although some of the charged groups on saccharides are detected after hydrolysis, hydrazinolysis, β-elimination, and limited degradation, some portions may be lost during these reactions. Therefore, the reproduction of the original composition is almost impossible. Instead, the amount of charged groups in glycan chains can be estimated by ion chromatography combined with combustion flask method [11]. The amount of S converted to SO_4^{2-} is estimated and the value is converted to a percentage if the mass of initial Q-mucin is known. Residual free SO_4^{2-} in samples after dialysis must be taken into consideration (sometimes its concentration is comparable with S atoms on glycan chains). An effective method is to compare the concentration of SO_4^{2-} in the solution after combustion and the Q-mucin solution simply dissolved in distilled water. We divide one sample into two of 5–10 mg. One is used for combustion and the other is for preparing a simple aqueous solution. The total amount of AEPs and phosphates can be estimated indirectly from the P (PO_4^{3-}) concentration. The mass measurement of initial Q-mucin has a large error because the moiety of the sample is not easy to control. Instead, we can estimate the mass of initial Q-mucin from total N amount (NO_2^- and NO_3^-) (also *see* **Note 11**).

1. Combustion flask: We use 1-L combustion flask equipped with a platinum mesh underneath the glass cap.
2. Ion chromatograph with a suppressing gel: To obtain sufficient sensitivity, an ion chromatograph with a suppressing gel is recommended. We use IC-2000 from Tosoh Corp.
3. 10% H_2O_2 solution: Dilute 30% H_2O_2 solution with distilled water.
4. O_2 line.
5. Filter paper: To reduce the background, a thin filter paper is used. Cut it into a shape convenient for wrapping the Q-mucin sample.
6. Vacuum desiccator.

2.12 Molecular Mass Estimation

Since the count of TR in the entire Q-mucin is variable, molecular mass (MM) of Q-mucin shows a wide distribution. Size exclusion chromatography (SEC) technique can be used for daily qualification from sample to sample. However, since no suitable standard materials are available for Q-mucin (*see* **Note 12**), SEC provides only a relative value of molecular mass of Q-mucin. Instead, we can use the MM values obtained by matrix-associated laser desorption/ionization (MALDI) with time-of-flight mass spectrometry (TOF-MS) as the absolute MM.

1. HPLC system: Any HPLC system can be used. No programmable gradient pump is needed. Both UV detectors and refractive index (RI) detectors can be used. When SP Q-mucin is

observed, dual beam UV observation (215 and 275 nm) is valuable. In our laboratory, the GL-7400 series from GL Sciences Inc. is used.

2. Column: We use TSKgel® G50000PWXL (7.8 × 300 mm) from Tosoh Corp. whose spec is following.
   (a) Average particle size in diameter: 10 μm.
   (b) Average pore size in diameter: 100 nm.
   (c) Molecular-mass exclusion limit: 2 MDa.
3. Mobile phase: Distilled water or ultrapure water. The solvent is degassed with a vacuum pump before its use.

3 Methods

3.1 Catching and Storage of JFs

Although the JF (dead or alive) body is stable in seawater in which microbial activities are low and the gravity is reduced by buoyant force, it degrades quickly in the atmosphere owing to microbial decomposition, self-decomposition by its own enzymes, and sol-gel transition induced by gravity. Therefore, use of fresh JF is indispensable in Q-mucin extraction. Storage in freezers is possible for 3–5 years. Note that 95–99 wt% of JFs is seawater (salt and water) and the rest (<5 wt%) is organic materials.

1. Collect appropriate JF species from a gulf or a pier using a soft brail net or a dipper so as not to injure the JF body. It is difficult to collect undamaged raw JFs from sandy beaches and rock reefs.
2. Keep the JF body in seawater if the transportation period is shorter than 24 h. If it is too long, JFs secrete a mucosal liquid that inactivates themselves, their bodies get thinner owing to their metabolic activities which are finally lost.
3. At the laboratory, wash the JF body with tap water and remove the liquid portion with a plastic sieve. Measure the wet weight at this stage.
4. When Q-mucin cannot be extracted immediately after the catch, store the caught JFs in a freezer as follows. Place the JFs in a zippered pack and add an antiseptic solution (1 mL/kg). Two or more zippered packs are recommended for preparing a duplicate package because the edge of ice may break the zippered pack. If we need to suppress the degradation of ECM jelly, add about 1/10 volume of wet JFs of 100 mM EDTA solution (pH 7). If the solution part exhibits acidity (such as pH 4–5), addition of NaOH solution is recommended to neutralize the sample for storage. JFs can be stored under frozen conditions (~ −20 °C) for several years. Storage at lower temperatures significantly delays the degradation of ECM.

3.2 Extraction of SP Q-mucin

1. When frozen JFs are used, thaw them in a stream of tap water.
2. Crush the whole unfrozen or raw JF bodies (including liquid from unfrozen JFs) in a blender for about 30 s. After crushing, the liquid part may exhibit acidity again because of some constituents in ECM. If so, we can add NaOH solution at this stage. pH control on extraction affects the kinds of counter cations in the final solid Q-mucin.
3. Remove the solid or jelly portion using a large-volume ultracentrifuge (12,000 × *g* for 15 min at 4 °C).
4. Collect the supernatant (liquid) and transfer it to a glass bottle and place the bottle in a showcase refrigerator.
5. Start stirring the supernatant in the glass bottle kept in the refrigerator and add ethanol (here we use IPA contaminating alcohol) of three times the volume of the supernatant dropwise. Initially, the liquid may separate into two phases. However, owing to the surfactant nature of mucin, the liquid will gradually become uniform. After dropwise addition is over, settle the liquid in the refrigerator overnight.
6. Centrifuge the liquid in a large volume ultracentrifuge (12,000 × *g* for 15 min at 4 °C) and collect the resulting precipitate. Measure the weight of the precipitate in wet condition.
7. Place the precipitate in a beaker and add ultrapure water (2 mL for 1 g wet precipitate) into it. Since the precipitate is pressed into paste form, crush it by a spoon, and start stirring the beaker. After some moment, the precipitate will be well dispersed.
8. After 2 h of stirring, centrifuge the supernatant by a small volume ultracentrifuge (12,000 × *g* for 15 min at 4 °C) and collect the supernatant.
9. Dialyze the obtained supernatant with a membrane tube in 10 L of distilled water. Change the water several times until the digital ion meter indicates 0.00% for the liquid inside the tube.
10. Repeat changing water once to remove free ions (such as SO_4^{2-}) completely.
11. Transfer the dialyzed solution to an eggplant-shaped flask and place the flask to freeze on a prefreezer. After the solution is completely frozen on the wall of the flask, the flask can be stored in the freezer until the freeze dryer is ready.
12. Start freeze-drying.
13. Cotton-like SP Q-mucin will appear on the wall of the eggplant flask. Sometimes Q-mucin may appear colored depending on the species used as the source.

3.3 Anion Exchange Chromatography (AEXC)

3.3.1 AEXC with Continuous Ion Gradient Program (Including Fractionation)

1. 1 mg of SP Q-mucin is dissolved in 1 mL of distilled water. Samples are shaken overnight to dissolve completely.
2. HPLC conditions that we use in our laboratory.
   (a) Sample volume: 20 μL for analysis (300 μL for fractionation)
   (b) Flow rate: 0.5 mL/min for analysis (4.0 mL/min for fractionation)
   (c) Scan period: 50 min
   (d) Column for analysis: TSK-gel® DEAE-5PW from Tosoh Corp. (7.5 × 75 mm)
   (e) Column for fractionation: TSK-gel® DEAE-5PW from Tosoh Corp. (21.5 × 150 mm)
   (f) Column temperature: 27 °C
   (g) Wavelength of UV absorption observation: 215 nm and 275 nm
3. Examples of the continuous gradient program are indicated in Fig. 3a–c.
4. Examples of the obtained chromatograms are also shown in Fig. 3a–c.
5. Identify Q-mucin components. Select the most prominent peak indicating strong UV absorption at 215 nm with no intensity at 275 nm (Marked with "Q" in Fig. 3a, b). Normally Q-mucin peak appears as a single peak around [NaCl] = 0.13 M as shown in Fig. 3a for Q-mucin from scyphozoan species. (The delay volume from the mixing pump to the injection port must be considered for correction.)
6. Check other impurity peaks. The peak of an uncharged component appears at [NaCl] = 0 M. Several aligned sharp peaks corresponding to polyglutamic acid (PGA) may appear at late retention times (marked with PGA in Fig. 3b) in high-ion-strength region.
7. Although fractionation may be possible with a column for analysis (7.5 × 75 mm length), the use of a larger column optimized for fractionation (21.5 × 150 mm length) is more practical. However, the chromatogram may thoroughly change and severe overlap of Q-mucin and impurity peaks may be found.
8. Collect all the fractions of Q-mucin (shaded retention time regions in Fig. 3a, b) obtained from repeated AEXC, and the mixture is dialyzed and freeze-dried to obtain purified Q-mucin.
9. The quality of purified Q-mucin can be checked by another round of AEXC.

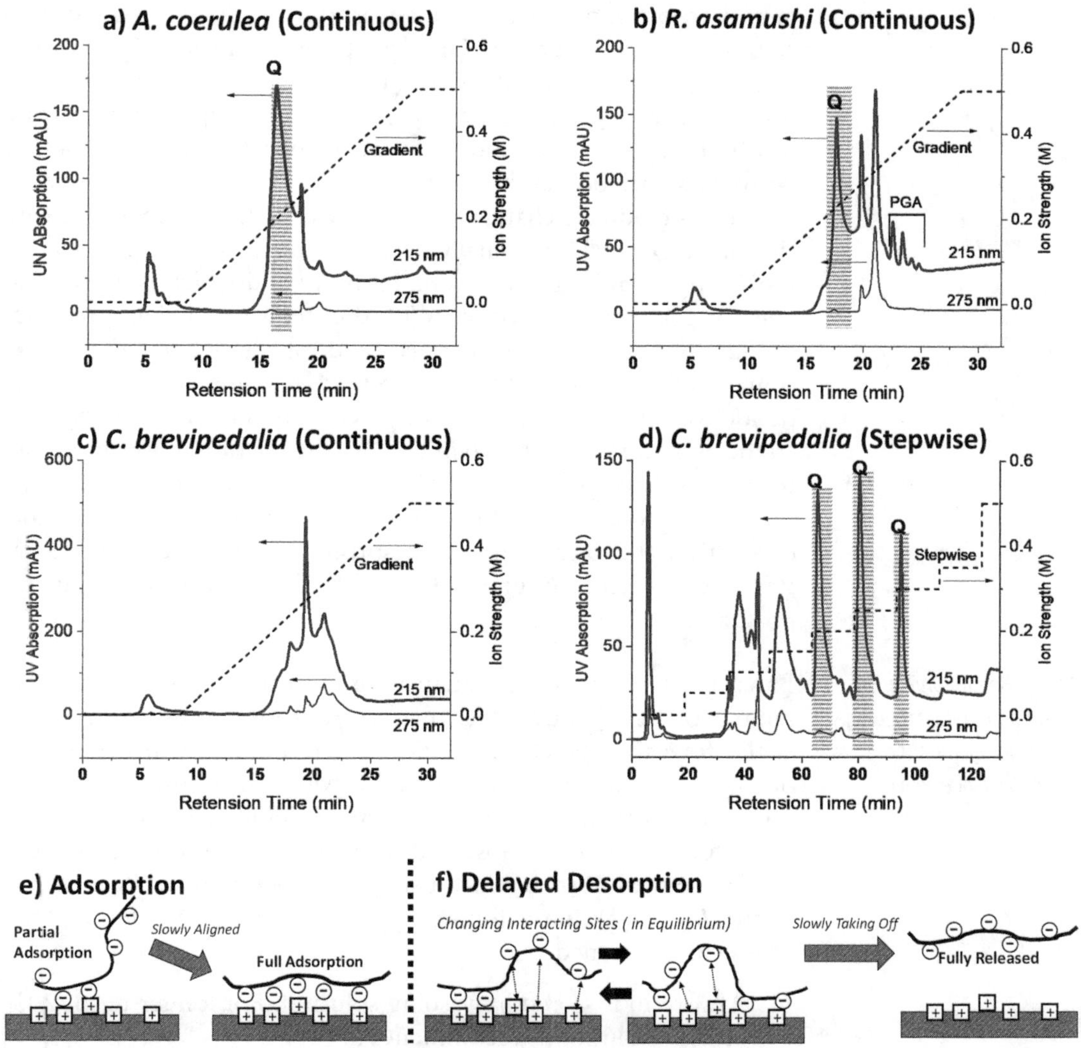

Fig. 3 Results of anion exchange chromatography used for purification. UV absorption was monitored at 215 nm and 275 nm. (**a**) *A. coerulea*, (**b**) *R. asamushi*, (**c**) *C. brevipedalia* (with continuous change in ion strengths), (**d**) *C. brevipedalia* (with stepwise change in ion strengths). Schematics of mucin molecule showing (**e**) adsorption and (**f**) delayed desorption on anion exchange resin

3.3.2 AEXC with Stepwise Ion Gradient Program (Including Fractionation)

In some species, the overlap between the peaks for Q-mucin and impurities is severe in AEXC with the continuous gradient program as shown in Fig. 3c. In such a case, the stepwise gradient program is effective as shown in Fig. 3d, where the peaks corresponding to Q-mucin and impurities are well separated. We consider that the adsorption of Q-mucin on the anion exchange resin is essentially a simple adsorption–desorption process controlled by ion strength. Q-mucin has many negative charges, which interact with amino groups on the AEX resin. Therefore, on adsorption, Q-mucin is pinned on the resin surface similarly to a stitch on the cloth (Fig. 3e). Although the adsorption may gradually proceed from partial adsorption to complete adsorption, it does not affect the resulting chromatograms. On desorption at an appropriate ion strength, every single interaction point starts to depart from the resin surface (Fig. 3f). Initially, however, each departed point can return to the surface, and they are involved in an equilibrium between association and dissociation. The distribution of interacting points gradually changes around the critical ion strength, then, the entire Q-mucin will finally depart from the resin surface to the mobile phase after a long equilibrating period which is indispensable for complete desorption. With the stepwise gradient program in Fig. 3d, the ion strength is sustained for 5 min and the desorption corresponding to the ion strength is completed within the period forming an isolated peak in each step in the chromatogram. In the present sample shown in Fig. 3d, the fractions corresponding to three isolated peaks were collected as purified Q-mucin.

Because of the equilibration period necessary to complete the desorption, the ion strength for desorption tends to be smaller with stepwise gradient program than with the continuous gradient program.

3.3.3 Batch Purification Based on the Adsorption and Desorption on the Anion Exchange Resin

Batch purification is convenient for the treatment of a large-scale sample of SP Q-mucin. As mentioned above, the ion strength values for each operation should be determined from the results of anion exchange chromatography with the stepwise gradient program. If the ion strength of ith step is indicated as $I(i)$ and the peaks for Q-mucin are observed from the jth to the kth steps, we can define the ion strengths for stripping operations as $IS_1 = I(j-1)$ and $IS_2 = I(k)$. In the following protocol, the use of 50 mg of SP Q-mucin is assumed.

1. AEX resin is stored in washing solution. For longer term, 20% ethanol solution is recommended.
2. Separate 5 g of AEX resin from the washing buffer by filtration.
3. Wash AEX resin with distilled water three times.
4. Wash AEX resin with the basic buffer.
5. Completely dissolve 50 mg of SP Q-mucin in 25 mL of the basic buffer.
6. Put 5 g of AEX resin into **step 5** and shake it for 30 min.
7. Separate the solution from the AEX resin by filtration. The solution contains some amount of residual Q-mucin that cannot be adsorbed on AEX resin.
8. Put the AEX resin into 25 mL of stripping buffer #1 (e.g., [NaCl] = 0.05 M) and shake the mixture for 10 min. (This process may be skipped so as not to decrease the yield.)
9. Separate the solution from the AEX resin by filtration. Together with impurities, the solution contains some amount of Q-mucin weakly adsorbed on AEX resin.

10. Put the AEX resin into 25 mL of stripping buffer #2 (e.g., [NaCl] = 0.12 M) and shake the mixture for 10 min.
11. Separate the solution from the AEX resin. The solution obtained as this step is mainly composed of purified Q-mucin.
12. Dialyze and freeze-dry the final solution to obtain the solid of purified Q-mucin.
13. Wash AEX resin with the washing solution for three times.
14. Store AEX resin in the washing solution. For longer term, use specified solvent for AEX resin by the provider.

3.4 Amino Acid Composition Analysis (AACA)

1. Connect a clean 8 ϕ glass test tube to a clean greaseless female connector by glasswork and cool it. (The shapes are shown in Fig. 2b #1.)
2. Put 6 M HCl (containing 1% phenol) from the ampules to the test tube (Fig. 2b #1).
3. Put 0.2–1.0 mg of purified or SP Q-mucin into a glass insert.
4. Place the glass insert gently at the bottom of the test tube (Fig. 2b #2).
5. Reduce the diameter of the test tube at about the height of 80 mm (Fig. 2b #3).
6. Connect the test tube with the vacuum line already started, freeze the 6 M HCl solution using liquid N_2 (Fig. 2b #4), then open the greaseless cock to pump out the gas inside the test tube (Fig. 2b #5).
7. After completely pumping out the air inside, close the greaseless cock, and thaw the solution. Some foams may appear (Fig. 2b #6).
8. Again, freeze the 6 M HCl solution using liquid N_2, then open the greaseless cock to pump out the residual gas inside the test tube. (Back to Fig. 2b #4.)
9. Repeat **steps** 7 and **8** (Fig. 2b #4–#6) twice or more (Freeze-pump-thaw cycling).
10. Keep the solution frozen (do not remove the Dewar pot) and close the shrink point of the test tube with a hand burner (Fig. 2b #7).
11. If the test tube is sealed successfully, leave it at room temperature to warm up (Fig. 2b #8).
12. Wrap the test tube with aluminum foil avoiding its explosion caused by pinholes that may be accidentally generated during glassworks (**steps 1**, **5**, and **10**).
13. Incubate the test tube at 110 °C for 20 h. The solution containing HCl remains.

Table 2
Examples of AACA results of SP Q-mucin and purified Q-mucin (*Aurelia coerulea*)

	Amino acid contents (mol%)						
Sample	**Val**	**Ile**	**Glu**	**Thr**	**Ala**	**Pro**	**Total**
SP Q-mucin	13.7	3.0	11.3	15.9	19.9	12.9	76.7
Purified Q-mucin	19.1	5.3	15.4	21.7	20.8	11.4	93.7

14. Cut the test tube, transfer the solution to a 1.5 mL plastic microtube, and remove residual HCl using a centrifuge concentrator.
15. Completely dissolve the dried solid in 100 μL of 0.02 M HCl, ultracentrifuge the solution, and remove the resulting precipitate. The supernatant is a mixture of amino acids after hydrolysis.
16. Set the sample on an HPLC system for amino acid composition analysis.
17. Examples of AACA results for SP Q-mucin and purified Q-mucin is shown in Table 2. In this case, we found that the content of amino acids involved in the TR increased from 76.7 to 93.7 mol%. These values can be used as a metric indicating the quality of purified Q-mucin.

3.5 Glycan Chain Analysis

For the glycan analysis of Q-mucin, we need to use several methods to reproduce the original glycan compositions because all methods are destructive and imperfect, having both advantages and disadvantages.

Monosaccharide analysis by HPLC after hydrolysis can distinguish the isomers of hexose and pentose using standard samples. In this analysis, however, various ester moieties such as sulfates are detached from saccharides while AEP esters are rather persistent. HPLC with UV detection after hydrazinolysis can distinguish various oligosaccharides from their retention times, if their standard samples are available (usually they are not). HPLC-MS methods cannot distinguish the isomers of saccharides (MS-MS analysis may distinguish some of the isomers.). Since hydrolysis, hydrazinolysis, β-elimination, limited degradation by proteinases, and ionization before MS are destructive, we cannot determine the amounts of sulfates, phosphates, and AEPs lost during these operations. We perform total sulfur analysis by ion chromatography because it is the most reliable method to determine the original sulfate contents in Q-mucin.

Usually, major glycans in Q-mucin are limited to four to five kinds (GalNAc, AEP-GalNAc, SO_3^--GalNAc, and SO_3^--Gal-GalNAc). However, we have identified more than 20 oligosaccharides whose amounts are extremely small. At present, we cannot conclude that they are not from some impurities or fragment products of some side reactions.

3.5.1 Monosaccharide Analysis After Hydrolysis

1. Dissolve 0.5 mg of purified Q-mucin into 10 μL of ultrapure water in a screw-capped microtube.
2. Add 10 μL of 8 M trifluoroacetic acid to the solution, seal the microtube, and agitate it.
3. Allow the liquid to settle at the bottom of the microtube using an ultracentrifuge (500 × *g*).
4. Heat the microtube (at 100 °C for 3 h) in a block heater, and then, cool it to room temperature.
5. Remove the liquid portion using a centrifugal concentrator (70 × *g* at 35 °C).
6. Add 40 μL of 2-propanol to the sample, agitate, allow the liquid to settle at the bottom of the microtube using an ultracentrifuge (500 × *g*), and then, remove the liquid portion again using a centrifugal concentrator (70 × *g* at 35 °C).
7. (N-acetylation) Add 200 μL of 10 mM ammonium acetate and 10 μL of 2,4,6-trimethylpyridine, agitate, and lower the liquid (500 × *g*).
8. Add 10 μL of acetic anhydride, agitate, and allow the liquid to settle to the bottom.
9. After blocking the light with aluminum foil, allow the sample to stand at −20 °C for more than 8 h to freeze.
10. Freeze-dry the sample with a vacuum line.
11. (ABEE labeling) Add 10 μL of ultrapure water and 40 μL of ABEE reaction reagent to the sample, agitate, lower the liquid, and heat the sample at 70 °C for 1 h.
12. After cooling to room temperature, add 200 μL of chloroform and 200 μL of ultrapure water to the sample, agitate, centrifuge (1500 × *g*, 10 min), and collect the water layer (the upper layer of two layers).
13. Add 200 μL of ultrapure water again and repeat the same extraction protocol as in **step 12**.
14. Collect the two portions of extraction liquid to obtain the labeled sample for HPLC.
15. HPLC conditions that we use in our laboratory.
    (a) Sample volume: 10 μL
    (b) Flow rate: 1.0 mL/min
    (c) Scan period: 85 min
    (d) Gradient program: Eluent C:Eluent D = 100: 0 (0–60 min), 0:100 (60.1–65 min), 100:0 (75.1–85)

(e) Column: Honenpack C18 (4.6 × 75 mm) from MGC Corp.

(f) Column temperature: 30 °C

(g) Wavelength for UV absorption observation: 305 nm

(h) Wavelength for fluorescence observation: Excitation at 305 nm observation at 360 nm

3.5.2 Glycan Chain Analysis After Hydrazinolysis

1. Put 0.5 mg of purified Q-mucin into a 1.5 mL screw-capped microtube and dry up completely overnight in a centrifugal concentrator (70 × *g* at 35 °C).
2. Synchronously with **step 1**, dry up the reaction vessel completely in a heater (120 °C, 12 h).
3. Put 1.5 mL screw-capped microtubes of Q-mucin samples into the reaction vessel, seal the vessel, and vacuum out the air in the vessel.
4. Attach a syringe needle to the top of the reaction vessel as shown in Fig. 2c. Inject the syringe needle to a hydrazine ampule.
5. Open the cock, introduce 2 mL of hydrazine anhydrous through it from the ampule. (Hydrazine is injected via pressure difference.)
6. Heat the reaction vessel at 100 °C for 4 h.
7. Pump out residual hydrazine from the vessel with a vacuum pump equipped with a cold trap.
8. Trapped hydrazine must be decomposed by acetic acid.
9. (N-acetylation) Carry out **steps 7–10** described in Subheading 3.5.1.
10. (ABEE labeling) Carry out **steps 11–14** described in Subheading 3.5.1.
11. HPLC conditions (with UV absorption and fluorescence detectors) that we use in our laboratory.

    (a) Sample volume: 10 μL.

    (b) Flow rate: 0.8 mL/min.

    (c) Scan period: 100 min.

    (d) Gradient program: Eluent E: Eluent F = 82:18 (0 min), 77.5:22.5(67.5 min) 0:100 (72.5–82.5 min), 82:18 (82.6–100 min).

    (e) Column: ULTRON® VX-ODS 5 mm (4.6 × 150 mm) from Shinwa Chemical Industries Ltd.

    (f) Column temperature: 40 °C.

    (g) Wavelength for UV absorption observation: 305 nm.

    (h) Wavelength for fluorescence observation: Excitation at 305 nm observation at 360 nm.

12. HPLC-ESI-MS conditions (with UV absorption detector) that we use in our laboratory.
    (a) Sample volume: 10 μL.
    (b) Flow rate: 0.8 mL/min.
    (c) Scan period: 90 min.
    (d) Gradient Program: Eluent G: Eluent F = 100:0 (−10–0 min), 94.5:5.5 (67.5 min), 0:100 (72.5–82.5), 100:0 (82.5–90).
    (e) Column: ULTRON® VX-ODS 5 mm (4.6 × 150 mm) from Shinwa Chemical Industries Ltd.
    (f) Column temperature: 40 °C.
    (g) Wavelength for UV absorption observation: 305 nm.
    (h) Spray voltage and conditions: 2.00 kV at 300 °C in negative mode.
    (i) MS range: $m/z = 150–1500$ or 100–1500.

3.5.3 Glycan Chain Analysis After β-elimination

1. Place 400 μL of 0.5 M NaOHaq containing 1 M $NaBH_4$ to 1.0 mg of purified Q-mucin into a 1.5 mL screw-capped microtube.
2. Seal the microtube, agitate it, allow the liquid to settle at the bottom of the microtube, and incubate at 45 °C for 15 h.
3. After cooling to room temperature, add 4 M aqueous solution of acetic acid until bubbling stops, indicating neutralization.
4. Apply the entire solution to a cation exchange cartridge to obtain the filtrate after washing it with 500 mL of ultrapure water twice.
5. Collect the filtrate and two portions of wash fluid and remove the liquid portion using a centrifugal concentrator (70 × *g* at 35 °C).
6. Add 100 μL of 1% (w/w) acetic acid/methanol solution to remove borate salts using a centrifugal concentrator (70 × *g* at 35 °C). Repeat this operation two more times.
7. Add 120 mL of ultrapure water to the sample, and collect the supernatant after ultracentrifugation (12,000 × *g* at 35 °C for 15 min).
8. ESI-MS conditions that we use in our laboratory.
    (a) Sample volume: 40 μL.
    (b) Flow rate: 0.2 mL/min.
    (c) Wavelength for UV absorption observation: 305 nm.
    (d) Spray voltage and conditions: 2.00 kV at 300 °C in negative mode.
    (e) MS range: $m/z = 150–1500$.

Table 3
Summary of glycan chains detected by ESI-MS after hydrazinolysis or β-elimination

Category		Detected glycans
Unsubstituted	Monosaccharides	Hex, **HexNAc***, Pen
	Disaccharides	$(Hex)_2$, $(Pen)_2$, Hex + HexNAc, Pen + Hex, Pen + HexNAc
	Trisaccharides	$(Hex)_3$, $(Hex)_2$ + HexNAc, $(HexNAc)_3$, Pen+Hex+HexNAc, $(Pen)_2$ + HexNAc
AEP (phosphonate)-substituted	Monosaccharides	AEP(Ac) + Hex, AEP + Hex, **AEP(Ac) ± HexNAc, AEP ± HexNAc*** (with one AEP) $(AEP(Ac))_2$ + HexNAc, AEP + AEP(Ac) + HexNAc (with two AEPs)
	Disaccharides	AEP(Ac) + $(Hex)_2$, AEP(Ac) + Hex+HexNAc (with one AEP) $(AEP(Ac))_2$ + Hex+HexNAc, AEP + AEP(Ac) + Hex +HexNAc (with two AEPs)
	Trisaccharides	$(AEP)_3$ + $(HexNAc)_3$** (with three AEPs)
	Tetrasaccharides	$(AEP)_2$ + Pen+$(Hex)_2$ + $(HexNAc)_2$** (with two AEPs)
Sulfate-substituted	Monosaccharides	SO_3H + Hex, **SO_3H ± HexNAc***
	Disaccharides	**SO_3H ± Hex ± HexNAc***
	Trisaccharides	SO_3H+$(Hex)_2$ + HexNAc, SO_3H + Hex+$(HexNAc)_2$*
	Tetrasaccharides	SO_3H + Pen+Hex+$(HexNAc)_2$**
Phosphate-substituted	Monosaccharides	PO_3H_2 + HexNAc
	Disaccharides	PO_3H_2 + Pen+HexNAc
	Pentasaccharides	PO_3H_2 + $(Hex)_3$ + $(HexNAc)_2$**

*Detected after either hydrazinolysis or β-elimination
**Detected after β-elimination only
Other unmarked glycans are detected after hydrazinolysis only
Most of Hex and HexNAc are Gal and GaNAc, respectively, and most of Pen is Ara
Five major glycans constituting large part of the entire glycans are indicated with boldface and undelines.

3.5.4 Summary of Glycan Analysis

We show some results of glycan analyses for various species as examples. In Table 3, all detected glycans after hydrazinolysis or β-elimination of all species are listed including minor glycans which may originate from some impurities or fragments formed on pretreatments or ESI on MS. There are five major glycans constituting large part of the entire glycans (indicated with boldface and underlines). In Table 4, the compositions of five main glycans determined by UV absorption are summarized as examples. The results seem to depend not only on species but also on other conditions of raw JFs used as source. Numbers may also change irregularly from sample to sample.

3.6 Limited Degradation of the Main Peptide Chain

1. Dissolve 1.0 mg of purified or SP Q-mucin in a microtube containing 500 μL of 10 mM phosphate buffer (pH 7) or 10 mM ammonium acetate buffer (pH 4).
2. Add 5 μL of Glu-C stock solution to the microtube and incubate the solution at 37 °C for 24 h.

Table 4
Varieties of compositions of five major glycans determined from UV absorption of HPLC after hydrazinolysis (examples)

Class	Species	unsubstituted	AEP substituted	Sulfate substituted		
		GalNAc	AEP(+Ac) + GalNAc	SO_3H + GalNAc	SO_3H + Gal+GalNAc	SO_3H + GalNAc+GalNAc
Scyphozoa	*Aurelia coerulea*	23.6	74.0	0.9	1.5	(N.D.)
	Aurelia limbate	20.0	78.0	0.4	1.7	(N.D.)
	Chrysaora pacifica	56.4	39.8	2.6	1.7	(N.D.)
	Chrysaora sp.	57.4	16.6	4.9	21.1	(N.D.)
	Chrysaora fuscescens	9.9	33.4	11.8	44.9	(N.D.)
	Chrysaora melanaster	38.4	29.5	8.9	23.2	(N.D.)
	Cyanea capilata	63.3	21.2	6.8	8.6	(N.D.)
	Mastigias albipunctata	31.6	61.8	1.8	4.8	(N.D.)
	Rheopilema asamushi	10.9	15.4	8.3	65.4	(N.D.)
Cubozoa	*Carybdea brevipedalia*	52.9	12.0	11.2	(N.D.)	23.8
	Chironex yamaguchi	15.7	31.0	16.9	(N.D.)	36.5
	Morbakka virulenta	85.9	7.1	2.9	(N.D.)	4.2

The results seem to depend not only on classes and species but also on other various JF conditions. Numbers may also fluctuate irregularly from sample to sample. Addition of EDTA on sampling may provide a large effect on the results(*see* **Note 13**)
(N.D.) indicates that the content is smaller than 0.05%

3. Remove the components whose molecular masses are larger than 3 kDa using a centrifugal filter device in an ultracentrifuge and collect the filtrates as the degraded samples.
4. Measure the samples using the appropriate HPLC-ESI-MS instrument.
5. HPLC-ESI-MS (with a UV absorption detector) conditions that we use in our laboratory.
   (a) Sample volume: 40 μL.
   (b) Flow rate: 1 mL/min.
   (c) Scan period: 35 min.
   (d) Gradient program: Eluent I:Eluent J = 98:2 (0–5 min), 30:70 (20–35 min).
   (e) Column: inertsil ODS-3 (4.6 × 75 mm) from GL Sciences Inc.
   (f) Column temperature: 40 °C.
   (g) Wavelength for UV absorption observation: 305 nm.
   (h) Spray voltage and conditions: 2.00 kV at 300 °C in positive and negative mode.
   (i) MS range: $m/z = 150–1500$ or 100–1500.
6. Check the results.

Previously identified adducts are summarized in Table 5. Some adducts are detected in both positive and negative modes. Double-ionized fragments were identified in some adducts. Several adducts show more than one peak in the chromatograms, indicating the existence of isomers whose structure is unknown. Owing to the detachment of glycans during enzyme reaction or ESI-MS, the observed adducts include decomposed fragments. Note that two threonine residues exist in a single TR and each identified fragment may include several different isomers.

3.7 Quantitation of Sulfate Groups and Other Acidic Moieties

1. Take two portions of about 5–10 mg each from one sample of Q-mucin. One is used for combustion and the other is used as for the reference of free ions.
2. Quickly weigh both portions precisely to 0.01 mg digit. The measurement results may be disturbed by moisture involved in the samples.
3. Dissolve one portion in 100 mL of ultrapure water for reference.
4. Tightly wrap the other portion with a piece of filter paper.
5. Roll the wrapped sample in the platinum mesh attached to the glass cap of the combustion flask. Prepare the mesh to maintain its tightness so that the sample will not fall off before complete combustion.

Table 5
Summary of previously identified adducts from the results of limited degradation experiments using endoproteinase Glu-C for Q-mucin

Category	Identified fragments	+1	−1	+2	−2
TR only	TR1(VVETTAAP), TR2(AIETTAAP)	o	o		
One saccharide	TR1 + HexNAc, TR2 + HexNAc	o	o		
	TR1 + HexNAc + AEP, TR2 + HexNAc + AEP	o	o		
Two saccharides	TR1 + HexNAc × 2, TR2 + HexNAc × 2	o	o		
	TR1 + HexNAc + Pen	o	o		
	TR1 + HexNAc × 2 + AEP, TR2 + HexNAc × 2 + AEP	o	o	o	o
	TR1 + HexNAc × 2 + AEP × 2	o	o	o	o
	TR2 + HexNAc + DeoxyHex	o	o		
Three saccharides	TR1 + HexNAc × 3 + AEPAc		o		
	TR2 + HexNAc × 3 + AEP		o		
	TR1 + HexNAc + Hex × 2, TR2 + HexNAc + Hex × 2,	o	o	o	o
	TR1 + HexNAc×2 + Hex, TR2 + HexNAc × 2 + Hex	o	o	o	o
	TR1 + HexNAc × 2 + Pen	o			
	TR1 + HexNAc × 2 + DeoxyHex		o		o
	TR1 + HexNAc + Hex × 2 + AEP	o	o		o
	TR1 + HexNAc × 2 + Hex + AEP		o		o
	TR1 + HexNAc × 2 + Hex + SO_3^-	o	o		o

6. Place approximately 80 mL of 5% H_2O_2 solution in the combustion flask.
7. Allow oxygen to flow into the flask for 1 min.
8. Covering the top of the flask with one hand, ignite the sample wrapped in the platinum mesh with a standing flame or a cigarette lighter outside the flask, and introduce the glass cap with the mesh quickly but gently into the flask.
9. If the flask is sufficiently filled with oxygen, the sample burns brightly until the wrapped sample disappears completely. Block the glass cap tightly with a hand because the gas inside the flask expands during the combustion.
10. The operation may fail when some residual ash falls into the solution.
11. Cool the flask to room temperature. There may be mist filling inside the flask.

12. Shake the flask to dissolve soluble gases into the 5% H_2O_2 solution.
13. Transfer the solution to a 100 mL volumetric flask and wash the combustion flask twice. Washing liquid is also poured into the same volumetric flask.
14. Add ultrapure water up to a final volume of exactly 100 mL.
15. With this procedure, the combusted sample from **step 14** and the reference sample from **step 3** contain free anions.
16. Estimate the concentrations of SO_4^{2-}, PO_4^{3-}, NO_2^-, and NO_3^- in the two samples by ion chromatography.
17. Calculate the contents of S, P, and N in the two samples from the results of ion chromatography.
18. Since the combusted sample involves both the atoms in Q-mucin and the atoms in residual free ions that cannot be removed by dialysis, we must subtract the concentrations of residual free ions that is estimated from the reference sample from the concentrations in the combusted sample.

3.8 Estimation of Molecular Mass

Chromatographic conditions that we use for SEC in our laboratory.

(a) Sample volume: 10–20 μL.
(b) Flow rate: 0.5 mL/min.
(c) Scan period: 30 min.
(d) Column: TSKgel® G50000PWXL (7.8 × 300 mm) from Tosoh Corp.
(e) Column temperature: 25 °C.

4 Notes

1. Since Q-mucin is not a constituent of mucus, we also use another category name mucin-type glycoprotein (MTGP). However, with such an expanded definition, proteoglycans (PGs) may be included as an MTGP.
2. We consider that the pKa values of sulfates, phosphonates, the amino groups in AEPs, and carboxyl groups are <1, ~4, ~10, and ~5, respectively, as determined from the titration results (*see* supporting information of ref. [5]).
3. Some simple glycoproteins similar to Q-mucin exist whose amino acid composition is different from VVETTAAP or VIETTAAP: *Proboscidactyla flavicirrata* and *Turritopsis* spp.
4. Q-mucin from the *Aurelia* genus shows a single isolated corresponding peak in AEXC in the present protocol. A large portion of glycan chains are modified by AEPs.

5. A large swarm of JFs normally rides on a tide, then moves more than several 10 km between full and low tides. Sometimes, it stays deep in the ocean. Accordingly, the appearance of JFs is irregular to us who are waiting at somewhere on the sea surface.
6. Since EDTA inhibits enzymic activity, matrix metalloproteinase (MMP) may be a candidate.
7. When the sea is highly polluted, we need to consider the inclusion of heavy metal ions some of which are toxic. Microelement analysis using ICP may be necessary.
8. In sequence analysis with Edman degradation, it sometimes happens that the N-terminus starts from Val. In such a case, the sequence analysis of the TR is to be successful. However, unwanted hydrolysis which is temporally uncontrollable tends to occur in purified samples [2].
9. Using cross-peaks between ^{13}C signals from peptide groups and ^{1}H signals from alpha protons on an HMBC 2D chart, we can follow the periodic sequence of TRs [2].
10. To distinguish collagens, analyses of HydPro and HydLys are desired. However, from our experience, these two amino acids are not present in JFs in large amount [5].
11. There is another method using ATR-FTIR to detect the existence of sulfates and phosphates on saccharides. This method is not destructive, and we can conserve precious samples. Free SO_4^{2-} and bound -O-SO_3^- are distinguishable from the lineshape and intensity.
12. We have an experience of using a set of analytical standards of pullulan. However, the values were several times different from those obtained from MALDI-TOF.
13. The effect of EDTA addition is complicated and we cannot explain all phenomena. There is a tendency that the elimination of sulfate groups is enhanced occasionally during the extraction even when we control the pH of the bulk solution. We speculate that protons generated in the local acidity around -SO_3H act as a catalyst for hydrolysis to enhance the self-dissociation of sulfate groups. In heterogeneous solutions of polymers, the deviation between the bulk pH and the local pH may be difficult to elucidate.

Acknowledgments

The authors thank Dr. Hiromasa Morishita, Riken, for his sustained advice for this project. One of the authors (KU) dedicates this chapter to the late professor Hiroyuki Hatano who invented the automatic amino acid analyzer and introduced him to the analyses of biomaterials. This research is partly supported by Japan Society for the Promotion of Science (JSPS) KAKENHI, No. 21550030 and No. 24650282.

References

1. Ushida K, Murata T (2013) Materials science and engineering of mucin: a new aspect of mucin chemistry. In: Rahman A (ed) Studies in natural products chemistry, vol 39. Elsevier, Amsterdam, pp 115–159. https://doi.org/10.1016/B978-0-444-62615-8.00004-7
2. Masuda A, Baba T, Dohmae N et al (2007) Mucin (Qniumucin), a glycoprotein from jellyfish and determination of its main chain structure. J Nat Prod 70:1089–1092. https://doi.org/10.1021/np060341b
3. Uzawa J, Urai M, Baba T et al (2009) NMR study on a novel mucin from jellyfish in natural abundance, qniumucin from Aurelia aurita. J Nat Prod 72:818–823. https://doi.org/10.1021/np800601j
4. Urai M, Nakamura T, Uzawa J et al (2009) Structural analysis of O-glycans of mucin from jellyfish (Aurelia aurita) containing 2-aminoethylphosphonate. Carbohydr Res 344:2182–2187. https://doi.org/10.1016/j.carres.2009.08.001
5. Ushida K, Sato R, Momma T et al (2022) Jellyfish mucin (qniumucin) extracted with a modified protocol indicated its existence as a constituent of the extracellular matrix. Biochim Biophys Acta Gen Subj 1866:130189. https://doi.org/10.1016/j.bbagen.2022.130189
6. Arai MNA (1997) Functional biology of scyphozoa. Chapman & Hall, London
7. Kameyama A, Yamakoshi K, Watanabe A (2019) A rapid separation and characterization of mucins from mouse submandibular glands by supported molecular matrix electrophoresis. Biochim Biophys Acta Proteins Proteom 1867: 76–81. https://doi.org/10.1016/j.bbapap.2018.05.006
8. Miller GH, Hare PE (1980) In: Hare PE, Hoering TC, King K Jr (eds) Biogeochemistry of amino acids. Wiley, New York, pp 415–443
9. Yasuno S, Murata T, Kokubo K et al (1997) Two-mode analysis by high-performance liquid chromatography of ρ-aminobenzoic ethyl ester-derivatized monosaccharides. Biosci Biotechnol Biochem 61:1944–1946. https://doi.org/10.1271/bbb.61.1944
10. Yasuno S, Kokubo K, Kamei M (1999) New method for determining the sugar composition of glycoproteins, glycolipids, and oligosaccharides by high-performance liquid chromatography. Biosci Biotechnol Biochem 63:1353–1359. https://doi.org/10.1271/bbb.63.1353
11. Macdonald AMG (1961) The oxygen flask method. A review. Analyst 86(1018):3–12. https://doi.org/10.1039/AN9618600003

Chapter 2

Extraction of Mucin from Rodent Feces and Determination of *O*-Linked Oligosaccharide Chain Equivalent Derived from Fecal Mucin

Naomichi Nishimura, Hiroki Tanabe, and Tatsuya Morita

Abstract

Since the core protein of mucin in the digesta of the stomach and small intestine, which is less affected by bacteria, remains intact, mucin content can be measured by enzyme-linked immunosorbent assay (ELISA). However, the mucin core protein in bacteria-rich colon digesta and feces is partially hydrolyzed by bacterial enzymes and not fully recognized by mucin antibodies, so mucin cannot be accurately quantified by ELISA. This method quantifies the glycan content linked to the mucin core protein and expresses mucin content in the colon digesta and feces as the equivalent of *O*-linked oligosaccharide chain. Although mucin glycans are also hydrolyzed by colonic bacteria, this method is a more accurate and simple way to measure mucin content in the digesta of the large intestine and feces than the ELISA method.

Key words Fecal mucin, Feces, *O*-linked oligosaccharide chain equivalent, Rat

1 Introduction

In the human gastrointestinal tract, mucins with the core proteins, MUC2 (small and large intestine), MUC5AC (stomach and large intestine), MUC5B (saliva), MUC6 (stomach and large intestine), and MUC7 (saliva) as secretory mucins are produced and secreted by goblet cells in the intestine, stomach surface, stomach glands, and salivary glands [1]. Mucins not only serve as a lubricant, but also protect the gastrointestinal mucosa from being broken down by stomach acid, digestive enzymes, and bile acids. In addition, the gastrointestinal tract is protected from bacteria by membrane-bound mucins, which generate an epithelial cell apical glycocalyx, and by secreted mucins, which form mucus of different shapes [1]. Recently, it has been also reported that the mucin glycans are used as an energy source by colonic bacteria and decide the composition of colonic bacteria [2]. These phenomena indicate that mucins have a significant impact on the physiology of the

Akihiko Kameyama (ed.), *Mucins: Methods and Protocols*, Methods in Molecular Biology, vol. 2763,
https://doi.org/10.1007/978-1-0716-3670-1_2,

gastrointestinal tract. Fecal mucin is the most readily available of the mucins secreted into the gastrointestinal tract, and simple and accurate measurements of it can provide insight into gastrointestinal mucins.

Mucins are glycoproteins, and because of the diversity of their constituent sugar chains, it is difficult to quantify mucin content from their glycan structures. Enzyme-linked immunosorbent assay (ELISA) targeting the mucin core protein, e.g., MUC2 protein, is used for the measurement of mucin but the core protein is partially degraded by bacteria in the colon, so the mucin content in the colon contents and feces cannot be measured correctly. Mucin oligosaccharide chains are *O*-glycosidically linked to hydroxy groups on the side chains of threonine and serine, the predominant amino acids of the mucin core protein, i.e., mucins have a large number of *O*-glycans. In the crude mucin fraction purified from the contents of the small intestine, the *O*-linked oligosaccharide chain equivalent is positively correlated with the amount of periodic acid-Schiff regent-reactive substances such as mucins, and sialic acid [3]. In addition, the *O*-linked oligosaccharide chain equivalent is correlated with the amount of mucin measured by the ELISA method. Although mucin glycans are also degraded by colonic bacteria, as long as even one *N*-acetylgalactosamine (GalNAc) that binds to the core protein remains, it is determined as the number of glycans. Therefore, measurement of *O*-linked oligosaccharide chain equivalent, targeting mucin glycans, is a suitable method for simple estimation of mucin content in feces.

2 Materials

Prepare all solutions using distilled water or equivalent and analytical grade reagents. Store all reagents at the temperature indicated (unless indicated otherwise).

2.1 Mucin Extraction

1. Phosphate-buffered saline (PBS, pH 7,4): Add 900 mL of distilled water to a 1-L beaker. Weigh 8.006 g of NaCl, 2.922 g of $Na_2HPO_4 \cdot 12H_2O$, 0.2 g of KH_2PO_4 and 0.1977 g of KCl, transfer to the 1-L beaker, and dissolve. Mix and adjust pH to 7.4 with 0.325 mol/L HCl. Make up to 1 L with distilled water and store at 4 °C.
2. 0.15 mol/L NaCl: Add 1.7532 g of NaCl and about 180 mL of distilled water to an appropriate beaker and dissolve. Make up to 200 mL with distilled water.
3. Polypropylene tube of 15 mL volume.
4. Polytron homogenizer with generator shaft of outer diameter, 12 mmφ.

5. Plastic cup, large enough to keep 50 mL polypropylene tube on ice.
6. Boiling water bath.
7. Shaking water bath.
8. 10 mL Glass centrifuge tube with a scale and a screw cap.
9. Refrigerated centrifuge.
10. Ethanol.
11. Lyophilizer.

2.2 Fecal Mucin Assay

1. 0.6 mol/L 2-cyanoacetamide (CNA): Dissolve 504.1 mg of 2-cyanoacetamide (MW 84.08, purity ≥98%) in 10 mL of distilled water in a 15 mL polypropylene tube.
2. 0.15 mol/L NaOH: Immediately weigh 0.6 g of NaOH (MW 40.00) and transfer to a 100 mL beaker. Add about 80 mL of distilled water to the beaker, stir, and dissolve. Make up to 100 mL with distilled water in a 100 mL graduated cylinder. If heat is generated, let it cool to room temperature before filling up. Wear safety gloves and glasses when preparing.
3. 0.6 mol/L borate buffer (pH 8.0): Weigh 7.4196 g of boric acid (MW 61.83) and transfer to a 200 mL beaker. Add about 150 mL of distilled water to the beaker, stir and completely dissolve boric acid by dropping 1 mol/L NaOH or 30% NaOH. Once completely dissolved, adjust pH to 8.0 with 1 mol/L NaOH. Make up to 200 mL with distilled water.
4. *N*-acetylgalactosamine standard solution series (GalNAc Std.): 10, 5, 2.5, 1.25, 0.625, 0.313, and 0.156 μg/mL *N*-acetylgalactosamine, weigh ca. 10 mg of *N*-acetylgalactosamine and transfer to 15 mL polypropylene tube, vortex it, and dissolve (1 mg/mL GalNAc). Dilute 100 μL of 1 mg/mL GalNAc with 900 μL of distilled water to 100 μg/mL GalNAc. Moreover, dilute 100 μL of 100 μg/mL GalNAc with 900 μL of distilled water to 10 μg/mL GalNAc. Then, repeat the twofold dilution to prepare a dilution series up to 0.156 μg/mL.
5. Alkaline CNA solution [*Prepare just before use!*]: 33.3 mmol/L CNA, 0.125 mol/L NaOH: Add 1 mL of the 0.6 mol/L borate buffer to the microtube and vortex it.
6. Boiling water bath.
7. Fluorometer.

3 Methods

Carry out all procedures at room temperature unless specified otherwise.

3.1 Extraction of Mucins from Rodent Feces

After killing fecal bacteria and deactivating bacterial enzymes by heating, the mucin in feces is solubilized in water and the mucin fraction in feces is extracted. The mucin fraction is then purified by ethanol precipitation to remove pigments derived from bile and low molecular weight compounds.

1. Collect rodent feces daily for 3 days, weigh collected feces, and store at −30 °C.
2. Lyophilize stored feces, weigh, and mill.
3. Store in a desiccator to avoid humidity after milling of dried feces.
4. Weigh 200 mg of powdered feces and transfer it to a 50 mL glass tube with a screw cap.
5. Add 30 volume of PBS to the 50 mL glass tube. If 200 mg of powdered feces sample is used, add 6 mL of PBS (*see* **Note 1**).
6. After tightening the screw cap loosely, boil for 10 min (*see* **Note 2**). Stir gently once after 5 min of heating.
7. After boiling, tightly close the screw cap, transfer the 50 mL glass tube to a shaking water bath at 37 °C, and shake at 125 rpm for 1.5 h.
8. Homogenize the slurry with a Polytron® homogenizer: 11,000 rpm, 20 s × 3, 4 °C.
9. Transfer the homogenate to two centrifugation tubes. Don't wash the homogenate into a centrifuge tube with PBS or distilled water.
10. Centrifuge the homogenate at 16,580 × *g* for 30 min at 4 °C.
11. Transfer the supernatant to a test tube. Dispense the supernatant equivalent to half the volume of PBS used into a pre-weighed 15 mL polypropylene tube (*A* mg).
12. Add three volumes of ethanol to the supernatant. If 3 mL of the supernatant is used, add 9 mL of ethanol.
13. After mixing by inversion, stand at −30 °C overnight to promote the precipitation of mucin (*see* **Note 3**).
14. Centrifuge at 2330 × *g* for 10 min at 4 °C. After removing the supernatant by decanting, rinse the precipitate with 0.15 mol/L NaCl (*see* **Note 4**).
15. Add 3 mL of 0.15 mol/L NaCl (*see* **Note 4**) and vortex vigorously to completely dissolve the precipitate.

16. Add 9 mL of ethanol to the sample solution. After mixing by inversion, stand at −30 °C overnight to promote the precipitation of mucin.
17. Centrifuge at 2330 × *g* for 10 min at 4 °C. After removing the supernatant by decanting, rinse the precipitate with distilled water.
18. Add 5 mL of distilled water and completely dissolve the precipitate.
19. Freeze the sample solution at −20 °C to −30 °C.
20. After freezing, lyophilize the sample and then measure the total weight (*B* mg).
21. Add 1 mL of distilled water to the resulting dried sample and dissolve completely. Make up to 3 mL with water.
22. Divide the solution including fecal mucin into small portion and store at −30 °C.

3.2 Determination of Fecal Mucin

The glycans in mucin bind *O*-glycosidically to the hydroxy group of the serine and threonine side chains of the core protein, while the glycans in peptide glycans bind *N*-glycosidically to the amino groups of the aspartic acid side chains of the core protein. Unlike *N*-linked glycoproteins, *O*-linked glycoproteins undergo β-elimination under alkaline conditions, so that oligosaccharides with reducing ends are liberated from mucin by alkaline treatment. The liberated oligosaccharides are determined (Fig. 1).

3.2.1 Dilution of the Sample Prepared from Feces

Dilute 100 μL of the sample prepared from feces with 900 μL of distilled water to 1:9 diluted sample. If the equivalents of the *O*-linked oligosaccharide chain in the sample prepared from digesta in the stomach, the small intestine, and the cecum are measured, prepare 1:99, 1:49, and 1:9 diluted samples, respectively.

3.2.2 Determination of O-Linked Oligosaccharide Chain Equivalent

This method takes advantage of the difference in resistance to degradation of *N*-glycosidic and *O*-glycosidic linkages to dilute alkaline conditions. Although *N*-glycosidic linkages are stable under dilute alkaline conditions, *O*-glycosidic linkages undergo β-elimination under the same conditions, resulting in the reducing end of the mucin glycan (Fig. 1). The aldehyde group at the reducing end dehydrates and condenses with the active methylene group of 2-cyanoacetamide to form fluorescent compounds such as pyrrolidone and pyridone derivatives [4, 5]. The *O*-linked oligosaccharide chain equivalent can be determined by measuring the intensity of fluorescence (Fig. 1).

1. Dispense 100 μL of the diluted sample, standard solution, or distilled water (blank) to a 1.5 mL microtube.

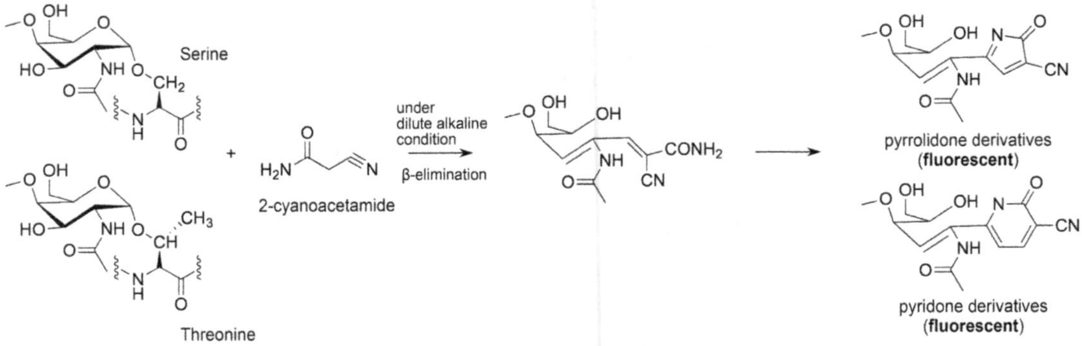

Fig. 1 β-Elimination of *O*-linked glycans from mucins by alkaline condition and production of fluorescent compounds by reducing sugar chains and 2-cyanoacetoamide.

2. Add 120 μL of the alkaline CAN solution to the 1.5 mL microtube. If multiple samples and standards are measured, finish dispensing the solution within 5 min. The solution must be heated to 100 °C within 10 min after the alkaline CAN solution are added since CAN is degraded under alkaline condition.
3. Vortex the microtube. If there is liquid on the inside of the lid, shake it down.
4. Lock the lid with a microtube cap lock to avoid the tube lid from opening during heating at 100 °C.
5. Heat the locked tube in boiling water (100 °C) for 30 min.
6. After heating, allow to cool at room temperature for 30 s to 1 min until the lid can be touched.
7. Add 1 mL of the 0.6 mol/L borate buffer to the microtube and vortex it.
8. Place the microtube in the dark because the fluorescent compounds produced are unstable until measurement.
9. After inverting the microtube and mixing the solution, transfer the solution to a quarts cuvette for the fluorometer and measure at excitation 336 nm/emission 383 nm. Record the fluorescent intensity 10 s after the cuvette is set and the fluorometer door is closed.

3.3 Calculation of Fecal Mucin Concentration Expressed as the Equivalent of O-Linked Oligosaccharide Chain

Subtract the fluorescence intensity of the blank from that of the standard solution and sample and calculate the true intensity of the standard solution and sample. As shown in Fig. 2, create a calibration curve with the GalNAc concentration on the horizontal axis and the true fluorescence intensity of the standard solution on the vertical axis. The fluorescence intensity detected by a fluorometer is expressed as the product of the fluorescence intensity of the sample itself and the wavelength specific to the fluorometer's light source, so the fluorescence intensity detected will be different for the same

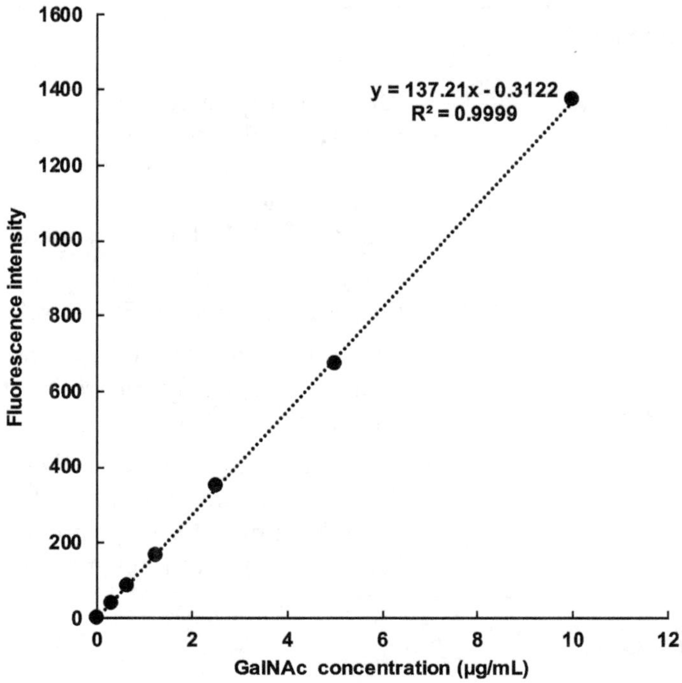

Fig. 2 Typical calibration curve between *N*-acetylgalactosamine concentration and fluorescence intensity

concentration of a standard solution if the fluorometer used is different. The equivalent of an *O*-linked oligosaccharide chain:

First, the equivalent (Eqv[diluted]; μg/mL of 1:9 diluted sample) of *O*-linked oligosaccharide chain in the 1:9 diluted samples is calculated from the true fluorescence intensity of the 1:9 diluted sample using the linear regression equation obtained from the calibration curve.

Next, the equivalent of the *O*-linked oligosaccharide chain (μmol/g dry feces) is calculated.

The equivalent of the *O*-linked oligosaccharide chain (μmol/g dry feces)

$$= \text{Eqv[diluted]}\ (\mu\text{g/mL}) \times 10 \times 3\ (\text{mL}) \times \frac{6\ (\text{mL})}{3\ (\text{mL})}$$

$$\times \frac{1}{221.1\ (\text{g GalNAc/mol})} \times \frac{1}{\text{Weight of fecal sample (g)}}$$

The equivalent of the *O*-linked oligosaccharide chain (μmol/mg mucin)

$$= \text{Eqv[diluted]}\ (\mu\text{g/mL}) \times 10 \times 3\ (\text{mL})$$

$$\times \frac{1}{221.1\ (\text{g GalNAc/mol})} \times \frac{1}{B - A\ (\text{mg})}$$

$B - A$, weight of mucin crudely purified from feces.

4 Notes

1. Although it is stated to add 30 volumes of PBS for the weight of fecal sample, it is more workable to add 6 mL of PBS for all fecal samples less than 200 mg and about 200 mg of fecal sample. There is no problem with quantitation even if the volume of PBS added is standardized to 6 mL.
2. Do not tighten the screw cap tightly, as heating may increase the pressure inside the 50 mL tube and cause the tube to burst.
3. To ensure precipitation of mucin, −30 °C is strongly recommended. Mucin may not precipitate sufficiently at temperatures higher than −30 °C.
4. Be sure to rinse with 0.15 mol/L NaCl (**step 14** in Subheading 3.1) and remember to add 3 mL of 0.15 mol/L NaCl (**step 15** in Subheading 3.1). These processes should never be forgotten because salting out facilitates the precipitation of mucin.

References

1. Hansson GC (2020) Mucins and the Microbiome. Annu Rev Biochem 89:769–793. https://doi.org/10.1146/annurev-biochem-011520-105053
2. Hino S, Mizushima T, Kaneko K, Kawai E, Kondo T, Genda T, Yamada T, Hase K, Nishimura N, Morita T (2020) Mucin-Derived O-Glycans Act as Endogenous Fiber and Sustain Mucosal Immune Homeostasis via Short-Chain Fatty Acid Production in Rat Cecum. J Nutr 150(10):2656–2665. https://doi.org/10.1093/jn/nxaa097
3. Tanabe H, Sugiyama K, Matsuda T, Kiriyama S, Morita T (2005) Small intestinal mucins are secreted in proportion to the settling volume in water of dietary indigestible components in rats. J Nutr 135(10):2431–2437. https://doi.org/10.1093/jn/135.10.2431
4. Crowther RS, Wetmore RF (1987) Fluorometric assay of O-linked glycoproteins by reaction with 2-cyanoacetamide. Anal Biochem 163(1): 170–174. https://doi.org/10.1016/0003-2697(87)90108-4
5. Honda S, Matsuda Y, Takahashi M, Kakehi K, Ganno S (1980) Fluorimetric determination of reducing carbohydrates with 2-cyanoacetamide and application to automated analysis of carbohydrates as borate complexes. Anal Chem 52(7): 1079–1082

Chapter 3

Preparation of Soluble Mucin Solutions from the Salivary Glands

Takanori Sugiura and Akihiko Kameyama

Abstract

Studying salivary gland mucins is important for elucidating the pathogenesis of salivary gland diseases, including tumors and xerostomia, and developing diagnostic methods for them. Classic methods for isolating mucins from salivary glands require sacrificing several animals to obtain sufficient quantities of mucin and are time-consuming. Supported molecular matrix electrophoresis (SMME) was used to characterize mucins and their glycans. With this method, mucins can be analyzed within 2 days using less than 100 mg of tissue and without using expensive equipment, such as an ultracentrifuge. This chapter describes a method for preparing mucin solutions for SMME analysis of salivary gland mucins.

Key words Mucin, *O*-linked glycan, Submaxillary gland, Sublingual gland, Submandibular gland, Ethanol precipitation, Supported molecular matrix electrophoresis

1 Introduction

Mucins, which are one of the major components of saliva, are produced by the salivary glands and function as lubricants in the oral cavity as well as adhere to microbial adhesion molecules, thereby contributing to the maintenance of oral hygiene. In elderly individuals with xerostomia, salivary viscosity increases and oral hygiene deteriorates. With aging, alterations in the physicochemical properties of mucin may be responsible for changes in their function. Mucoepidermoid carcinoma (MEC) is the most prevalent malignancy glands. MEC is distinguished by its ability to produce mucin and abnormally express Mucin 1 (MUC1) [1], a highly glycosylated, high molecular weight glycoprotein. Alterations from glycosylation are often observed in tumors. This suggests that the expression of glycans on mucins and mucins themselves may be altered by oral diseases, especially pathological conditions of the submandibular and sublingual glands. Thus, studying salivary gland mucins is crucial for elucidating the pathogenesis of salivary

Akihiko Kameyama (ed.), *Mucins: Methods and Protocols*, Methods in Molecular Biology, vol. 2763,
https://doi.org/10.1007/978-1-0716-3670-1_3,

gland malignancies and xerostomia as well as developing diagnostic methods for such malignancies.

Conventional methods for isolating mucins from salivary glands include cetylammonium bromide precipitation of the water-soluble fraction followed by dialysis [2], ultracentrifugation [3], gel filtration, and polyacrylamide gel electrophoresis [4]. These methods involve sacrificing many animals to obtain sufficient quantities of mucin and are considerably time-consuming. We investigated a simple and rapid method known as supported molecular matrix electrophoresis (SMME) for analyzing mucins and published it in 2009 [5]. This method can analyze mucins within 2 days using less than 100 mg of tissue and without using special equipment. This chapter describes a method for preparing soluble mucin fractions for the analysis of salivary gland mucins via SMME.

The general flow of the procedure is depicted in Fig. 1a [6]. In brief, salivary glands are homogenized in acetone to produce acetone powder and then phosphate-buffered saline (PBS) is used to extract soluble proteins. A protease inhibitor mixture is introduced at this time to prevent protein degradation. The solution of soluble protein is treated with calcium acetate and then ethanol is precipitated. Calcium ions have a high affinity for mucins, which contain numerous glycans containing sialic acids, and are an effective auxiliary agent for ethanol precipitation. Proteins including mucins are extracted from the precipitate with 2 M urea. The extract is again precipitated with ethanol, and the precipitate is dissolved in a small amount of 8 M urea to complete the sample.

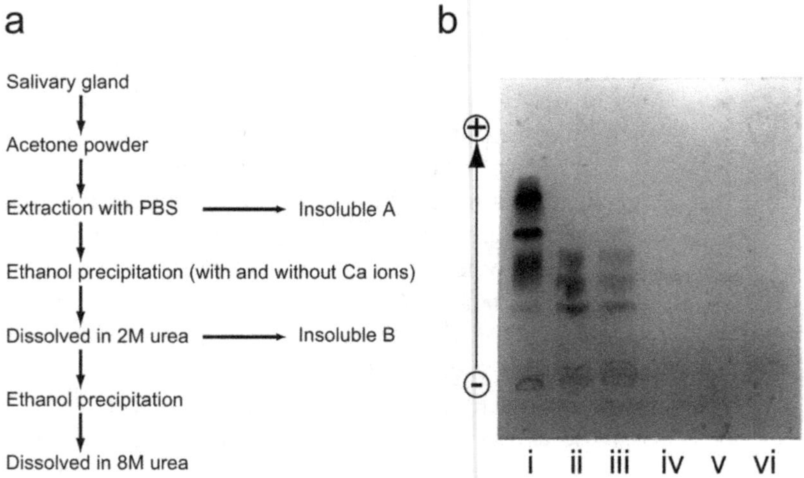

Fig. 1 Overview of the mucin enrichment procedure from salivary glands and supported molecular matrix electrophoresis (SMME) of the fractions obtained via the procedure. (**a**) Overview of the procedure. (**b**) SMME of the obtained fractions and reference mucin: porcine gastric mucin (i), 8 M urea solution with (ii) and without (iii) Ca ion, insoluble B with (iv) and without (v) Ca ion, and insoluble A (vi). The SMME membrane was stained with Alcian blue (pH 4.0). Three bands in lane (ii) are darker than those in lane (iii), and no bands are observed in the other lanes. The porcine submaxillary gland was used as an example of salivary glands in this experiment. (Reproduced from [6] with permission. Copyright 2018 Elsevier B.V.)

Using this method, we have demonstrated that salivary gland mucin and its *O*-glycans change with aging in mice [7, 8] and that Bmi-1, an oncogene, suppresses submandibular mucin levels and regulates its *O*-glycosylation [9]. We have also demonstrated that mucoepidermoid carcinoma, a human salivary gland tumor, expresses core 2 type sialoglycans [10, 11].

2 Materials

Prepare all solutions using ultrapure water and analytical grade reagents. Store all reagents at room temperature (unless indicated otherwise).

1. Salivary gland (*see* **Note 1**).
2. Disposable micro tissue homogenizer (*see* **Note 2**).
3. PBS (pH 7.4) containing protease inhibitor mix: Add 10 μL of a commercially available protease inhibitor mix into 1 mL of PBS (pH 7.4) and mix well (*see* **Note 3**).
4. 2 M urea in PBS (pH 7.4): Weigh 120 mg of urea and transfer it to a 1.5 mL microcentrifuge tube. Add 910 μL of PBS (pH 7.4) and dissolve the urea in the tube.
5. 8 M urea in PBS (pH 7.4): Weigh 480 mg of urea and transfer it to a 1.5 mL microcentrifuge tube. Add 620 μL of PBS (pH 7.4) and dissolve the urea in the tube.
6. Saturated calcium acetate solution: Weigh 350 mg of calcium acetate monohydrate and transfer it to a 1.5 mL microcentrifuge tube. Add 1 mL of water and dissolve the calcium acetate in the tube. Transfer the supernatant into a new tube.
7. Acetone.
8. Ethanol.
9. Refrigerated microcentrifuge.
10. Centrifugal evaporator.
11. Vortex mixer.

3 Methods

Unless otherwise specified, perform all procedures at room temperature. Care should be taken not to disturb the pellet when discarding or transferring the supernatant. Preparation of mucin using this method has been successful in mouse submandibular glands, porcine submandibular glands, and human submandibular and sublingual glands. However, homogenization may be difficult in aged tissues with fibrosis and hardening.

1. Weigh an empty tube of the disposable microtissue homogenizer kit and transfer the tissue specimen (approximately 50 mg) into the tube. Then, weigh the tube containing the tissue again and calculate the wet-weight of the tissue specimen (*see* **Notes 1** and **4**).
2. Add 500 μL of acetone to the tube containing the tissue specimen and homogenize using a homogenization pestle.
3. Centrifuge the solution at 16,000 × *g* for 10 min at 4 °C and discard the supernatant.
4. Add 500 μL of acetone to the tube and homogenize with the pestle again. Then, centrifuge the solution at 16,000 × *g* for 10 min at 4 °C and discard the supernatant. Repeat the addition of acetone, homogenization, centrifugation, and discarding of the supernatant at least twice until the tissue becomes a fine homogenate.
5. Evaporate the acetone in the homogenate in the tube using a centrifugal evaporator to obtain powder.
6. Add 500 μL of PBS containing the protease inhibitor mix to the tube and homogenize at 4 °C again.
7. Centrifuge the solution at 16,000 × *g* for 10 min at 4 °C and transfer the supernatant equally to two new 1.5 mL microcentrifuge tubes (*see* **Note 5**).

   Perform the following operations on each tube.
8. Add a quarter of the supernatant volume of saturated calcium acetate solution to the tube and mix well using a vortex mixer (*see* **Note 6**).
9. Add three times the volume of ethanol to the supernatant containing calcium in the tube and mix well using a vortex mixer. Then, incubate the solution at −80 °C for at least 1 h (*see* **Note 7**).
10. Centrifuge the solution at 16,000 × *g* for 10 min at 4 °C and discard the supernatant.
11. Add 100 μL of 2 M urea in PBS to the precipitate in the tube and mix well using a vortex mixer.
12. Centrifuge the solution at 16,000 × *g* for 10 min at 4 °C.

    The next step is to combine the contents of the two separate tubes into a single tube.
13. Transfer the supernatants of the two tubes into a new 1.5 mL microcentrifuge tube.
14. Add three times the volume of ethanol to the supernatant in the tube and mix well using a vortex mixer. Then, incubate the tube at −80 °C for at least 1 h (*see* **Note 8**).

15. Centrifuge the solution at 16,000 × *g* for 10 min at 4 °C and discard the supernatant.
16. Add the same volume of 8 M urea in PBS as the initial tissue piece used and dissolve the pellet in the tube (*see* **Note 9**).
17. Store the solution at 4 °C until use (*see* **Note 10**).

4 Notes

1. Use fresh or frozen samples that have been thawed. Formalin-fixed tissue cannot be used.
2. We use the BioMasher II (Nippi Incorporated, Japan).
3. We use Protease Inhibitor Mix (Cytiva) as a protease inhibitor. If using a protease inhibitor from another provider, follow their instructions.
4. Mouse submandibular glands can be used as whole tissues, but pig and human salivary glands are larger and should be cut into appropriate sizes (50–100 mg).
5. Since the volume will increase in the subsequent steps, divide it into two parts at this stage.
6. For example, if the supernatant volume is 200 μL, add 50 μL of calcium acetate solution.
7. For example, if the supernatant containing calcium is 250 μL, add 750 μL of ethanol. Adding ethanol will make the solution cloudy.
8. For example, if the supernatant is 200 μL, add 600 μL of ethanol.
9. For example, if the wet weight of the tissue used is 50 mg, add 50 μL of 8 M urea in PBS.
10. Can be stored at 4 °C for 1 year.

References

1. Robinson L, van Heerden MB, Ker-Fox JG et al (2021) Expression of mucins in salivary gland mucoepidermoid carcinoma. Head Neck Pathol 15:491–502
2. Tettamanti G, Pigman W (1968) Purification and characterization of bovine and ovine submaxillary mucins. Arch Biochem Biophys 124:41–50
3. Oemrawsingh I, Roukema PA (1974) Isolation, purification and chemical characterization of mucins from human submandibular glands. Arch Oral Biol 19:615–626
4. Denny PA, Denny PC, Jenkins K (1980) Purification and biochemical characterization of a mouse submandibular sialomucin. Carbohydr Res 87:265–274
5. Matsuno Y-K, Saito T, Gotoh M et al (2009) Supported molecular matrix electrophoresis: a new tool for characterization of glycoproteins. Anal Chem 81:3816–3823
6. Kameyama A, Yamakoshi K, Watanabe A (2019) A rapid separation and characterization of mucins from mouse submandibular glands by supported molecular matrix electrophoresis. Biochim Biophys Acta Proteins Proteom 1867:76–81
7. Iida M, Matsuno YK, Watanabe A et al (2019) A sialo-oligosaccharide-rich mucin-like

molecule specifically detected in the submandibular glands of aged mice. Arch Oral Biol 97: 52–58

8. Kameyama A, Thet Tin WW, Nishijima R et al (2021) Alteration of mucins in the submandibular gland during aging in mice. Arch Oral Biol 121:104967
9. Kameyama A, Nishijima R, Yamakoshi K (2021) Bmi-1 regulates mucin levels and mucin O-glycosylation in the submandibular gland of mice. PLoS One 16:e0245607
10. Isaka E, Sugiura T, Hashimoto K et al (2021) Characterization of tumor-associated MUC1 and its glycans expressed in mucoepidermoid carcinoma. Oncol Lett 22:702
11. Sugiura T, Hashimoto K, Kikuta K et al (2023) Expression and localisation of MUC1 modified with sialylated core-2 O-glycans in mucoepidermoid carcinoma. Sci Rep 13:5752

Chapter 4

Isolation of Membrane Bound Mucins from Human Bronchial Epithelial Cells

Jerome Carpenter and Mehmet Kesimer

Abstract

Membrane-bound mucins constitute a large portion of the periciliary layer of lung epithelial surfaces, and thus play an important role in many aspects of innate defense. The biophysical and biochemical properties of the membrane-bound mucins have important implications for mucociliary clearance, viral penetration, and potential therapeutics delivered to the airway surface. Hence, isolating them and determining these properties is important in understanding airways disease and ultimately in developing treatments. Here, we describe a method using isopycnic centrifugation to enrich and isolate shed membrane-bound mucins from the washings of human bronchial epithelial cell cultures.

Key words Membrane-bound mucins, Human bronchial epithelial cells, MUC1, MUC4, MUC16

1 Introduction

The airway epithelial surface can be described as a gel-on-brush model: the overlying mucus gel consisting of the major gel-forming mucins, MUC5B and MUC5AC, and their interacting proteins; and the "brush," the periciliary layer (PCL), consisting of the membrane-bound mucins (MBMs) and other glycoproteins on the epithelial surface [1]. Though the overlying mucus layer deservedly draws a lot of attention, the role of the PCL and its constituents cannot be overlooked. The PCL functions as a barrier; specifically, the membrane-bound mucins have been shown to exclude some pathogens from entering the PCL and entrapping others [2, 3]. Likewise, inhaled intracellular therapeutics must successfully penetrate the PCL barrier to reach their targets. Therefore, isolating and characterizing MBMs and determining their interactions with pathogens, as well as therapeutic vehicles and vectors, is an essential area of study. MBMs can be acquired from the washings of well-differentiated human bronchial epithelial (HBE) cell cultures. HBE cell cultures maintained under air–liquid

Akihiko Kameyama (ed.), *Mucins: Methods and Protocols*, Methods in Molecular Biology, vol. 2763,
https://doi.org/10.1007/978-1-0716-3670-1_4,

interface conditions shed MBMs apically along with cell secretions [4]. These MBMs can then be isolated from HBE cell culture washings [5–7]. Here, we describe a protocol for isolating membrane-bound mucins MUC1, MUC4, and MUC16 from human bronchial epithelial cell washings. We collect and pool the HBE washings and then use a "dissociative" isopycnic gradient to separate the MBMs from the gel-forming mucins and other non-interacting proteins. We then take advantage of the keratan sulfate glycosylation found on MBMs to identify the MBM-rich fractions and pool them together [8], so that they are ready for further purification and analysis.

2 Materials

2.1 Collecting and Storing Washings

1. Well-differentiated human bronchial epithelial (HBE) cultures.
2. Sterile phosphate-buffered saline (PBS).
3. Sterile 50 mL centrifuge tubes (with a screw cap).
4. 70% Ethanol.
5. Water bath.
6. Biosafety cabinet.

2.2 Cesium Chloride Isopycnic Density Gradient Centrifugation

1. Cesium chloride.
2. Guanidine hydrochloride (8 M) buffer: Prepare an 8 M guanidine HCl buffer (pH 8.0). 8 M guanidine hydrochloride, 0.1 M Tris–HCl, 5 mM EDTA. Prepare using ultrapure water and adjust pH to 8 using hydrochloric acid. Filter the final buffer twice through a 0.2 um cellulose nitrate membrane filter.
3. Guanidine hydrochloride (4 M) buffer. The same as the buffer prepared above but diluted 1:1 with de-ionized water.
4. Ultracentrifuge tubes (40 mL) (Polypropylene Quick-Seal centrifuge Tubes #342414, Beckman Coulter, Ca, USA) or an equivalent.
5. Ultracentrifuge and a fixed angle rotor capable of spinning 40 mL tubes at 145,500 $\times g_{avg}$.
6. Large plastic syringe (30 mL) and 18-gauge needles.

2.3 Unloading the Gradient and Running a Slot/Dot Blot

1. 5 mL syringes and 18-gauge needles.
2. Slot/dot blot apparatus.
3. 0.45 um Nitrocellulose blotting membrane cut to the dimensions of the slot/dot blot apparatus.
4. Grade 3MM chromatography filter paper cut to the dimensions of the slot/dot blot apparatus.

5. Plastic trays that accommodate the dimensions of the cut blotting membrane.
6. Tris-buffered saline and TWEEN (TBST) solution: Prepare a TBST buffer (pH 8.0). 10 mM Tris base, 150 mM NaCl, and 0.05% TWEEN. Prepare using ultrapure water and adjust the pH to 8 using hydrochloric acid.
7. Blocking solution: 1 g of non-fat dry milk mixed into 50 mL of TBST.
8. MUC5B antibody solution (HPA08246, MilliporeSigma): Prepare in TBST diluted 1:500.
9. Keratan sulfate antibody (MAB2022, MilliporeSigma): Prepare in TBST diluted 1:500.
10. Secondary antibodies for detection (matched to scanner).
11. Scanner/Biomolecular imager.

3 Methods

Carry out all procedures at room temperature unless specified otherwise.

3.1 Collecting HBE Cell Culture Washings

All of the techniques in this subheading will use sterile technique. Before placing items into the biosafety cabinet, they should be sprayed with 70% ethanol. All of the buffers, pipettes, and pipette tips should be sterile.

1. Determine the total amount of PBS needed to collect the culture washings. If using 12 mm diameter inserts, each insert will require 250 μL of PBS. If using 30 mm diameter inserts, each insert will require 1 mL of PBS. Using sterile technique, transfer the PBS into an appropriately sized centrifuge tube with a screw top lid.
2. Using a water bath, warm the sterile PBS to 37 °C.
3. Remove the PBS from the water bath and place the tube in a holder inside of a biosafety cabinet.
4. Remove the cells from the incubator and place the culture plates into the biosafety cabinet.
5. For each plate, open the plate and pipette PBS onto the apical side of the insert. For 12 mm inserts dispense 250 μL, for 30 mm inserts, dispense 1 mL. Close the plate.
6. Return the culture plates to the incubator and let them sit for 30 min.
7. Place the cultures plates back into the biosafety cabinet.

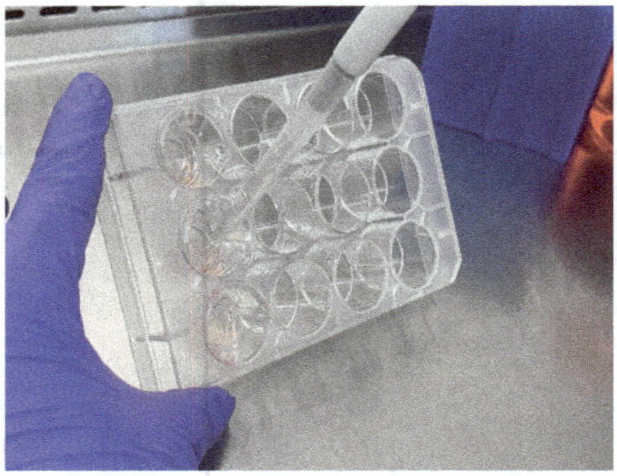

Fig. 1 Removing the apical washings from the culture. Tilt the culture plate toward you so that you can retrieve all of the apical fluid without damaging the cultures

8. Place a sterile 50 mL centrifuge tube into the bio safety cabinet into a test tube stand and remove the lid.
9. For each plate, open the plate and tilt it slightly toward you so that the apical liquid collects between the insert surface and insert wall (as shown in Fig. 1). Collect the apical liquid by gently pipetting the liquid up, and then pipetting it into the centrifuge tube.
10. Repeat this for all the plates. Once finished, return the plates to the incubator, cap the collection centrifuge tube, and store the tube at 4 °C (*see* **Note 1**).
11. Continue the collection once weekly, until there is enough material to run an isopycnic gradient (ideally at least 20 mL of material) (*see* **Note 2**).

3.2 Cesium Chloride Isopycnic Density Gradient Centrifugation

1. Using low speed centrifugation, spin the cell culture washings at 500 × *g* for 10 min.
2. Take the supernatant of the cell culture washings and mix 1:1 with the 8 M guanidine hydrochloride buffer. Discard the pellet. Mix using a large beaker with a stir bar (*see* **Notes 3** and **4**).
3. Determine how many centrifuge tubes will be needed: consider the current volume of material and that the added cesium chloride will account for ~13% of the final volume. The *final volume* should be the amount needed to fill this number of tubes +5% extra material for any mistakes.
4. Determine the amount of cesium chloride needed to bring the density of the solution to 1.45 g/mL *(at the final volume)* using the following formula.

$$\text{Mass of cesium chloride (g)} = \text{Final volume (mL)} * 0.47895$$

5. Measure out the cesium chloride and slowly add it to supernatant/guanidine mixture stirring in the large beaker (*see* **Note 5**).
6. Add 4 M guanidine hydrochloride buffer to reach the final intended volume.
7. Let the solution continue to mix until the cesium chloride is completely dissolved.
8. Let the solution mix for an additional 15 min.
9. Check the density of the solution by stopping the stir bar, pipetting 1 mL of material, and weighing the material in a weigh boat on a balance. The target mass for 1 mL is 1.45 g. If lower than this, adjust by adding cesium chloride to the solution, letting it stir for 5 min and rechecking the density. If higher, adjust by adding 4 M guanidine hydrochloride buffer, letting the solution stir for 5 min, then checking the density (*see* **Note 6**).
10. Once satisfied with the density, fill the ultracentrifuge tubes with the solution. Using a 30 mL or larger plastic syringe and a blunt tip syringe needle, pull the solution into the syringe. Slowly inject the solution into the top of the centrifuge tube. Fill the tubes stopping just short of the neck.
11. Check the weight of the tubes and ensure that they can be balanced. If the number of tubes can't be balanced, a blank tube can be quickly made by adding cesium chloride to 4 M guanidine hydrochloride buffer. As there is no protein, this can be rapidly mixed by placing into a 50 mL tube, capping the tube, and then mixing using a vortex mixer. Once mixed and the density has been checked, a blank ultracentrifuge tube can be prepared as in **step 10**.
12. Seal the tubes and load them into the rotor in a balanced configuration. Load the rotor into the centrifuge and run the centrifuge with the following parameters: 145,500 × *g*, 14 °C, and set the time to hold. (Final time can be anywhere from 60–96 h.)

3.3 Unloading the Gradient and Running a Slot/Dot Blot

1. Stop the centrifuge. Carefully remove the rotor from the centrifuge.
2. Open the rotor, and carefully remove the tubes from the rotor and place them onto a stable rack/stand.
3. For 40 mL centrifuge tubes, each tube will be fractionated from the top in 2 mL increments. Set aside 20 tubes for fractions that will be able to accommodate the volume of each fraction across all the loaded centrifuge tubes (2 mL * number of tubes loaded) and label the tubes 1–20.

4. Fractionating a tube: Start by gently using a sharp tip 18-gauge syringe needle to puncture a hole in the top of the centrifuge tube. Make sure that the puncture hole is complete, but there is no need to full insert the needle into the tube.
5. Make a second hole at the top of the centrifuge tube.
6. Attach a 5 mL syringe to the needle and remove the top 2 mL of fluid from the tube. Place this into the correspondingly numbered tube. Repeat this for the next five fractions (six fractions in all).
7. Using a razor blade, carefully remove the top of the centrifuge tube (approximately 5 mm above the liquid level after removing the top six fractions).
8. Use a pipette to remove fractions 7–19 in 2 mL increments.
9. Prior to collecting fraction 20, pipette vigorously to remove any material that may have pelleted to the bottom of the tube. Collect fraction 20.
10. Repeat **steps 4–9** for any remaining tubes.
11. Perform slot/dot blot antibody analysis on the 20 fractions. For each tube use 20 μL of material for the blot. After blocking with milk, probe the blots with MUC5B (#HPA008246, MilliporeSigma) and keratan sulfate (#MAB2022, MilliporeSigma) antibodies to determine their presence, distributions, and abundance across the fractions. In our lab, we image using an Odyssey scanner and the corresponding secondary antibodies (IRDYE 680LT and IRDYE 800LT) (Licor Biosciences, NE, USA) but any equivalent scanner/secondary antibody combination can be used (*see* **Notes 7** and **8**).
12. For each antibody, plot the signal vs. the fraction to generate an antibody concentration profile (as shown in Fig. 2).
13. Identify the keratan sulfate peak.
14. Pool the fractions at the keratan sulfate-rich peak together as enriched membrane-bound mucin pool. These fractions are dominated by the MBMs as compared to the MUC5B dominant fractions, as assessed by proteomic analysis in Fig. 3. Atomic force microscopy imaging of these fractions shows the presence of membrane-bound mucins (Fig. 4).

4 Notes

1. It is fine to collect samples over the course of several weeks, extending out toward a few months. We haven't seen evidence of degradation at that time length, provided that the washings are stored at 4 °C and the cultures are free of contamination.

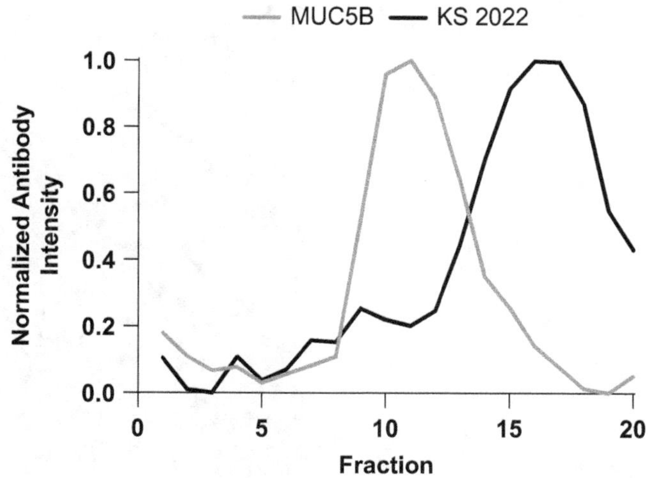

Fig. 2 Antibody intensity plot of the isopycnic fractions. In the plot above, the antibody signal for keratan sulfate (black) and MUC5B (gray) are plotted for each fraction. The values have been normalized to the peak value for each antibody, thus the peak has a value of 1. MUC5B has a peak in fraction 11, while KS peaks in fraction 17

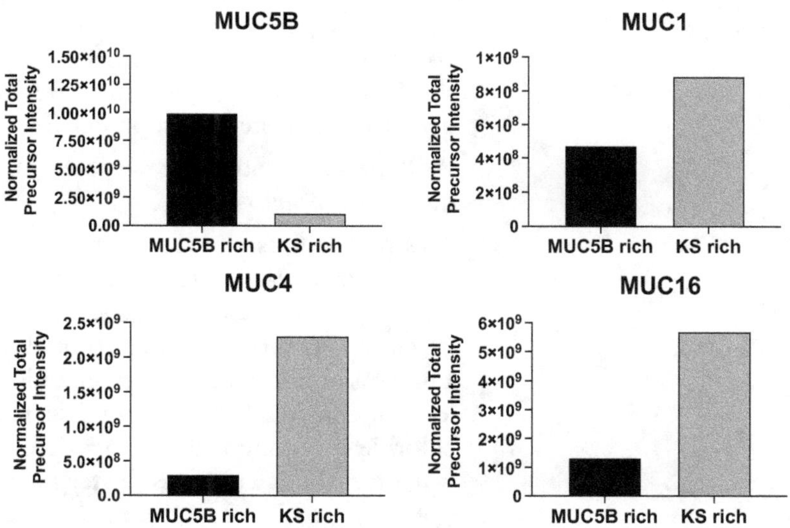

Fig. 3 Total precursor intensity comparison of MUC5B and the membrane-bound mucins in the MUC5B rich fraction vs. the keratan sulfate rich fraction. MUC5B is 10× more abundant in the MUC5B rich fraction as compared to the KS rich fraction. MUC1, MUC4, and MUC16 are more abundant in the KS rich fraction by factors of ~2×, 4×, and 4×, respectively

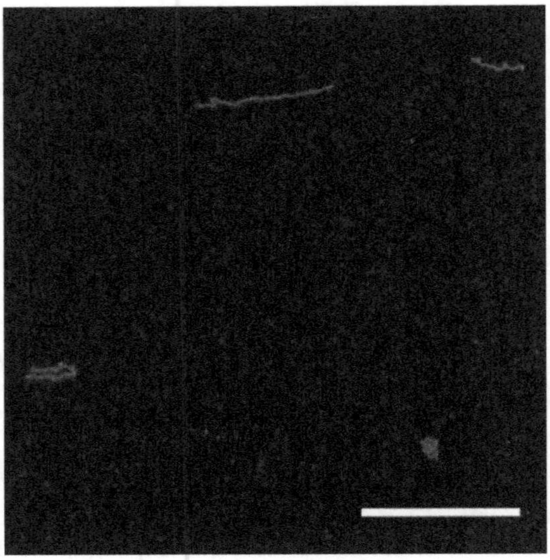

Fig. 4 Atomic force microscope image produced from a dilute preparation of the keratan sulfate rich peak showing a collection of membrane-bound mucins (scale bar = 1 um)

2. This protocol is designed around using 40 mL centrifuge tubes because we get better separation at this size than with smaller volumes. This procedure can be done for smaller (or larger) sized tubes, but the spin time, rate, and fraction sizes will all need to be adjusted appropriately.
3. We typically avoid using a vortex mixer when trying to mix mucin solutions.
4. This protocol uses a dissociative isopycnic gradient (by using the chaotropic agent guanidine chloride) to separate out interacting proteins and improve the separation between the mucins. This protocol can also be done associatively by using PBS as a buffer instead of guanidinium, but this will result in less enrichment. If making an associative buffer, more cesium chloride is required to get to a density of 1.45 g/mL, which will affect the calculations in **steps 3** and **4** in Subheading 3.2.
5. When mixing cesium chloride, it may help to use a wider beaker so that you can minimize having a density gradient if your solution isn't mixed well.
6. The density doesn't have to be exactly 1.45 g/mL, but it is a good goal. If the density is higher, then both the gel-forming mucins and MBMs will be closer to the top of the gradient, and vice versa if the density is lower. The further you get away from 1.45 g/mL in either direction, however, will limit your separation between the two mucin types.

7. The volume required to get a decent antibody signal will depend on the concentration of the material, but 20 μL is a good starting point for enriched cell culture washings. You may need to adjust downward if there is too much material.
8. For this protocol, we use the keratan sulfate antibody to identify the distribution of the membrane-bound mucins within the fractions. The primary reason for doing this is that the presence of keratan sulfate can reduce the activity of some of the antibodies specifically to MBMs. This reduced activity can vary greatly depending on the specific antibody and heterogeneity between cell cultures. That said, antibodies specific for each membrane-bound mucin can also be used as a probe to determine their distributions, but may not be reliable due to the presence of keratan sulfate.

Acknowledgments

We would like to thank members of the Kesimer lab as well as the Marsico Lung Institute for the help and support during the development of this protocol. The protocols presented here were developed for work supported by the NIH (P01HL164320-01) and Cystic Fibrosis Foundation (CFFCARPEN19I0).

References

1. Button B, Cai LH, Ehre C et al (2012) A periciliary brush promotes the lung health by separating the mucus layer from airway epithelia. Science 337(6097):937–941. https://doi.org/10.1126/science.1223012
2. Kesimer M, Ehre C, Burns KA et al (2013) Molecular organization of the mucins and glycocalyx underlying mucus transport over mucosal surfaces of the airways. Mucosal Immunol 6(2):379–392. https://doi.org/10.1038/mi.2012.81
3. Iverson E, Griswold K, Song D et al (2022) Membrane-tethered mucin 1 is stimulated by interferon and virus infection in multiple cell types and inhibits influenza A virus infection in human airway epithelium. mBio 13(4):e0105522. https://doi.org/10.1128/mbio.01055-22
4. Hattrup CL, Gendler SJ (2008) Structure and function of the cell surface (tethered) mucins. Annu Rev Physiol 70:431–457. https://doi.org/10.1146/annurev.physiol.70.113006.100659
5. Davies JR, Wickstrom C, Thornton DJ (2012) Gel-forming and cell-associated mucins: preparation for structural and functional studies. Methods Mol Biol 842:27–47. https://doi.org/10.1007/978-1-61779-513-8_2
6. Davies JR, Kirkham S, Svitacheva N et al (2007) MUC16 is produced in tracheal surface epithelium and submucosal glands and is present in secretions from normal human airway and cultured bronchial epithelial cells. Int J Biochem Cell Biol 39(10):1943–1954. https://doi.org/10.1016/j.biocel.2007.05.013
7. Kesimer M, Kirkham S, Pickles RJ et al (2009) Tracheobronchial air-liquid interface cell culture: a model for innate mucosal defense of the upper airways? Am J Physiol Lung Cell Mol Physiol 296(1):L92–L100. https://doi.org/10.1152/ajplung.90388.2008
8. Carpenter J, Kesimer M (2021) Membrane-bound mucins of the airway mucosal surfaces are densely decorated with keratan sulfate: revisiting their role in the Lung's innate defense. Glycobiology 31(4):436–443. https://doi.org/10.1093/glycob/cwaa089

Chapter 5

Extraction and Fractionation of Human Gastric Mucins from Gastric Juice

Rei Kawashima, Daigo Tsubokawa, Katsunori Iijima, and Takafumi Ichikawa

Abstract

Mucin, a major component of the mucus, is considered to be one of the principal factors in the physiological defense mechanism of the gastrointestinal mucosa. Measuring the mucin content of human gastric mucus is a useful tool for the assessment of *Helicobacter pylori* (*H. pylori*) eradication or the involvement of mucus secretion in various gastroduodenal diseases. Here, we describe a methodology for the isolation of the mucin fraction from human gastric juice and the quantification of mucin.

Key words Gastric mucin, Column chromatography, Phenol-sulfuric acid method

1 Introduction

Gastric mucus, an important defensive factor in the function of the gastric mucosa [1], can be classified into three phases in vivo: soluble mucus present in the luminal solution, an insoluble mucus gel layer adherent to the mucosal surfaces, and an intracellular secreted mucus, which stored in the vesicles of mucus-producing cells. Mucin is the largest and most characteristic component of gastric mucus. It is also a high molecular weight substance of which more than 80% consists of carbohydrates. Two generally accepted methods have been used to isolate this substance for characterization; one utilizes the high molecular weight nature of mucin for chromatographic separation [2] and the other depends upon the reduction of the solubility of mucin in a high concentration of ethanol or in acidic conditions [3–5]. The latter method is straightforward and simple to use but is unsatisfactory for distinguishing mucin from lower sized materials which include a variety of non-mucinous glycoproteins originating from intracellular materials, cell membranes, and the extracellular tissues [6, 7]. These procedures have been established as a simple method to measure mucin

Akihiko Kameyama (ed.), *Mucins: Methods and Protocols*, Methods in Molecular Biology, vol. 2763,
https://doi.org/10.1007/978-1-0716-3670-1_5,

content in human gastric juice. These can be utilized to assess the *H. pylori* eradication degree and the participation of mucus secretion in various gastroduodenal diseases [8, 9]. Here, we describe the procedure for extracting and fractionating human gastric mucins from gastric juice.

2 Materials

Prepare all solutions using ultrapure water (prepared by purifying deionized water, to attain a sensitivity of 18 MΩ·cm at 25 °C) and analytical grade reagents. Diligently follow all waste disposal regulations when disposing waste materials. Do not add sodium azide to the reagents. Prepare and store all reagents at room temperature (unless indicated otherwise).

2.1 Collection of Gastric Juice by Endoscopy

1. Pentagastrin.
2. Upper gastrointestinal endoscopy.

2.2 Extraction of Human Gastric Mucins

1. Human gastric juices (*see* **Note 1**).
2. Centrifuge (Angle rotor 50 mL round bottom).
3. Dialysis tube (Flat width 20–30 mm).
4. Freeze-dry machine.
5. 50 mM Tris buffer (pH 7.2): Add ~100 mL of water to a 1-L graduated cylinder. Weigh 6.06 g of Tris base and transfer to the cylinder. Mix with the water. Add more water until the volume of the mix reaches 800 mL. Mix and adjust the pH to 7.2 with HCl. Add more water until the volume reaches 1 L, and mix the solution.
6. Polytron homogenizer (*see* Subheading 2.1 in Chapter 18).
7. 20% (v/v) Triton X-100: Add 80 mL of water to a 100 mL graduated beaker (*see* **Note 2**). Add 20 mL Triton X-100 to the beaker, and mix the solution using a stirrer.
8. Shaking incubator.
9. Centrifuge tubes (50 mL).
10. Gel filtration resin (Bio-Gel® A-1.5 m, Bio-Rad Laboratories, Richmond, CA, USA) (*see* **Note 3**).
11. Open glass columns (2.5 × 100 cm).
12. Fraction collector.
13. Elution buffer: 5 mM Na_2 EDTA in 0.5 M NaCl. Add ~2500 mL of water to a 5-L beaker. Weigh 9.30 g of Na_2 EDTA and 147.75 g of NaCl, and transfer to the beaker. Mix with the water. Add more water until the volume reaches 5 L, and mix the solution.

14. Collection tubes.
15. 5% phenol: Add 5 mL of phenol to 100 mL of water (*see* **Notes 4** and **5**).
16. Sulfuric acid (*see* **Note 6**).
17. Ultraviolet spectrophotometer.
18. Glass *cuvette*.
19. Ethanol.

2.3 Fractionation of Human Gastric Mucins

1. Ion-exchange resin: Diethylaminoethyl cellulose (DEAE)-Sepharose CL-6B (Pharmacia Fine Chemicals, Sweden).
2. Open glass columns (2.5 × 100 cm).
3. Collection tubes.
4. Fraction collector.
5. 10 mM Tris–HCl buffer (pH 7.2): Add ~100 mL of water to a 1-L graduated cylinder. Weigh 1.21 g of Tris base and transfer to the cylinder. Mix with the water. Add more water until the volume of the mix reaches 800 mL. Mix and adjust the pH to 7.2 with HCl. Add more water until the volume reaches 1 L, and mix the solution.
6. Discontinuous gradient concentration of NaCl (0, 0.25, 0.5, 1.0 mol/L) in 10 mmol/L Tris–HCl (pH 7.2): Add 14.6 g (0.25 mol/L), 29.2 g (0.5 mol/L), and 58.4 g (1.0 mol/L) of NaCl to 1 L of 10 mM Tris–HCl buffer (pH 7.2).

3 Methods

Carry out all procedures at room temperature unless specified otherwise.

3.1 Gastric Aspiration with the EGT (Endoscopic Gastrin Test) Technique

1. Inject intramuscularly with pentagastrin at a dose of 6 μg/kg about 15 min before the endoscopy (*see* **Note 7**).
2. On entering the stomach with the endoscope, aspirate the gastric fluid pooled in the stomach and discard.
3. Aspirate and collect gastric juice newly secreted between 20 and 30 min after the pentagastrin injection (*see* **Note 8**).
4. After the collection of gastric juice, remove the endoscope.
5. Freeze immediately and store at −80 °C until use.
6. Record the volume of gastric juice collected in the 10-min period.
7. Determine the H^+ concentration of the gastric juice by titration (*see* **Notes 9** and **10**).

3.2 Extraction and Measurement of Mucin in Gastric Juice

1. Centrifuge the collected gastric juice at 1500 × *g* for 30 min at room temperature (*see* **Note 11**).
2. The mucin in gastric juice samples can be quantified by the following two methods.

3.2.1 Method I-Gel Filtration Method

1. Dialyze 50 mL aliquot of the collected gastric juice supernatant against running water (*see* **Note 12**).
2. Lyophilize the dialyzed solution, and store at 4 °C.
3. Add a solution of 50 mM Tris–HCl (pH 7.2) to each sample at a ratio of 2 mL/100 mg of lyophilized weight.
4. Homogenize the suspension with polytron homogenizer and add Triton X-100 to a final concentration of 2% (*see* **Note 13**).
5. Incubate the mixture obtained at 37 °C for 1 h (*see* **Note 14**).
6. Centrifuge at 5000 × *g* for 30 min at 4 °C.
7. Obtain the homogenate supernatant.
8. Carry out the extraction procedure using the Triton-containing buffer twice (*see* **Note 15**).
9. Apply 2 mL of supernatant to a gel-filtration column (bed volume 100 mL) (*see* **Note 16**), elute with 80 mL of elution buffer, and collect the effluent in collection tubes using a fraction collector (2 mL each).
10. Determine the hexose content in 0.5 mL of each as described below.
11. Measure the hexose content by the phenol-sulfuric acid method.
12. Add phenol-sulfuric acid solution to 0.5 mL column elution sample and mix.
13. Measure with a spectrophotometer (490 nm) (*see* **Note 17**).
14. Plot an elution curve and confirm the three peaks appeared on each elution curve (Fig. 1a).
15. Collect the first, which elutes near the void volume, and designate Fr. 1 (*see* **Note 18**).
16. Determine the mucin content of each sample by dividing the hexose content of Fr. 1 (*see* **Note 19**).

3.2.2 Method II-EtOH Precipitation Method (see Note 20)

1. Add absolute ethanol (6 mL) to 2 mL of the supernatant obtained from the gastric juice (*see* **Note 21**).
2. Maintain the resultant suspension at 4 °C overnight (*see* **Note 22**).
3. Collect the precipitate by centrifugation (5000 × *g* for 30 min at 4 °C).

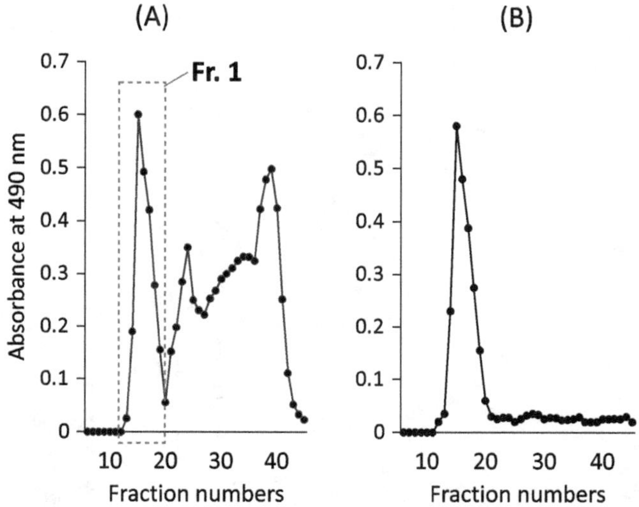

Fig. 1 Elution profiles of Bio-Gel A-1.5 m column chromatography monitored by hexose measurement. (**a**) Elution pattern of one of the extracts from the gastric juice sample obtained from healthy volunteers. (**b**) Elution pattern of the material isolated from the same sample by ethanol precipitation

4. Solubilize the obtained pellet in 2 mL of 0.9% NaCl and 0.2 mL of 20% Triton X-100 solution and ethanol precipitation (*see* **Note 23**).
5. Repeat this precipitation-solubilization procedure twice.
6. Dissolve the final pellet in distilled water (2 mL).
7. Measure its hexose content by the phenol-sulfuric acid method (*see* **Notes 24** and **25**).
8. The mucin content of the gastric juice at the start of the precipitation method is expressed as the hexose content (μg/mL) of the final solution (Fig. 2).

3.3 Fractionation of Human Gastric Mucin (see Note 26)

1. Digest the mucin solution obtained by Method II with 1 mg/mL chymotrypsin.
2. Dialyze extensively against deionized water.
3. Lyophilize the dialyzed solution, and store at 4 °C.
4. Redissolve in 10 mmol/L Tris–HCl, pH 7.2.
5. Apply the sample to a 5 × 50 mm ion-exchange column containing DEAE-Sepharose CL-6B.
6. Elute at 1 mL/min with a discontinuous gradient concentration of NaCl (0, 0.25, 0.5, 1.0 mol/L) in 10 mmol/L TrisHCl, pH 7.2.
7. The effluent solution collected from the DEAE-Sepharose column is checked by hexose measurement (*see* **Notes 27** and **28**) and the mucin recovery is found to be approximately 90% (*see* **Note 29**).

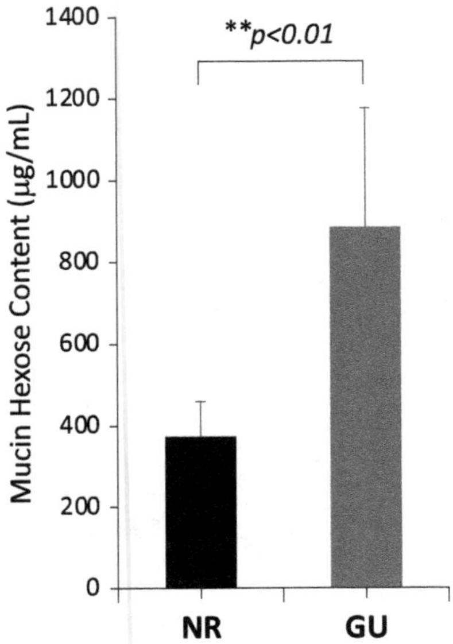

Fig. 2 Purification of gastric juice from healthy volunteers and patients with chronic gastric ulcer by Method II. Mean gastric luminal mucin contents in the normal control group (NR) ($n = 17$) and chronic gastric ulcers (GU) ($n = 21$). The mucin hexose contents of gastric ulcer patients were significantly higher than normal control

4 Notes

1. Frozen gastric juice should be thawed under running water before use.
2. Triton X-100 should be added to beaker following water to prevent bubbling. Pre-heating of water can dissolve Triton X-100 faster.
3. A commercially available gel filtration agarose resin (fractionation range: 10–1500 kDa) can be used.
4. Phenol should be measured after it has completely dissolved in a hot water bath.
5. Keep the containers warm beforehand.
6. Avoid direct contact with skin.
7. Procedures should be performed in accordance with institutional policies and ethical standards.
8. During the examination, the subjects are asked not to swallow their saliva.
9. Calculate acid output in the 10-min period by multiplying the volume by the H^+ concentration, and the EGT value is expressed in units of mEq/10 min.

10. EGT levels are more stable with pentagastrin administration than without pentagastrin.
11. Contaminated debris can be removed.
12. Unnecessary small molecules can be removed.
13. The suspension with 2% Triton X-100 is adjusted by adding prepared 20% (v/v) Triton X-100.
14. Mucin solubilization is facilitated.
15. The same buffer as **step 4** in Subheading 3.2.1.
16. A gel-filtration column should be equilibrated with three bed volumes of elution buffer before a sample applying. Elute with the same buffer solution as that used for extraction.
17. The spectrophotometer may be degraded by phenol, so wipe the inside of the device with a dry cloth after use.
18. Most of the hexose-containing materials in Fr. 1 had already been characterized as mucin [8, 10].
19. Express the initial volume of the gastric juice sample as micrograms of mucin hexose per milliliter of gastric juice.
20. Method II results in less sample loss than method I when this is followed by a charge-dependent mucin-type separation procedure [10–12].
21. The final ethanol concentration is 75%.
22. Complete the precipitation.
23. The same method as the **step 1** in Subheading 3.2.2.
24. The degree of mucin purification may be confirmed by gel filtration (*see* Subheading 3.2.1, **steps 6–16**).
25. Method II is straightforward and simple to use but is unsatisfactory for distinguishing mucin from lower sized materials which include a variety of non-mucinous glycoproteins originating from intracellular materials, cell membranes, and the extracellular tissues [6, 7]. (*see* Fig. 1b).
26. Charge-dependent mucin-type separation procedure.
27. The same method as **steps 10–13** in Subheading 3.2.1.
28. Figure 3 shows a typical DEAE-Sepharose chromatography elution pattern of the mucins obtained from the gastric juice sample of a patient before and after eradication of the *H. pylori*. The gastric luminal mucins were divided into three fractions (Fr. I, Fr. II, and Fr. III) by ion-exchange chromatography using a discontinuous salt gradient with three salt steps. These mucin Frs. I–III were further studied in order to determine whether they represented compositionally distinct mucin species using gas–liquid chromatography. The carbohydrate of Fr. III was composed of hexosamine, hexose, and sialic acid in an average ratio of 1.00:1.69:1.70, expressed as moles per mole

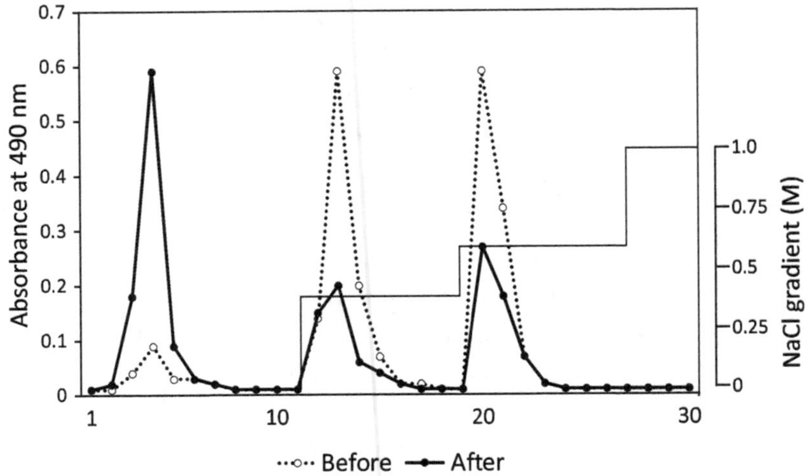

Fig. 3 DEAE-Sepharose chromatography of mucins obtained from the gastric juice sample of a patient before and after eradication of the *H. Pylori*

of hexosamine. In contrast, the proportion of sialic acid to each carbohydrate in Frs. I and II was much less than that in Fr. III (hexosamine/hexose/sialic acid = 1.00:1.88:0.00 for Fr. I, 1.00:2.05:0.21 for Fr. II) [13].

29. The sugar composition of the fractionated mucin can be determined by the method using a Shimadzu GC7A gas chromatograph described by Sweeley et al. [9].

Acknowledgments

The authors express our sincere appreciation to S. Sugawara for her technical assistance.

References

1. Alle A, Hutton D, Leonard A et al (1986) The role of mucus in the protection of the gastroduodenal mucosa. Scand J Gastroenterol 21 (Suppl. 125):71–77
2. Azuumi Y, Ohara S, Ishihara K et al (1980) Correlation of quantitative changes of gastric mucosal glycoproteins with aspirin-induced gastric damage in rats. Gut 21:533–536
3. Tettamanti G, Pigman W (1968) Purification and characterization of bovine and ovine submaxillary mucins. Arch Biochem Biophys 124: 41–50
4. Pamer T, Glass G, Horowitz M (1968) Purification and characterization of sulfated glycoproteins and hyaluronidase-resistant mucopolysaccharides. Biochemistry 7:3821–3829
5. Dekanski J, Macdonald A, Sacra P (1975) Effects of fasting, stress and drugs on gastric glycoprotein synthesis in the rat. Br J Pharmacol 55:387–392
6. Hunter AC, Allen A, Garner A (1989) Studies on mucus biosynthesis in gastrointestinal tract. Symp Soc Exp Biol 43(27):27–36
7. Yoshimura K, Kraus ER, Shimakura S et al (1992) Role of prostaglandin E2 in cholinergic-mediated glycoprotein synthesis in canine antrum. Dig Dis Sci 37(7):1045–1050
8. Ohara S, Ishihara K, Hotta K (1986) Comparative study on mucus glycoproteins in rat

stomach and duodenum. Comp Biochem Physiol 83B:273–275

9. Sweeley C, Bentley R, Makita M et al (1963) Gas-liquid chromatography of trimethylsilyl derivatives of sugars and related substances. J Am Chem Soc 85:2497–2507
10. Azzumi Y, Ichikawa T, Ishihara K et al (1993) The validity of the ethanol precipitation method for the measurement of mucin content in human gastric juices and its possible relationship to gastroduodenal diseases. Clin Chim Acta 221:219–225
11. Ijima K, Ara N, Abe Y et al (2012) Association of gastric acid and mucus secretion level with low-dose aspirin-induced gastropathy. J Gastroenterol 47(2):150–158
12. Ikezawa T, Ichikawa T, Adachi K et al (2005) Analysis of mucin composition in gastric juices of chronic rheumatic patients with upper gastrointestinal damage. Biomed Res 26(4): 147–151
13. Ichikawa T, Koizumi W, Takeuchi H et al (2001) A new strategy for evaluation of the eradication treatment of *Helicobacter pylori*: validation of the analysis of mucin composition in human gastric juice. Clin Chim Acta 303: 69–73

Chapter 6

Extraction of Mucins from the Mammalian Intestinal Tract

Shota Okamoto, Mugen Taniguchi, and Ryu Okumura

Abstract

In the intestine, mucus covering the mucosa plays a critical role in maintaining gut homeostasis by protecting the mucosa from invasion by commensal bacteria. The gut mucus is composed primarily of MUC2 mucin secreted by goblet cells. MUC2 is highly *O*-glycosylated, and *O*-glycans are necessary for the function and polymer structure of MUC2. In addition, recent evidence revealed that several glycan modifications, such as sialylation and sulfation, confer resistance of mucins to proteolysis and affect the viscosity and lubricity of mucus. Therefore, characterizing glycan structures of mucins is required to understand their functions fully. In this chapter, we describe how to purify secreted mucins from the mammalian intestine for analysis of their glycan structures. This description includes the extraction of MUC2 mucin from the mucosal surface of the mouse colon and colon explants.

Key words Mucin extraction, Chaotropic agent, Urea, DL-dithiothreitol, Iodoacetamide

1 Introduction

Intestinal mucosal barriers constructed by intestinal epithelial cells are essential for maintaining intestinal homeostasis. The integrity of mucosal barriers contributes to the spatial segregation of gut microbiota and intestinal epithelia (Fig. 1). In the large intestine, which carries higher bacterial loads than the small intestine, these physical barriers contribute mainly to host immunity to avoid unnecessary immune responses to gut microbiota [1]. Mucins are a major component of mucosal barriers that can be classified into two types: cell-surface mucins and gel-forming mucins [2]. Among several mucins, Mucin2 (MUC2) is the predominant gel-forming mucin in the intestine. The goblet cell-secreted MUC2 constructs the mucus layer, which is divided into the two layers: the inner firm and the outer loose mucus layer [3]. The inner mucus layer is characterized as being insoluble with a stratified appearance. The polymeric net-like structure anchored to the intestinal epithelia prevents bacterial adhesion to epithelial cells and thus ensures the inner mucus layer remains essentially sterile. In contrast, the outer

Akihiko Kameyama (ed.), *Mucins: Methods and Protocols*, Methods in Molecular Biology, vol. 2763,
https://doi.org/10.1007/978-1-0716-3670-1_6,

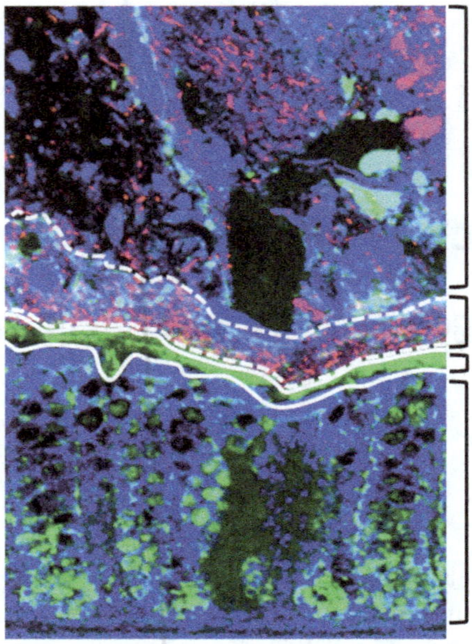

Fig. 1 Fluorescence in situ hybridization and MUC2 staining of the mouse normal colon. The mucus layers contribute to the spatial segregation of gut microbiota and intestinal epithelia. In contrast to the outer mucus layer, the inner mucus layer is firmly attached to epithelia and impenetrable to bacteria. By the scraping method in this protocol, the gel-forming mucins covering epithelia are extracted as "insoluble" fraction. Green: MUC2; red: bacteria; blue: 4′,6-diamidino-2-phenylindole (DAPI)

mucus layer is a loosely attached and guanidinium chloride-soluble gel. This layer hosts a distinct intestinal microbial niche. Proteolytic cleavage of polymerized MUC2 by bacterial proteases facilitates the expansion of the net-like structure, resulting in the conversion of the inner mucus layer into the outer mucus layer. As such, these rheological and structural characteristics establish beneficial symbiotic relationships with their host [4]. Mucins are large glycoproteins (~2 MDa) with a high carbohydrate content (50–90% of total molecular weight) that is attributed to their large number of *O*-linked glycans. Mucin glycosylation protects the protein core from endogenous protease degradation and represents a reliable energy source for commensal mucus-residing bacteria through their glycosidase enzymes. Bacteria produce several metabolites, such as short-chain fatty acids which regulate intestinal epithelial and immune cells. The glycan profile of mucins is regulated by complex mechanisms. Glycosyltransferases and sulfotransferases expressed by intestinal epithelial cells are responsible for mucin glycosylation and sulfation. Growth factors and inflammatory cytokines can alter the expression level of these genes. Commensal bacteria directly or indirectly modulate mucin glycosylation patterns through the upregulation of interleukin (IL)-22 and IL-23, in immune cells of the lamina propria [5]. Notably, mucin dynamics attributed to glycan alterations are implicated in the development of diseases. Genetic or epigenetic alterations reduce the mucus layer, impair host–microbe interactions and mucosal immunity, and ultimately contribute to inflammatory bowel disease pathogenesis [6]. Additionally, the regulation of mucin expression has been reported to be disrupted in colorectal cancer, and structural alterations of mucin *O*-linked glycans have also been reported [7, 8]. The analysis of mucins and their glycans should provide deeper insights into disease pathogenesis because mucins are intimately involved in inflammation and cancer.

In this chapter, we describe methods for extracting and purifying mucins from the mammalian intestinal tract (*see* **Note 1**). Two different extraction methods can be selected depending on the target mucins: from the surface epithelium or the whole intestinal tissue. The protocols are mainly aimed at the analysis of mucin glycans. Urea, used as a chaotropic agent, may degrade to isocyanic acid, which causes N-terminal carbamylation of the resultant mucins [9]. Thus, this reaction should be considered in mucin polypeptide analysis. Moreover, other mucosal components, such as nonspecific antimicrobials and specific antimicrobial immunoglobulins, are not suitable for analysis using these protocols.

2 Materials

2.1 Extraction of Mucins with Chaotropic Agents

1. 8 M urea solution: Dissolve 24 g of urea in double-distilled water up to 47 mL. Add 0.5 mL of 0.5 M ethylenediaminetetraacetic acid (EDTA) disodium salt solution (pH 8.0) and 2.5 mL of 10 mM sodium phosphate buffer. Store at 4 °C. Immediately prior to extraction, add *N*-ethylmaleimide (NEM) and diisopropyl fluorophosphate (DFP) to the solution to give final concentrations of 5 mM NEM and 1 mM DFP (*see* **Note 2**).
2. 8 M urea reduction buffer: Dissolve 24 g of urea in double-distilled water up to 44.5 mL. Add 0.5 mL of 0.5 M EDTA disodium salt solution (pH 8.0) and 5 mL of 1 M Tris-HCl (pH 8.0). Store at room temperature. Immediately prior to reduction, add DL-dithiothreitol (DTT) to the solution to give a final concentration of 10 mM DTT (*see* **Note 3**).
3. 0.5 M iodoacetamide (IAA) solution: Dissolve 96 mg of IAA in 1038 μL double-distilled water (*see* **Note 4**).
4. Tube rotator (*see* **Note 5**).
5. Ultracentrifuge (*see* **Note 5**).
6. Fixed angle rotor (*see* **Note 5**).
7. Curved tweezer or glass microscope slide.

8. 1.5 mL microcentrifuge tube.
9. Scissors.

2.2 Protein Measurement and Identification

1. Bicinchoninic acid (BCA) protein assay kit.
2. Microplate reader (*see* **Note 5**).
3. Rabbit anti-MUC2 antibody (GTX100664, GeneTex, California, USA).
4. Alcian blue stain solution pH 2.5 (37154-15, Nacalai Tesque, Kyoto, Japan).

3 Methods

If the analysis target is mucins covering the surface epithelium, *see* Subheading 3.1, whereas if it is the whole intestinal tissue containing mucin vesicles before exocytosis, *see* Subheading 3.2. Carry out all procedures at room temperature unless specified otherwise. The following protocol uses mice as the animal model (*see* **Note 6**).

3.1 Extraction of Mucins with Chaotropic Agents from the Surface Epithelium

1. Resect the colon and cut it open longitudinally. Gently remove the fecal contents (do not wash the mucus on the surface).
2. Scrape mucins covering the surface epithelium using a curved tweezer or a glass microscope slide.
3. Place the scrapings in a 1.5 mL plastic sterile microcentrifuge tube containing 500 μL of the 8 M urea solution. As a recommendation, use one tube per mouse.
4. Repeat **step 2** about ten times and pool the scrapings.
5. After mixing thoroughly by pipetting up and down, incubate the sample overnight on a rotator at 4 °C.
6. Place the sample in the fixed angle rotor and centrifuge at 24,000 × *g* for 45 min at 4 °C.
7. Carefully transfer the supernatant to a new 1.5 mL tube and store at −80 °C as the "soluble" mucins sample (*see* **Note 7**).
8. If necessary, add 500 μL of the 8 M urea solution to the pellet and repeat **steps 5–7** two more times and pool the extracts.
9. Re-suspend the pellet from the tube (**step 7**) in 100 μL of the 8 M urea reduction buffer and incubate for 5 h at 37 °C (*see* **Note 8**).
10. If multiple samples are processed, combine them into one tube.
11. Add 0.5 M IAA to give a final concentration of 25 mM and incubate overnight in darkness at room temperature.
12. Place the tube in the fixed angle rotor and centrifuge at 24,000 × *g* for 45 min at 4 °C.

13. Carefully transfer the supernatant to a new 1.5 mL tube and store at −80 °C as the "insoluble" mucins sample (*see* **Note 7**).

3.2 Extraction of Mucins with Chaotropic Agents from the Whole Intestinal Tissue

1. Resect the colon and cut it open longitudinally. Gently remove the fecal contents (do not wash the mucus on the surface).
2. Place the tissue into a 1.5 mL tube and snap-freeze in liquid nitrogen. As a recommendation, use one tube per mouse.
3. Cut the frozen tissue into small pieces using scissors.
4. Mix the powdered tissue with 500 μL of the 8 M urea solution and incubate the sample overnight on a rotator at 4 °C.
5. Place the tube in the fixed angle rotor and centrifuge at 24,000 × *g* for 45 min at 4 °C.
6. Carefully transfer the supernatant to a new 1.5 mL tube and store at −80 °C as the "soluble" mucins sample (*see* **Note 7**).
7. If necessary, add 500 μL of the 8 M urea solution to the pellet and repeat **steps 4–6** two more times and pool the extracts.
8. Re-suspend the pellet from the tube (**step 6**) in 100 μL of the 8 M urea reduction buffer and incubate for 5 h at 37 °C (*see* **Note 8**).
9. If multiple samples are processed, combine them into one tube.
10. Add 0.5 M IAA to give a final concentration of 25 mM and incubate overnight in darkness at room temperature.
11. Place the tube in the fixed angle rotor and centrifuge at 24,000 × *g* for 45 min at 4 °C.
12. Carefully transfer the supernatant to a new 1.5 mL tube and store at −80 °C as the "insoluble" mucins sample (*see* **Note 7**).

3.3 Protein Measurement and Identification

1. Measure the concentration of extracted mucins using BCA Protein Assay Kit and microplate reader.
2. Determine extracted mucins by western blotting using anti-MUC2 antibody and Alcian Blue staining (Fig. 2). If further separation for the glycan analysis is needed, supported molecular matrix electrophoresis (SMME) is recommended (*see* Chapter 7), (*see* **Note 9**).

4 Notes

1. As discussed previously [10], research using commercially available mucins (e.g., porcine gastric mucin, bovine submaxillary mucin) has limitations because their biochemical structures differ from intestinal mucins and do not form a gel.
2. Guanidine hydrochloride is more likely to be used as a chaotropic agent, although this agent can inhibit the β-elimination

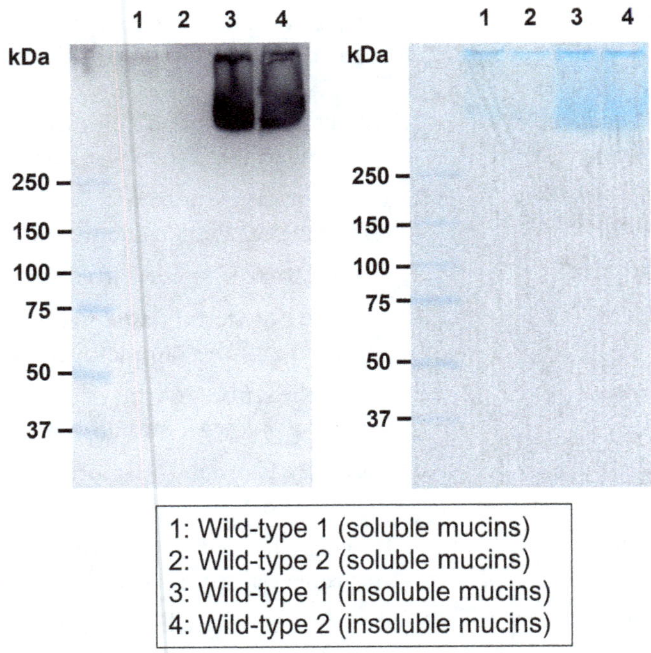

Fig. 2 Sodium dodecyl sulfate-polyacrylamide gel electrophoresis analysis of soluble or insoluble colonic mucins from wild-type mice. Western blotting using anti-MUC2 antibody (left panel) and gel staining with Alcian blue (right panel). Protein loading was 20 μg/lane and 30 μg/lane, respectively. MUC2 is hardly detectable in soluble mucins. In contrast, insoluble mucins are rich in MUC2, which is the high molecular weight glycoprotein (over 250 kDa) negatively charged with acidic residues. Wild-type 1: mucins extracted from three 10-week-old female wild-type mice; Wild-type 2: mucins extracted from three 11-week-old female wild-type mice

of glycans from mucins. In such cases, urea is recommended. EDTA is a metalloprotease inhibitor. NEM is a cysteine protease inhibitor and also prevents the formation of new disulfide crosslinks. DFP is a serine protease inhibitor, and because this chemical is an acetylcholinesterase inhibitor, it should be handled with extreme care.

3. DTT is a reagent for reducing disulfide bonds between and within mucins. DTT treatment converts mucin dimers to monomers and facilitates further solubilization.
4. IAA is an alkylating reagent that modifies protein sulfhydryls. Permanent modification of protein sulfhydryls prevents protein aggregation, ensuring the sample remains soluble.
5. We use Biospin MBS-1 (Tokyo Rikakikai, Tokyo, Japan) as a tube rotator, Optima Max XP ultracentrifuge (Beckman Coulter, California, USA) as an ultracentrifuge, TLA-55 rotor (Beckman Coulter, California, USA) as a fixed angle rotor,

and iMark Plate Reader (Bio-Rad, California, USA) as a microplate reader.

6. These protocols are available for surgically removed human tissues.
7. This step should be performed on ice. Do not leave the sample for too long after centrifugation as doing so may disturb the pellet.
8. A previous paper reported that the "guanidinium chloride insoluble" fraction is the pellet from the centrifugation step after three rounds of extraction with 6 M guanidinium chloride extraction buffer [10]. However, MUC2 mucins are resistant to 6 M guanidinium chloride extraction buffer and are only solubilized after reduction and alkylation. Similarly, in our protocols, the definition of "insoluble" fraction is the pellet following centrifugation in Subheading 3.1, **step** 7 or Subheading 3.2, **step 6**. MUC2 mucins are solubilized only after reduction and alkylation following 8 M urea extraction (*see* Fig. 2, above).
9. The glycans of extracted mucins can be analyzed by mass spectrometry.

Acknowledgments

This work was supported by grants from the Ministry of Education, Culture, Sports, Science and Technology of Japan: JP18K15187, Uehara Memorial Foundation and Takeda Science Foundation. We thank Edanz (https://jp.edanz.com/ac) for editing a draft of this manuscript.

References

1. Okumura R, Takeda K (2018) Maintenance of intestinal homeostasis by mucosal barriers. Inflamm Regen 38:5
2. McGuckin MA, Lindén SK, Sutton P et al (2011) Mucin dynamics and enteric pathogens. Nat Rev Microbiol 9:265–278
3. Johansson ME, Larsson JM, Hansson GC (2011) The two mucus layers of colon are organized by the MUC2 mucin, whereas the outer layer is a legislator of host-microbial interactions. Proc Natl Acad Sci U S A 108: 4659–4665
4. Paone P, Cani PD (2020) Mucus barrier, mucins and gut microbiota: the expected slimy partners? Gut 69:2232–2243
5. Pickard JM, Chervonsky AV (2015) Intestinal fucose as a mediator of host-microbe symbiosis. J Immunol 194:5588–5593
6. Kudelka MR, Stowell SR, Cummings RD et al (2020) Intestinal epithelial glycosylation in homeostasis and gut microbiota interactions in IBD. Nat Rev Gastroenterol Hepatol 17: 597–617
7. Kufe DW (2009) Mucins in cancer: function, prognosis and therapy. Nat Rev Cancer 9:874–885
8. Brockhausen I (2006) Mucin-type *O*-glycans in human colon and breast cancer: glycodynamics and functions. EMBO Rep 7:599–604

9. Kesimer M, Sheehan JK (2012) Mass spectrometric analysis of mucin core proteins. Methods Mol Biol 842:67–79
10. Davies JR, Wickström C, Thornton DJ (2012) Gel-forming and cell-associated mucins: preparation for structural and functional studies. Methods Mol Biol 842:27–47

Chapter 7

Supported Molecular Matrix Electrophoresis

Akihiko Kameyama

Abstract

Distinct bands of mucins cannot be banded using a gel electrophoresis based on a molecular sieving effect due to their very large molecular weight and remarkable diversity in glycosylation. In contrast, membrane electrophoresis can separate mucins as round bands. Here, we present an analysis of mucin separation via membrane electrophoresis using a porous polyvinylidene difluoride membrane, which is highly stable against chemical modifications and various organic solvents. The separated mucins can not only be stained with dyes but also with antibodies and lectins, and glycans can be released from the excised bands and analyzed.

Key words Polyvinylidene difluoride membrane, Membrane electrophoresis, Polyvinylalcohol, Glycans, Alcian blue, Lectin

1 Introduction

Mucins have a tremendous potential as novel clinical biomarkers for various malignancies and are characterized by their large molecular weight (approximately 2 MDa) and high glycan content (50–90% of the total molecular weight), which is a result of the presence of numerous *O*-linked glycans. Dysregulation of mucin expression and structural changes in their *O*-linked glycans have been reported in several tumors [1, 2]. Different methods for protein electrophoresis have been developed for decades; however, most of them cannot be applied to mucins due to their large size and heterogeneity. Mucins do not migrate through the separation gel during sodium dodecyl sulfate polyacrylamide gel electrophoresis (SDS-PAGE), due to their large molecular size, and even if they do, they do not converge into round bands due to high glycan heterogeneity, resulting in smears. Moreover, transfer from the gel to the polyvinylidene difluoride (PVDF) membrane is essential for glycan analysis [3, 4]. Recovering transferred mucins for quantitative evaluation is often difficult. Recently, we reported a new membrane electrophoresis method called "supported molecule matrix

Akihiko Kameyama (ed.), *Mucins: Methods and Protocols*, Methods in Molecular Biology, vol. 2763,
https://doi.org/10.1007/978-1-0716-3670-1_7,

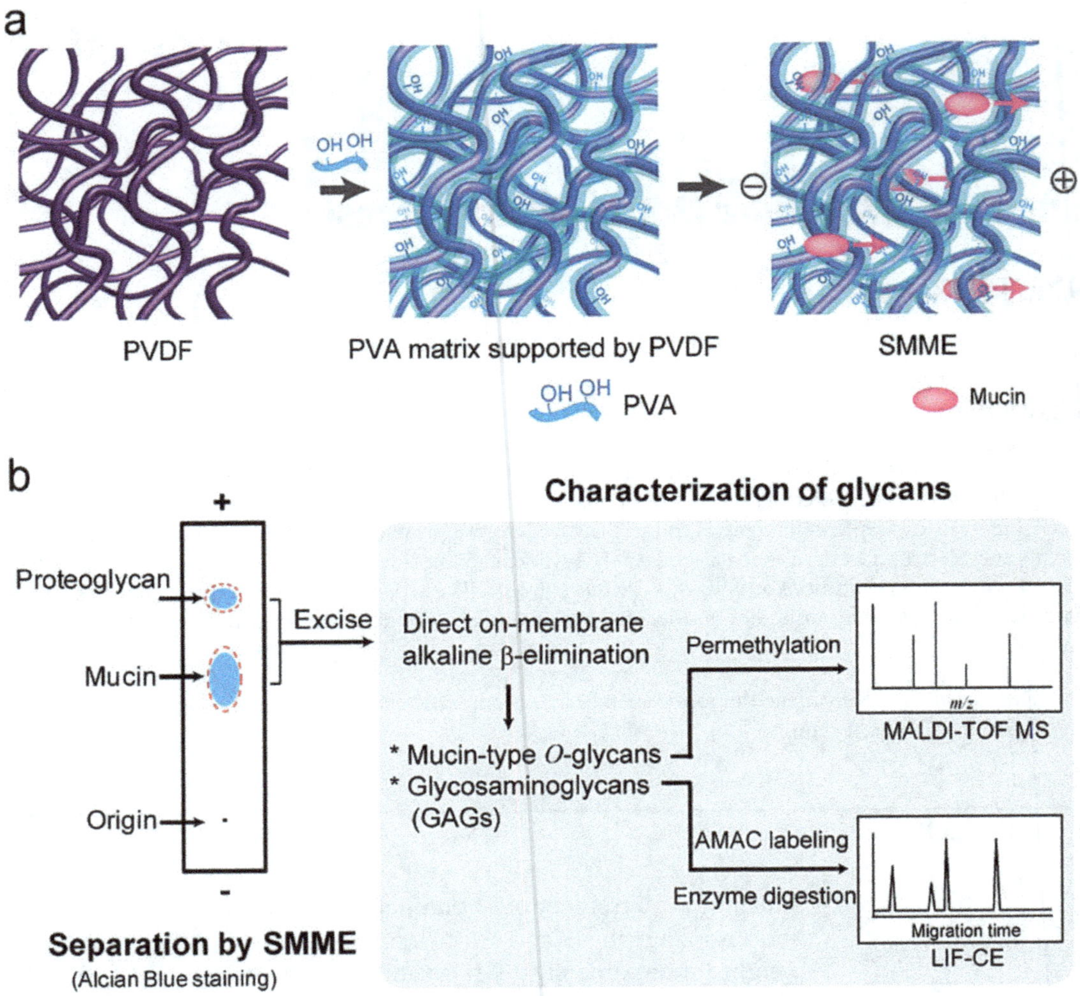

Fig. 1 Supported molecular matrix electrophoresis (SMME). (**a**) A conceptual overview of the SMME technique. (**b**) A schematic diagram illustrating the characterization of glycoproteins using the SMME technique. (Reproduced from [5] with permission. Copyright 2009 American Chemical Society)

electrophoresis" (SMME), in which hydrophilic polymers, such as polyvinyl alcohol (PVA), are used as separation media in a porous PVDF membrane filter (Fig. 1a) [5]. SMME is a novel method for separating mucins and proteins with multiple glycans, such as proteoglycans. These molecules can be visualized on the membrane using Alcian blue staining [6]. Mucins present on the membrane after separation can also be visualized with monoclonal antibodies [7, 8] or lectins by adjusting the composition of the hydrophilic polymer of the separation medium. Since the proteins are already on the PVDF membrane, the SMME procedure does not require a transfer phase [9]. A significant advantage of the SMME method is that it can analyze the glycans in the separated spots (Fig. 1b). Conventional membrane electrophoresis methods use filter paper or cellulose acetate membranes, which are degraded to glucose oligomers during the glycan release reaction under alkaline conditions. These glucose oligomers pose a substantial barrier to glycan analysis; the PVDF filter and hydrophilic polymer of the SMME membrane are stable under the same conditions and do not generate interfering substances. Here, we describe mucin separation, visualization with Alcian blue and lectin, and glycan analysis using the SMME method.

2 Materials

Prepare all solutions with ultrapure water and reagents of analytical grade. No sodium azide should be added to the reagents. Unless otherwise specified, prepare and store all reagents at ambient temperature.

2.1 Preparation of Mucin Samples for SMME

1. Mucin from porcine stomach (Type III, partially purified powder, Sigma-Aldrich) (*see* **Note 1**).
2. Mucin sample extracted from biomaterials (*see* **Note 2**).
3. 0.1 M Tris–HCl buffer (pH 8.6): Add approximately 100 mL of water to a 1 L graduated cylinder or to a glass beaker. Weigh 12.1 g of Tris and transfer it to the cylinder. Dissolve the Tris and dilute the solution to a final volume of 900 mL. Adjust the pH to 8.6 with HCl and then adjust the final volume to 1 L.
4. Denaturing buffer: Prepare a solution of 20 mM DTT/8 M urea in 0.1 M Tris−HCl buffer (pH 8.6). Weigh 480 mg of urea and 3.1 mg of DTT, and transfer to a 1.5 mL microcentrifuge tube. Add 0.7 mL of the 0.1 M Tris−HCl buffer (pH 8.6) and dissolve the DTT and the urea. Use the solution immediately after preparation.
5. Iodoacetamide solution (250 mM): Weigh 9.2 mg of iodoacetamide and transfer it to a 500 μL microcentrifuge tube. Add 200 μL of water and dissolve the iodoacetamide. Use the solution immediately after preparation.
6. Trypsin solution: Dissolve 1 mg of trypsin from porcine pancreas in 1 mM HCl. Use the solution immediately after preparation.
7. 8 M Urea solution: Weigh 480 mg of urea and transfer it to a 1.5 mL microcentrifuge tube. Add 0.7 mL of the 0.1 M Tris−HCl buffer (pH 8.6) and dissolve the urea.
8. Mini dialysis kit, 8 kDa cut-off (250 μL) (*see* **Note 3**).
9. Ultrafiltration device (0.5 mL, 10 kDa cutoff) (*see* **Note 4**).

10. Magnetic stirrer.
11. Magnetic stirrer bar.
12. Refrigerate microcentrifuge.

2.2 SMME Membrane

Hydrophilic polymer solution 1 for Alcian blue staining and glycan analysis, and hydrophilic polymer solution 2 for antibody and lectin staining should be prepared.

1. PVDF filter membrane for protein blotting (Immobilon-P, Millipore, Billerica, MA, USA) (*see* **Note 5**).
2. Methanol.
3. Hydrophilic polymer solution 1: Dissolve 500 mg of PVA (88% saponified, average MW 22,000) in 200 mL of water to obtain a solution with a 0.25% (w/v) concentration (*see* **Note 6**).
4. Hydrophilic polymer solution 2: Dissolve 100 mg of PVA (88% saponified, average MW 22,000) and 400 mg of polyvinyl pyrrolidone (PVP, average MW 20,000) in 200 mL of water to obtain a 0.25% (w/v) concentration solution (*see* **Note 6**).
5. Plastic disposal tray (100 × 70 mm, depth:13 mm) (*see* **Note 7**).
6. Glass Petri dish (diameter 10 cm) (*see* **Note 7**).
7. Lint-free paper for the clean room (*see* **Note 8**).

2.3 Membrane Electrophoresis

1. SMME membrane.
2. Pencil (*see* **Note 9**).
3. Plastic disposal tray (100 × 70 mm, depth: 13 mm) (*see* **Note 7**).
4. Glass Petri dish (diameter: 10 cm) (*see* **Note 7**).
5. Running buffer: 0.1 M Pyridine-formic acid buffer (pH 4.0). Add approximately 100 mL of water to a 1-L graduated cylinder or to a glass beaker. Add 7.8 mL of pyridine to the cylinder and mix it with the water. Add water to make up a volume of 900 mL. Add formic acid to adjust the pH to 4.0. Make up the final volume to 1 L with water. Store at room temperature or at 4 °C (*see* **Note 10**).
6. Electrode filter papers: Cut a qualitative filter paper (Advantec grade 131, Toyo Roshi, Tokyo, Japan) to an appropriate size (length = 8 cm, width = 1 cm broader than the width of the SMME membrane) (*see* **Note 11**).
7. Membrane electrophoresis chamber (EPC105AA-type, Advantec, Tokyo, Japan) (*see* **Note 12**).
8. Power supply for electrophoresis.

2.4 Alcian Blue Staining

1. Alcian blue solution: Weigh 100 mg of Alcian blue 8GX and transfer it to a 200 mL glass bottle. Dissolve it in 100 mL of 0.1% acetic acid. Store at 4 °C (*see* **Note 13**).
2. Plastic disposal tray (100 × 70 mm, depth: 13 mm) (*see* **Note 7**).
3. Glass Petri dish (diameter: 10 cm) (*see* **Note 7**).
4. Methanol.
5. 30% Acetic acid in methanol.
6. Diethylene glycol.
7. Oriented polypropylene (OPP) film pocket sheet (*see* **Note 14**).
8. Heat block.
9. White plate.
10. Gel imaging system (*see* **Note 15**).

2.5 Lectin Staining

1. Phosphate-buffered saline containing 0.05% Tween-20 (PBS-T).
2. Lectin solution: Dissolved lyophilized biotin-labeled lectins (e.g., Sambucus sieboldiana agglutinin [SSA], Aleuria aurantia lectin [AAL], Maackia amrensis lectin II [MAL-II], and *N*-terminal domain of BC2L-C lectin derived from Burkholderia cenocepacia [BC2LCN]) with distilled water to prepare stock lectin solutions (1 mg/mL). Dilute the stock lectin solutions 500 times with PBS-T. Store at 4 °C (*see* **Note 16**).
3. HRP-conjugated streptavidin solution: Dilute a commercially available HRP-conjugated streptavidin stock solution (1 mg/mL) 4000 times with PBS-T. Store at 4 °C (*see* **Note 16**).
4. Blocking solution: Weigh 500 mg of bovine serum albumin and transfer it to a 15 mL conical tube. Dissolve it into 10 mL of PBS-T (*see* **Note 17**).
5. Plastic disposal tray (100 × 70 mm, depth: 13 mm) (*see* **Note 7**).
6. Methanol.
7. Acetone.
8. Glassine paper.
9. Glass fiber filter paper.
10. Clothes iron without steam (*see* **Note 18**).
11. Glass Petri dish.
12. Hybridization bag.
13. Heat sealer.
14. Plastic wrap.

15. Chemiluminescent peroxidase substrate kit (*see* **Note 19**).
16. Gel imaging system with CCD camera for chemiluminescence (*see* **Note 20**).

2.6 Glycan Analysis (see Note 21)

1. Excised mucin bands from the SMME membrane.
2. Surgical scissors. The cutting operation must be performed in 1.5 mL microtubes.
3. 1 M $NaBH_4$ in 0.1 M NaOH: Weigh 38 mg of $NaBH_4$ and transfer it to a 1.5 mL microtube. Add 1 mL of 0.1 M NaOH and dissolve $NaBH_4$ in the tube. Use the solution immediately after preparation (*see* **Note 22**).
4. Glacial acetic acid.
5. Methanol (*see* **Note 23**).
6. Cation exchange resin (H^+ form) (*see* **Note 24**).
7. Acetic acid in methanol, 1% (v/v). Use immediately after preparation.
8. Dimethyl sulfoxide (DMSO) containing 1% (v/v) distilled water: Add 10 μL of water into 1 mL of DMSO. Use immediately after preparation (*see* **Note 25**).
9. NaOH (granules) (*see* **Note 26**).
10. Methyl iodide.
11. Sep-Pak C18 Vac Cartridge (1 cc, 50 mg, Waters) (*see* **Note 27**).
12. 1% Acetic acid.
13. 50% Acetonitrile: Mix 25 mL acetonitrile (LC-MS grade) and water (*see* **Note 28**).
14. 2,5-Dihydroxybenzoic acid (2,5-DHB) solution: Weigh 10 mg of 2,5-DHB and transfer it to a 1.5 mL microcentrifuge tube. Add 300 μL of ethanol to dissolve the 2,5-DHB, and then add 670 μL of water and 30 μL of 0.1 M NaOH into the tube. Use the solution within 3 days following the preparation unless frozen (*see* **Note 29**).
15. Mass spectrometer: Matrix-assisted laser desorption/ionization (MALDI) time-of-flight (TOF) mass spectrometer.
16. Centrifugal evaporator.
17. Large orifice tip (200 μL) (*see* **Note 30**).
18. Disposable micro tissue homogenizer (*see* **Note 31**).
19. Capillary micropipette (*see* **Note 32**).
20. Vortex mixer with tube holder.
21. Gravity flow manifold (*see* **Note 33**).
22. Luer Stop Valve (PTFE) (*see* **Note 33**).
23. Oil free air pump (*see* **Note 34**).

3 Methods

This section describes sample preparation methods for SMME, beginning with the preparation of porcine gastric mucin (PGM) as a reference material (*see* Subheading 3.1.1) and intestinal mucin (*see* Subheading 3.1.2). *See* Chapter 3 for the preparation of water-soluble mucins for SMME. After describing the preparation of SMME membranes (*see* Subheading 3.2) and electrophoresis using these membranes (*see* Subheading 3.3), Alcian blue staining of the mucins present in the membranes after electrophoresis is described (*see* Subheading 3.4). Alcian blue cannot stain stomach mucins and other substances that lack acidic glycans. Please refer to Chapter 9 in such cases. After electrophoresis, the membranes can also be stained with lectins or antibodies. This section describes the method of staining with lectin (*see* Subheading 3.5), but the same method can also be used to stain with antibodies. We conclude by discussing glycan analysis (*see* Subheading 3.6). Glycan analysis begins with the release of *O*-glycans from mucins on the membrane (*see* Subheading 3.6.1), followed by permethylation of the released glycans (*see* Subheading 3.6.2) and measurement by mass spectrometry. The overall flow is depicted in Fig. 2.

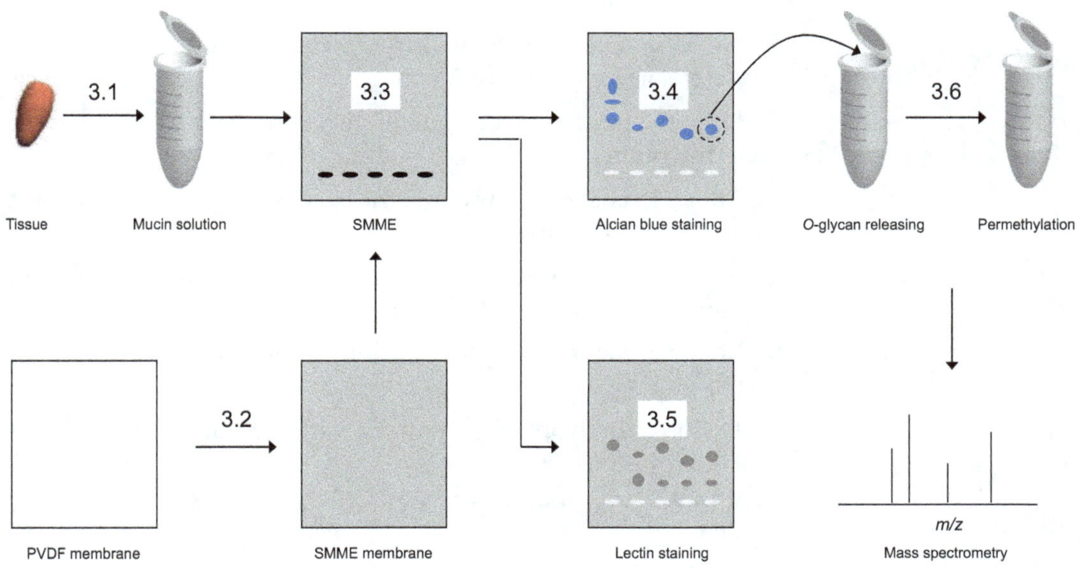

Fig. 2 Overall outline of the method

3.1 Preparation of Mucin Samples for SMME

3.1.1 Porcine Stomach Mucin Sample as a Reference Material for SMME

1. Weigh 11 mg of porcine stomach mucin and transfer it to a 1.5 mL microcentrifuge tube. Add 1 mL of the denaturing buffer and dissolve the mucin.
2. Allow the solution to remain at room temperature for 3 h.
3. Add 120 μL of the iodoacetamide solution to the mixture and mix it thoroughly. Then keep the tube in a dark environment at room temperature for 3 h (*see* **Note 35**).

3.1.2 Intestinal Mucin Sample for SMME

Mucin extracted from biological samples is processed for SMME. In the case of intestinal mucin, heavily glycosylated regions of the mucins are subjected to the SMME, whereas other regions are digested by trypsin and removed through ultrafiltration. When protease inhibitors are present in the mucin extracts, they are eliminated through the following **steps 1–9**. This method can also be used for MUC1 sample preparation.

1. Transfer 100 μL of the intestinal mucin extract into a dialysis tube (a component of mini dialysis kit).
2. Cap the dialysis tube with a dialysis membrane (a component of mini dialysis kit).
3. Insert the tube into a float adapter (a component of mini dialysis kit).
4. Add 300 mL of water and a magnetic stirrer bar to a 500 mL glass beaker.
5. Float the dialysis tube with the membrane side touching the water surface (*see* **Note 36**).
6. Place the beaker on a magnetic stirrer and gently stir the water in the beaker at 4 °C.
7. After 8 h, change the water in the beaker and gently stir again at 4 °C (*see* **Note 36**).
8. After 8 h, remove the dialysis tube from the float adapter and centrifuge it at 100 × *g* for 3 min.
9. Remove the membrane cap and lyophilize the solution in the tube.
10. Dissolve the lyophilized powder with 100 μL of 0.1 M Tris-HCl (pH 8.6).
11. Add 10 μL of the trypsin solution and incubate at 37 °C for 18 h.
12. Transfer the trypsinized solution into the ultrafiltration device.
13. Wash the tube used for trypsinization with 200 μL of water and combine the washing with the trypsinized solution in the ultrafiltration device.
14. Centrifuge the ultrafiltration device at 14,000 × *g* for 15 min. Discard the filtrates (*see* **Note 37**).

15. Add 200 μL of water to the ultrafiltration device and centrifuge at 14,000 × *g* for 15 min again. Discard the filtrates (*see* **Note 37**).
16. Add 40 μL of 8 M urea to the ultrafiltration device and invert the device into a clean tube.
17. Centrifuge the ultrafiltration device at 1000 × *g* for 2 min. The recovered solution can be used as the mucin sample for SMME (*see* **Note 38**).

3.2 Preparation of the SMME Membrane

1. Cut a PVDF membrane to an appropriate size, with a standard length of 6 cm and a width depending on the number of samples (1 cm per sample) (*see* **Note 39**).
2. Add approximately 5 mL of methanol to a glass Petri dish and dip the membrane in methanol (*see* **Note 40**).
3. Add approximately 20 mL of the hydrophilic polymer solution to a plastic disposal tray. Transfer the membrane from the methanol dish into the tray with the hydrophilic polymer solution 1 or 2. Submerge the membrane into the solution using tweezers (*see* **Note 41**).
4. Gently shake the container for 30 min.
5. Take out the membranes and place them on a lint-free paper to remove excess solution from the membrane surface.
6. Transfer the membranes onto a new, lint-free piece of paper once again (*see* **Note 42**).
7. Place the membranes between two new sheets of lint-free paper and allow them to dry at room temperature for 16–24 h.
8. Store the membranes in a box at room temperature until use (*see* **Note 43**).

3.3 Membrane Electrophoresis

1. Use a pencil to make a 7 mm-wide mark at 1.5 cm from the bottom edge of the membrane. This mark will serve as a marker for applying the samples.
2. In a plastic disposal tray, add approximately 20 mL of the running buffer.
3. Take a glass Petri dish and add approximately 5 mL of methanol. Dip the SMME membrane into the methanol in the dish and then quickly transfer the membrane into the running buffer in the disposal tray. Submerge the membrane in the solution using tweezers and gently shake it for 30 min.
4. Add approximately 100 mL of running buffer to the buffer tanks on both sides of the membrane electrophoresis chamber.
5. Wet the electrode filter papers with the buffer and place them in the chamber according to the instructions shown in Fig. 3.

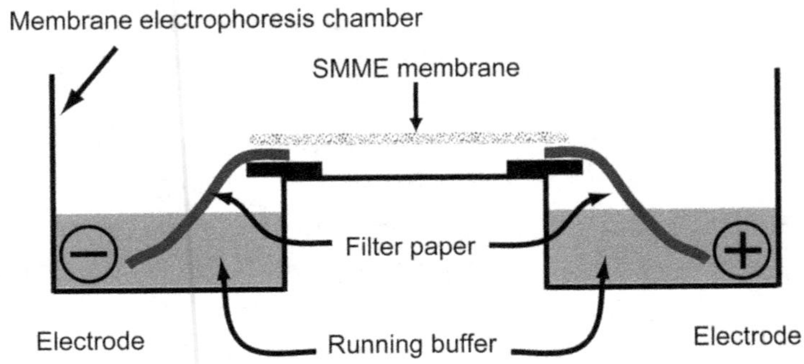

Fig. 3 Membrane electrophoresis assembly for SMME, utilizing an apparatus for cellulose acetate membrane electrophoresis. (Reproduced from [9] with permission. Copyright 2015 Springer Nature)

6. Position the SMME membrane to bridge the wetted electrode filter papers on each side (*see* **Note 44**).
7. Apply electrical current to the membrane using the power supply at a rate of 1.0 mA/cm for 5–10 min (*see* **Note 45**).
8. Turn off the electricity and then load the samples (typically 1 μL) to the marked position of the membrane. Note that a maximum of 10 μg of protein can be applied (*see* **Note 46**).
9. Once the sample solution has completely soaked into the membrane, perform electrophoresis in the constant current mode, typically at 1.0 mA/cm for 30 min.
10. Turn off the electricity and carefully remove the membrane. Proceed immediately to the next step of the procedure (*see* Subheading 3.4 or 3.5).

3.4 Alcian Blue Staining

1. In a glass Petri dish, add approximately 20 mL of a solution consisting of 30% acetic acid in methanol. Transfer the electrophoresed membrane (*see* Subheading 3.3, **step 10**) into the solution, cover the Petri dish, and gently shake it for 20 min.
2. In a plastic disposal tray, add approximately 20 mL of the Alcian blue solution. Transfer the membrane from the 30% acetic acid in methanol into the Alcian blue solution and gently shake it for 20 min.
3. Remove the membrane from the Alcian blue solution and transfer it to a new plastic disposable tray. Wash the membrane several times with methanol for a few minutes to remove the background color (*see* **Note 47**).
4. Take the membrane out of the methanol, place it on a new plastic disposable tray or a glass Petri dish, and dry it by heating it on a heat block (*see* **Note 48**).
5. In a corner of the tray, place 1 mL of diethylene glycol. Slowly place the dry membrane over the diethylene glycol at the edge.

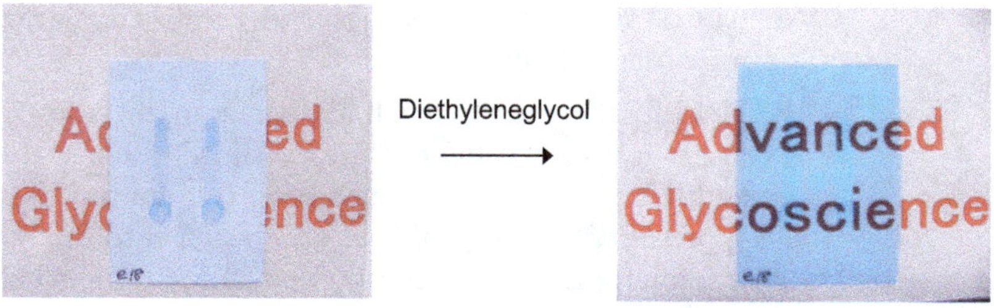

Fig. 4 Transparency of SMME membranes with diethylene glycol. Left: SMME membrane stained with Alcian blue. Right: SMME membrane soaked in diethylene glycol

As the diethylene glycol soaks into the membrane, it becomes transparent. Move the membrane slowly to ensure that the entire membrane is soaked with diethylene glycol (refer to Fig. 4) (*see* **Note 49**).

6. Remove the membrane from the tray and insert it into an OPP film pocket sheet, ensuring that any air bubbles are pushed out, if present.
7. Place the membrane in the OPP sheets on a white plate and acquire an image using a gel imaging system (*see* **Notes 15** and **50**).

3.5 Lectin Staining

To stain SMME membranes with lectin, first immobilize the membranes after electrophoresis, followed by blocking and processing as in normal western blotting. SMME membranes can be stained with antibodies using the same method. SMME membranes prepared using a PVA-PVP polymer blend should be used (*see* Subheading 2.2, **item 4**). Using SMME membranes prepared with PVA only may cause mucin release during incubation in a buffer solution containing Tween.

1. Add approximately 20 mL of acetone to a glass Petri dish. Transfer the electrophoresed membrane (*see* Subheading 3.3, **step 10**) into the acetone, cover the Petri dish, and gently shake it for 20 min (*see* **Note 51**).
2. Take out the membrane from acetone and dry it.
3. Sandwich the membrane between glassine papers and then sandwich them between glass fiber filter papers.
4. Press the sandwich at 150 °C for 5 min using a clothes iron without steam (*see* **Note 52**).
5. Add approximately 5 mL of methanol into a glass Petri dish. Pick up the membrane from the filter paper sandwich with a tweezer and dip the membrane in methanol for a few seconds. Then, quickly transfer the membrane into water in a plastic disposal tray. Submerge the membrane into the solution using tweezers and gently shake for 20 min.

6. Cut a hybridization bag several centimeter larger than the membrane edges.
7. Pick up the membrane from water in the disposal tray and insert it into the hybridization bag. Double seal the bottom and left and right sides of the bag using a thermo sealer. Add approximately 5 mL of blocking buffer into the bag using a 1 mL micropipette. Then, double seal the top of the bag.
8. Gently shake the bag for 60 min.
9. Cut the bag with scissors, remove the membrane from inside, and place it in the plastic disposal tray containing 20 mL of PBS-T. Gently shake it for 5 min, change PBS-T, and gently shake again. Repeat the washing procedure three times.
10. Pick up the membrane from PBS-T in the disposal tray and insert it into a newly cut hybridization bag (**step 6**). Double seal the bottom and left and right sides of the bag using the thermo sealer. Add approximately 5 mL of the lectin solution into the bag using a 1 mL micropipette. Then, double seal the top of the bag.
11. Gently shake the bag for 60 min.
12. Wash the membrane as described in **step 9**.
13. Pick up the membrane from PBS-T in the disposal tray and insert it into a newly cut hybridization bag (**step 6**). Double seal the bottom and left and right sides of the bag using the thermo sealer. Add approximately 5 mL of the HRP-conjugated streptavidin solution into the bag using a 1 mL micropipette. Then, double seal the top of the bag.
14. Gently shake the bag for 60 min.
15. Wash the membrane as described in **step 9**.
16. Spread the plastic wrap on a laboratory table, remove excess water from the washed membrane, and place it on the plastic wrap spread out on the table.
17. Cut the hybridization bag (**step 6**). Double seal the bottom side of the bag using the thermo sealer.
18. Pour a chemiluminescent peroxidase substrate onto the membrane on the plastic wrap according to the instructions in the chemiluminescent peroxidase substrate kit.
19. Insert the membrane into the hybridization bag (**step 17**). Double seal the top and left and right sides of the bag using the thermo sealer.
20. Acquire an image using the gel imaging system (*see* **Note 53**).

3.6 Glycan Analysis

Glycan analysis of mucin separated via SMME is performed by reductive β-elimination to release *O*-glycans (Subheading 3.6.1), which are then permethylated and analyzed via mass spectrometry (Subheading 3.6.2).

3.6.1 Releasing O-glycan from SMME Bands

1. Excise the stained mucin bands from the Alcian blue-stained SMME membrane (Subheading 3.4) and transfer them to a 1.5 mL microcentrifuge tube.
2. Add 500 μL of methanol to the tube and allow it to stand for 3 min. Then, aspirate the methanol using a micropipette and discard it. Add another 500 μL of methanol, allow it to stand for 3 min, and discard the methanol in the same manner. Repeat this procedure once more (*see* **Note 54**).
3. Add 500 μL of methanol to the tube and use surgical scissors to cut the stained mucin bands into approximately 1 mm squares in methanol (*see* **Note 55**).
4. Discard the methanol in the tube and add 25 μL of water, followed by 25 μL of 1 M $NaBH_4$ in 0.1 M NaOH into the tube. Cap the tube, mix the contents, and incubate at 45 °C for 16 h.
5. Add 5 μL of glacial acetic acid to neutralize and quench the reaction mixture (*see* **Note 56**).
6. Add 50 μL of cation exchange resin (H^+ form) to the tube and mix well.
7. Set up a conditioned Sep-Pak C18 Vac cartridge with a luer stop valve onto a gravity flow manifold.
8. Transfer the entire solution containing the resin onto the Sep-Pak C18 Vac cartridge using a micropipette with a large orifice tip. Wash the tube with 200 μL of water and transfer the washing onto the Sep-Pak C18 Vac cartridge. Open the luer valve and collect the flow-through fraction in a new 1.5 mL microcentrifuge tube.
9. Add 1 mL of water to the Sep-Pak C18 Vac cartridge and open the luer valve. Collect the washing in the 1.5 mL microcentrifuge tube containing the flow-through fraction.
10. Lyophilize the solution in the microcentrifuge tube using a centrifugal evaporator.
11. Add 100 μL of 1% (v/v) acetic acid in methanol to the microcentrifuge tube and dissolve the residues. Evaporate the solution using the centrifugal evaporator. Repeat this addition and evaporation procedure two more times (*see* **Note 57**).

3.6.2 Permethylation and MALDI MS

Iodomethane is a highly volatile methylating agent; **steps 3–8** must be performed in a fume hood.

1. Weigh approximately 80 mg of NaOH (granules) and transfer them into a disposable microtissue homogenizer tube. Add 1.0 mL of dimethyl sulfoxide (DMSO) containing 1% (v/v) distilled water per 80 mg NaOH to the tube and grind with a microhomogenizer pestle into a slurry (*see* **Note 58**).
2. Transfer approximately 50 μL of the slurry to the microcentrifuge tube containing glycans (*see* Subheading 3.6.1, **step 11**) (*see* **Note 59**).
3. In a fume hood, add 15 μL of methyl iodide to the tube using a micropipette with a glass capillary. Cap the tube securely and shake the mixture vigorously for 30–60 min (*see* **Note 60**).
4. Set up a conditioned Sep-Pak C18 Vac cartridge with a luer stop valve onto a gravity flow manifold.
5. Add 1 mL of 1% acetic acid to the reaction mixture **(step 3)** (*see* **Note 61**).
6. Transfer the mixture onto the Sep-Pak C18 Vac cartridge, open the luer valve, and collect the eluent in a waste pool.
7. Once all the reaction solution has flowed, add 1 mL of water onto the Sep-Pak cartridge and collect the washing in a waste pool. Repeat the washing procedure three times in the same manner.
8. With the valve open, push the liquid out of the Sep-Pak cartridge bed using an air pump and blow for an additional 3 min to remove any remaining methyl iodide.
9. Add 50 μL of MeOH to the Sep-Pak C18 Vac cartridge and allow it to soak into the Sep-Pak bed (*see* **Note 62**).
10. Add 1.5 mL of 50% acetonitrile into the Sep-Pak C18 Vac cartridge and collect the eluate in a new 1.5 mL microcentrifuge tube (*see* **Note 63**).
11. Evaporate the solution using a centrifugal evaporator until the sample is dry.
12. Add 20 μL of 50% acetonitrile to the tube containing the evaporated sample and dissolve the residue. Use the solution as a sample in the MALDI mass spectrometry (MS) analysis.
13. Deposit 0.5 μL of 2,5-DHB solution onto a MALDI target plate. Allow the spot to dry and then add 0.5 μL of the sample solution over the dried spot of 2,5-DHB. Allow the spot to dry again.
14. Acquire the MS spectrum in a positive-ion mode using a mass spectrometer (Fig. 5) (*see* **Note 64**).

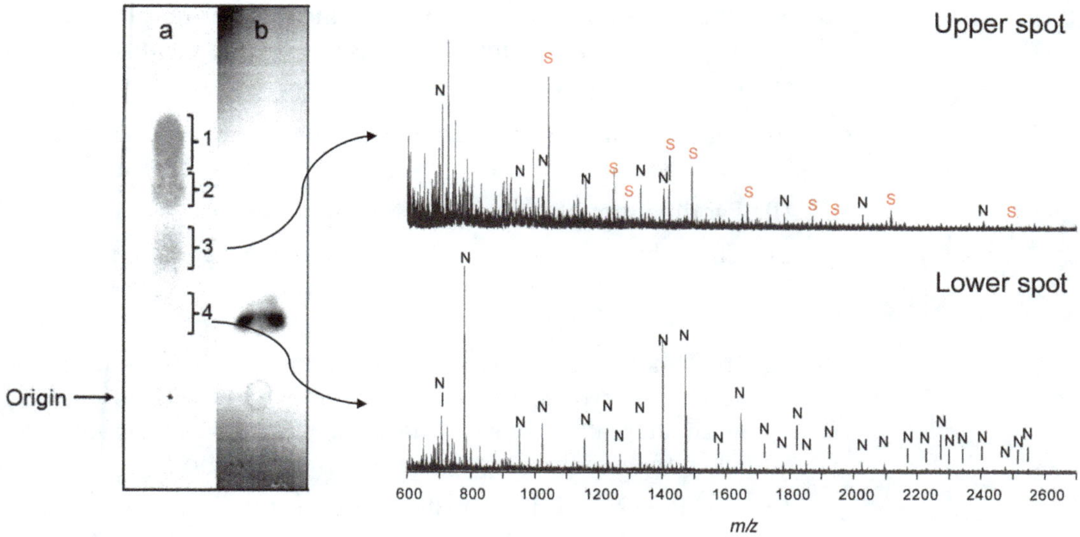

Fig. 5 MALDI-TOF MS analysis of *O*-glycans in PGM-glycoforms separated using the SMME technique. Left: crude mucin from porcine stomach was electrophoresed by SMME and stained with Alcian blue (**a**) and Pro-Q emerald (**b**). Right: mass spectra of permethylated glycans derived from the upper (band 3) and lower (band 4) spots. The capital letters "N" and "S" indicate neutral glycans and sulfated glycans, respectively. (Reproduced from [5] with permission. Copyright 2009 American Chemical Society)

4 Notes

1. This material contains hyaluronic acid, chondroitin sulfate, and porcine stomach mucins. These components are separated via SMME and can be used as reference markers.
2. Refer to Chapters 1, 2, 3, 4, 5, and 6 for the extraction of mucin. When performing SMME, avoid using detergents, such as SDS and Tween, during the extraction process.
3. We use the kit obtained from Cytiva.
4. We employ the Amicon Ultra-0.5 Centrifugal Filter Unit from Millipore.
5. A commercially available PVDF filter membrane for western blotting with a standard pore size of 0.45 μm (such as Immobilon-P, Hybond-P, or BioTrace) can be used. In our case, we use Immobilon-P from Millipore. It is important to consistently use membranes from the same provider, as the electrophoresis changes may vary slightly depending on the membrane used.
6. To dissolve the polymers, warm the beaker containing the polymer and water to 50 °C if necessary. Always use the same beaker since the polymer tends to adsorb onto it. Prepare polymer solutions just before use and ensure they are used only once.

7. Select a size that matches the cut membrane. In our case, we utilize 50 mm × 70 mm membranes, which is an ideal size.
8. Lint interferes with glycan analysis.
9. Lightly mark the SMME membrane using a pencil with a soft lead.
10. Pyridine should be used in a fume hood.
11. The provided filter paper is specifically designed for electrophoresis and has a low calcium content. However, other filter papers can be used as well.
12. If a membrane electrophoresis chamber is unavailable, a submarine-type agarose electrophoresis chamber can be used as an alternative. In such cases, the jig should be designed to allow the membranes to form a bridge (refer to Fig. 3).
13. Dividing 25 mL into smaller portions makes it easier to use. This amount is sufficient to stain five membranes (5 × 6 cm).
14. It is preferable to use a transparent sheet that allows the membrane to be sandwiched inside.
15. It should be possible to obtain an image of the gel stained with Coomassie brilliant blue.
16. The provided dilution ratio is suitable for initial studies. However, it is important to determine the optimum concentration for each lectin as needed.
17. Adjust the amount according to the number of membranes; 10 mL is sufficient to block two membranes (5 × 6 cm).
18. We utilize the TLC thermal blotter (AC-5970, ATTO, Tokyo, Japan). While a clothes iron can be used, avoid using steam.
19. We employ the ECL Prime Western Blotting Detection Reagents from Cytiva.
20. We utilize the ChemDoc XRS plus system from Bio-Rad Laboratories Inc.
21. The method described here is specifically for the analysis of *O*-glycans of mucins separated via SMME. For a general analysis of permethylated *O*-glycans, please refer to Chapter 17.
22. Begin by measuring out $NaBH_4$ and calculating the required amount of 0.1 M NaOH solution. Add the calculated amount of NaOH solution accordingly.
23. It is desirable to degas.
24. We use Dowex 50W ×8, 50–100 mesh (H).
25. The addition of water prevents the oxidation of hydroxyl groups during permethylation.
26. Use granules, not pellets.

27. Conditioning should be performed immediately before use according to the provider's instructions.
28. Always use LC–MS grade. Can be stored at room temperature for 1 month.
29. The addition of sodium allows the glycan signals to converge on the sodium-adducted ions during mass spectrometry. It is easy to use by dividing 100 μL into smaller portions and freezing until use.
30. If not available, the tip of a regular pipette tip can be cut with scissors.
31. We use BioMasher II (Nippi Incorporated, Japan).
32. We use Acura 846.025 (Socorex Isba SA, Switzerland).
33. Used for solid-phase extraction cartridges with luer stop valves.
34. Always use an oil-free pump to avoid contamination by oil vapor.
35. This solution is used as a reference sample and can be stored at 4 °C for 1 year.
36. Ensure that no air exists between the membrane and water.
37. The duration should be adjusted accordingly.
38. This solution can be stored at 4 °C for 1 year.
39. We usually make them 5-cm wide. This size allows for the application of five samples.
40. This treatment is used to make the membrane hydrophilic. The entire membrane should be wetted with methanol.
41. Quickly transfer the membrane to the hydrophilic polymer while it is still wet with methanol.
42. The objective is to remove excess hydrophilic polymer solutions. This is done until the surface of the membrane no longer glows with liquid.
43. Can be stored at room temperature for 1 year.
44. Keep the membrane wet with buffer supplied via the electrode filter papers on each side.
45. Excess buffer on the surface of the membrane will reflect light. Apply the current until this reflection completely disappears.
46. If the sample solution contains urea, the applied area becomes translucent, but there is no problem.
47. Alcian blue solution is collected and can be used approximately five times.
48. The purpose is to completely remove methanol and water from the membrane. Temperature and time should be adjusted accordingly.

49. Adjust the amount of diethylene glycol based on the size of the membrane (30 μL/cm^2).
50. We acquire images using the epi-white mode of the ChemDoc XRS Plus system (Bio-Rad Laboratories Inc.) with the white light conversion screen.
51. The membrane may swell slightly with acetone, but this does not pose a problem.
52. Skip this step for samples in which the antigen is heat denatured. Glycans are not affected.
53. Due to chemiluminescence quenching, control the timing between pouring the peroxidase substrate solution and capturing the image according to the instructions provided with the reagent kit.
54. This step removes diethylene glycol from the membrane and can be omitted if diethylene glycol is not used in Alcian blue staining (**step 5** in Subheading 3.4).
55. Cutting the membrane while it is dry can make it difficult to handle due to static electricity. Always work with the membrane submerged in methanol.
56. The addition of acid may result in the rapid release of hydrogen gas.
57. The objective is to methyl esterify and distill boric acid. As a result, little or no visible residue should remain in the tube.
58. Quickly measure out NaOH and calculate the amount of DMSO to be added based on the measured amount of NaOH. To grind with a pestle, stir the liquid so that the sodium hydroxide granules mix in the DMSO, and grind the granules with the side of the pestle to prevent hardening. Avoid crushing the granules with the tip of the pestle.
59. Stir the slurry well to ensure homogeneity before dispensing 50 μL, as NaOH may precipitate out.
60. The reaction mixture will turn into a white solid.
61. Dissolve the white solid.
62. If soaking is slow, gently apply air pressure. Do not force out methanol.
63. Tubes coated with low adsorption materials can interfere with mass spectrometry. Use only uncoated tubes. We recommend Eppendorf Safe-Lock Tubes (Eppendorf SE).
64. Glycans should be detected as sodium-adducted ions $[M + Na]^+$ of the corresponding permethylated alditols.

Acknowledgments

This work was supported by a Grant-in-Aid for Scientific Research (KAKENHI 23310154, 17H03808, 17K19627, and 21H02123) from the Japan Society for the Promotion of Science (JSPS).

References

1. Andrianifahanana M, Moniaux N, Batra SK (2006) Regulation of mucin expression: mechanistic aspects and implications for cancer and inflammatory diseases. Biochim Biophys Acta 1765:189–222
2. Brockhausen I (2006) Mucin-type *O*-glycans in human colon and breast cancer: glycodynamics and functions. EMBO Rep 7:599–604
3. Schulz BL, Packer NH, Karlsson NG (2002) Small-scale analysis of *O*-linked oligosaccharides from glycoproteins and mucins separated by gel electrophoresis. Anal Chem 74:6088–6097
4. Andersch-Björkman Y, Thomsson KA, Holmén Larsson JM et al (2007) Large scale identification of proteins, mucins, and their *O*-glycosylation in the endocervical mucus during the menstrual cycle. Mol Cell Proteomics 6: 708–716
5. Matsuno Y-K, Saito T, Gotoh M et al (2009) Supported molecular matrix electrophoresis: a new tool for characterization of glycoproteins. Anal Chem 81:3816–3823
6. Dong W, Matsuno Y-K, Kameyama A (2012) A procedure for Alcian blue staining of mucins on polyvinylidene difluoride membranes. Anal Chem 84:8461–8466
7. Matsuno Y-K, Dong W, Yokoyamam S et al (2011) Improved method for immunostaining of mucin separated by supported molecular matrix electrophoresis by optimizing the matrix composition and fixation procedure. Electrophoresis 32:1829–1836
8. Matsuno Y-K, Dong W, Yokoyama S et al (2013) Identification of mucins by using a method involving a combination of on-membrane chemical deglycosylation and immunostaining. J Immunol Methods 394: 125–130
9. Matsuno Y-K, Kameyama A (2015) Supported molecular matrix electrophoresis. In: Kurien BT, Scofield RH (eds) Detection of blotted proteins methods and protocols. Springer Nature, New York
10. Ciucanu I, Costello CE (2003) Elimination of oxidative degradation during the per-*O*-methylation of carbohydrates. J Am Chem Soc 125: 16213–16219

Part II

Staining, Detection, and Quantitation

Chapter 8

Supported Immunohistochemical Staining of Mucins

Michiyo Higashi

Abstract

Immunohistochemistry (IHC) staining is the most common method to detect the distribution and localization of biomarkers in different parts of a tissue. Antibodies for tandem repeat peptide of mucins are very popular, but antibodies for glycosylation or others are also used. IHC for mucin is the same protocol as IHC for others. This description includes IHC according to ABC method for processed formalin-fixed paraffin-embedded (FFPE) tissues. Protocol of in situ hybridization is also shown.

Key words Immunohistochemistry, Mucins, Indirect method, Avidin-Biotin conjugate, In situ hybridization

1 Introduction

Immunohistochemistry (IHC) staining is the most common application to detect proteins expressed in cells of tissue using antigens and antibodies interaction to visualize localization of the proteins, although less quantitative than assays such as western blotting or Enzyme-Linked immunp Sorbent Assay (ELISA). IHC staining is widely used not only in pathology labs but also for basic research to detect the distribution and localization of biomarkers in different parts of a tissue. IHC detection systems are the direct method and indirect method including peroxidase antiperoxidase (PAP), the avidin-biotin conjugate (ABC), and the polymer enzyme system [1]. As the IHC method has been evolved, we can obtain greater sensitivity and reducing procedure with a higher cost. I mainly describe the method of IHC according to ABC method for processed formalin-fixed paraffin-embedded (FFPE) tissues, which is one of the typical methods. At first, preparing tissue is required: tissue processing, fixation, and sectioning. At the same time or in parallel, suitable primary antibody is picked for the research. Primary antibody selection is the most important decision, since successful IHC experiment relies on the primary antibody specifically

Akihiko Kameyama (ed.), *Mucins: Methods and Protocols*, Methods in Molecular Biology, vol. 2763,
https://doi.org/10.1007/978-1-0716-3670-1_8,

binding the target antigen. Detection system is also selected: direct method, indirect method. Before the process of primary antibody, antigen retrieval and/or blocking proteins are done as necessary. Blocking endogenous enzymes are required for the ABC method. After washing off the chromogen and running water, a suitable counterstain is selected. It is important to use a counterstain, so that we could be able to see where the staining from the antibody is in relation to the cellular structures within the tissue. On adding mounting media and coverslips for specimen protection, preparation of optimal viewing is completely done. Many commercial kits are available, following the manufacturers' instructions from the web of manufactures: Vector Laboratories [2], abcam [3], Agilent [4], and Roche [5]. Careful points and pitfalls are noticed, and I also show the protocol of in situ hybridization briefly in our experiences [6].

2 Materials

2.1 Reagents for Immunohistochemistry

1. Antibody. An antibody that has been shown to work in IHC is preferable. It doesn't matter polyclonal or monoclonal (*see* **Note 1**).
2. Tissue samples. Tissue or cells. Control tissues (positive and negative) are also prepared, if possible.
3. 10% neutral buffered formalin (*see* **Note 2**).
4. Paraffin.
5. Vectastain Elite ABC kit (Vector Laboratories, Burlingame, CA, USA) (*see* **Note 3**).
6. Diluted normal blocking serum: 1.5% normal blocking serum in Phosphate-buffered saline (PBS).
7. PBS stock solution: 0.1 M PBS stock solution (pH 6.9). Weigh Na_2HPO_4 11.5 g, KH_2PO_4 2 g, KCl 2 g, NaCl 80 g, and add ~100 mL of water to a 1-L graduated cylinder or glass beaker. Mix with the water. Add more water until the volume of the mixture reaches 900 mL. Mix and adjust the pH to 6.9 with HCl. Add more water until the volume reaches 1 L. Store the stock solution at 4 °C. The stock solution can be stored for up to 1 month.
8. PBS working solution (0.01 M PBS): Mix 100 mL of PBS stock solution and 900 mL of distilled water, adjust pH to 7.5 (7.4–7.6) by HCl (1 N) or NaOH (1 N) before use. Store at room temperature and use within a week.
9. PBS-BSA for the dilution of primary antibodies: Dissolve 1 g of bovine serum albumin and 0.1 g of sodium azide (NaN_3) with PBS working solution. Store at 4 °C and use within 6 months.

10. DAB (Diaminobenzidine) stock solution: Add 0.1 g of DAB to 10 mL of distilled water and dissolve. Filter and aliquot the solution to 330 μL and freeze. Use DAB and solution containing DAB with keeping gloves on.
11. DAB solution for detection: Mix 150 mL of PBS, 1 mL of 1% DAB stock solution, and 150 μL of conc. H_2O_2.

2.2 Oher Solutions for Immunohistochemistry

1. Xylene.
2. Ethanol.
3. 0.3% H_2O_2 in 100% methanol.
4. Hematoxyline (*see* **Note 4**).
5. Mounting media.

2.3 Instruments and Others

1. Microtome (Fig.1a): for FFPE tissue.
2. Cryostat (Fig.1b, c): for frozen tissue, keep tissues under −20 °C.
3. Blades.
4. Slide glasses (*see* **Note 5**).
5. Coplin staining jar (Fig. 2).
6. Staining basket (Fig. 3): Convenient tool for handling many slides.
7. Incubation chamber (Fig. 4): Applying antibody or detection dye (*see* **Note 6**).
8. Wide-mouth wash bottle.
9. Task wipers.
10. Cover glass.
11. Microscope (*see* **Note 7**).
12. Dropper or Container with syringe.
13. Oven.

2.4 Materials for In Situ Hybridization

1. Digoxigenin-labeled probes. Radio-, fluorescent- or antigen-labeled bases also can be used.
2. Tris buffered saline (TBS) buffer.
3. Developing solution with NBT/BCIP (nitroblue tetrazolium solution and 5-bromo-4-chloro-3-indolylphosphate solution).
4. Fast red.
5. Alkaline phosphatase-conjugated antidigoxigenin antibody.
6. 0.1 N HCl.
7. Pepsin solution: Pepsin in 0.01 N HCl. Final concentration is 0.5–3.0 mg/mL.

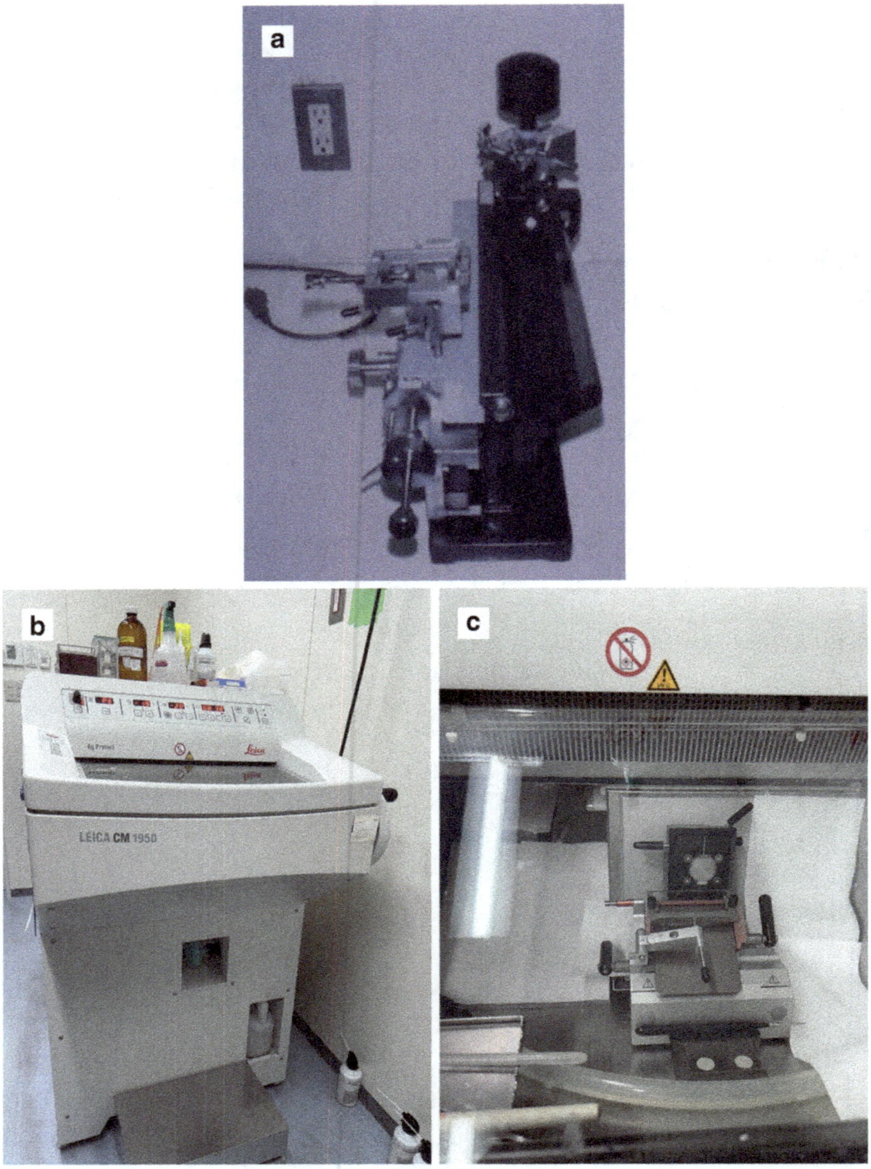

Fig. 1 Sectioning. (**a**) Microtome is used for formalin-fixed paraffin-embedded (FFPE) tissues. (**b**) Overview of cryostat. This is used for frozen tissues. (**c**) Inside of frozen chamber of cryostat. The room is kept under −20 °C

3 Methods

Carry out all procedures at room temperature unless specified otherwise. Keep wet the slides during procedures. You can keep the slides in 0.01 M PBS in order to suspend IHC. Use DAB, and solution including DAB with keeping gloves on.

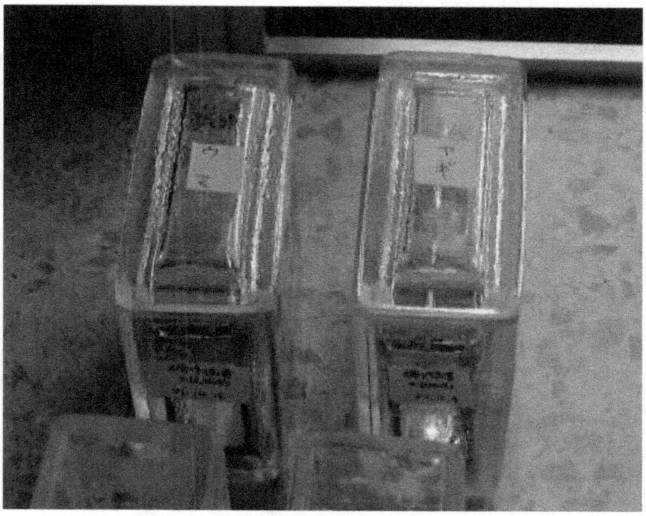

Fig. 2 Coplin stating jar. During the washing and DAB (Diaminobenzidine) processes, we can leave the slides in the jar

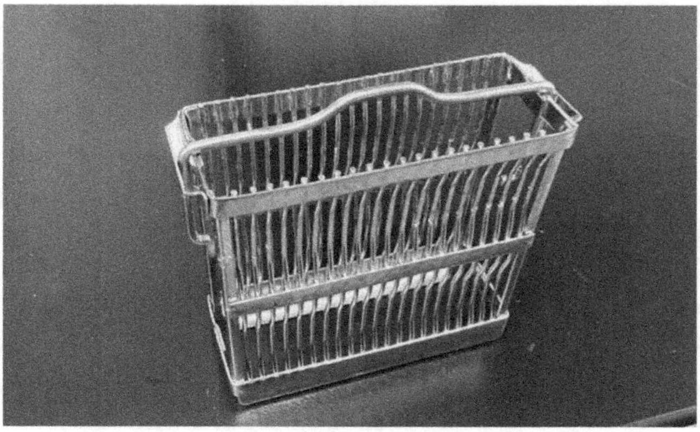

Fig. 3 Staining basket. You can shake this basket in a Coplin stating jar to wash well

3.1 Preparation of the Good Tissue Slides for IHC

1. Fixation: Dip the fresh tissue or cells in buffered 10% formalin for 6 h to 2 days (room temperature).
2. Paraffin embedding (complete dehydration).
3. Tissue sectioning used by microtome attached a blade and put on glass slide. Thickness is usually 4–5 μm. Cryostat is used for fresh and frozen section, fixtation and paraffin embedding are not necessary.
4. Complete adhesion (in 50 °C oven 2 h) to prevent detachment of tissue (*see* **Note 8**).

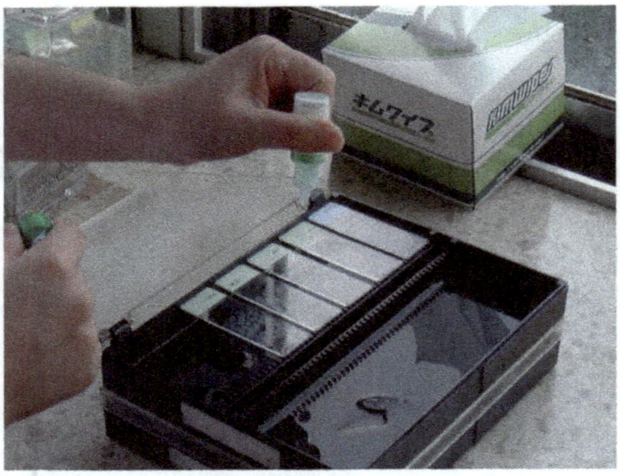

Fig. 4 Incubation chamber. During primary antibody incubation and detection, pour water in the bottom of the chamber and cover so as not to dry the slides

3.2 Deparaffinization, Rehydration and Blocking of Endogenous Peroxidase

1. Dip the slides into xylene (10 min × 2) and rehydration by ethanol gradient (100 to 70%).
2. Wash in distilled water for 5 min.
3. Incubating the slides in 0.3% H_2O_2 in 100% methanol at room temperature for 30 min to eliminate endogenous peroxidase (*see* **Note 9**).
4. Rinse for 5 min in tap water, and rinse with distilled water.
5. If antigen retrieval is required, perform the process continuously (*see* **Note 10**).

3.3 Primary Antibody Incubation

1. Incubate slides for 20 min with diluted normal blocking serum (1.5% in PBS) in order to block nonspecific staining. Wipe off excess serum from slides.
2. Incubate sections with diluted primary Antibody (typically 1 h at room temperature or overnight at 4 °C).
3. Wash slides in PBS for 3–4 times.

3.4 Detection

1. Prepare ABC complex (solution A and B each 1% in PBS) 30 min before use.
2. Incubate slides with ABC complex for 30 min.
3. Wash slides in PBS jet wash × 3–4 times.
4. Incubate slides in DAB solution (PBS 150 mL + 1%DAB in distilled water 1 mL + conc. H_2O_2 150 μL) for 10 min. Check color used by microscope with gloves (Fig. 5).
5. Wash slides with tap water in order to stop reaction.

3.5 Mounting and Examination

1. Counterstaining (hematoxylin) a few seconds.
2. Wash slides with running water for 10 to 15 min.

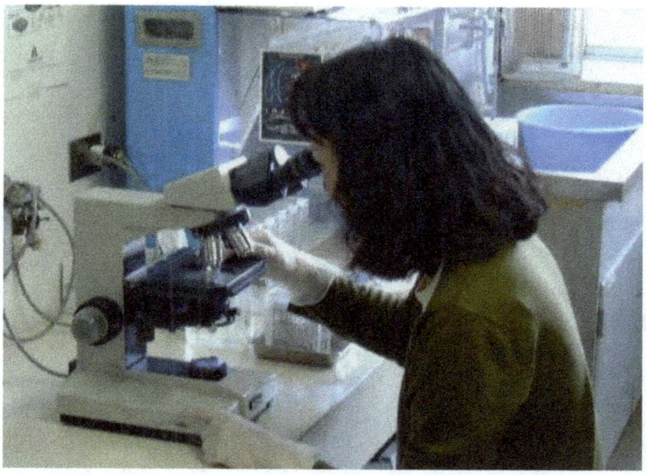

Fig. 5 Checking the colors of the slides used by microscope. Keep the gloves on while using DAB solutions

3. Incubate slides in 100% ethanol for 1 min × 3–4 times.
4. Incubate slides in xylene.
5. Add mounting media and coverslips for the specimen protection and optimal viewing.

3.6 In Situ Hybridization

1. Dip the slides into xylene (10 min × 2) and rehydration by ethanol gradient (100 to 70%).
2. Wash the slides with TBS, immersed in 0.1 N HCl, and digested in 0.01 N HCl containing pepsin for 30 min at 37 °C.
3. Wash the slides with TBS, and dehydration by ethanol gradient (70 to 100%) and air-dry.
4. Immerse the slides in prehybridization solution for 30 min at 37 °C.
5. Wash the slides with TBS, and dehydration by ethanol gradient (70 to 100%) and air-dry.
6. Heat the digoxigenin-labeled probes for 5 min at 100 °C and chill on ice. Mix in prehybridization solution.
7. Incubate the slides in mixture overnight at 37 °C.
8. Wash the slides three times in TBS at 37 °C.
9. Wash the slides twice in TBS at room temperature.
10. Incubate the slides in alkaline phosphatase-conjugated antidigoxigenin antibody for 30 min at room temperature.
11. Wash the slides three times in TBS at room temperature.
12. Incubate developing buffer pH 9.8 with NBT/BCIP at 37 °C.
13. Wash in tap water to stop reaction.

14. Counterstain with nuclear fast red.
15. Wash slides with running water for 10–15 min.
16. Incubate slides in 100% ethanol for 1 min × 3–4 times.
17. Incubate slides in xylene.
18. Add mounting media and coverslips for the specimen protection and optimal viewing.

4 Notes

1. We used antibodies shown in Table 1.
2. 10% formaldehyde, or 20% formaldehyde can be used, but 10% formaldehyde is preferred to preserve antigens, RNA, and DNA. The best way is the purchase of 10% neutral buffered formalin. Standard stock solution of formalin is typically 37–40% formaldehyde gas in aqueous solution and unbuffered. This is regarded as 100% formalin. 10% formalin represents 10% of the 37–40% stock solution.

   4% PFA (paraformaldehyde) can also be used: for cells or small tissue, such as biopsy.
3. A commercially available immunohistochemistry kit can be used.
4. The use for counterstain. Other dyes: Nuclear fast red or methyl green can also be used.
5. Silane-coated glass-slides would be preferred. Dip the uncoated slides into 1% silane diluted by acetone, and air-dry.
6. Commercial incubation chambers are available. We are reusing slide preservation cases.
7. Microscope for checking the color is preferred. If impossible, keep the gloves on.
8. Slides are used within 2 weeks to 1 month ideally, but you can stock the slide at room temperature for several years in an area

Table 1
List of antibodies for mucin staining

Antigen	Clone	Company	Dilution ratio
MUC1	DF3	Fujirebio	×10
MUC1-CORE	Ma695	Novocastra	×100
MUC2	CCP58	Novus biologicals	×200
MUC4	8G7	Santa Cruz biotechnology	×100
MUC4	1G8	Santa Cruz biotechnology	×500

out of direct sunlight and dusts. Reactivity of some antigens is weak or lost.

9. 3% H_2O_2 in 100% methanol at room temperature for 5 min can be available, but several antigens are lost sometimes. You can skip this step if your detection system is not ABC method.

10. Antigen retrieval methods break methylene bridges because of formaldehyde and expose antigenic sites, allowing antibodies. Heat-induced epitope retrieval and proteolytic-induced epitope retrieval are representative.

Acknowledgments

This work was supported by a Grant-in-Aid for Scientific Research (KAKENHI 21 K06889) from the Japan Society for the Promotion of Science (JSPS).

References

1. Cartum RW, Taylor CR, Dabbs DJ (2021) Techniques of immunohistochemistry. In: Dabbs DJ (ed) Diagnostic immunohistochemistry, 6th edn. Saunders Elsevier, Philadelphia
2. https://vectorlabs.com
3. https://www.abcam.com
4. https://www.agilent.com/ja-jp/product/immunohistochemistry
5. https://www.roche.com/
6. Yonezawa S, Sueyoshi K, Nomoto M et al (1997) MUC2 gene expression is found in non-invasive tumors but not in invasive tumors of the pancreas and liver: its close relationship with prognosis of the patients. Hum Pathol 28:344–352

Chapter 9

Succinylation-Alcian Blue Staining of Mucins on Polyvinylidene Difluoride Membrane

Weijie Dong and Akihiko Kameyama

Abstract

Mucins are often stained with the basic dye Alcian blue, but mucins with a low acidic glycan content cannot be stained with it. Succinylation-Alcian blue staining is a method that temporarily modifies glycans with succinic acid to visualize mucins with low acidic glycan content. This method can be used to stain mucins on polyvinylidene difluoride (PVDF) membranes separated via supported molecular matrix electrophoresis (SMME) and mucins blotted onto PVDF membranes from gel electrophoreses. The succinyl groups of the modified glycans can be easily and completely removed by releasing *O*-glycan from the stained mucin bands. Therefore, the glycans can be analyzed using the same methods as those used for mucins with a high acidic glycan content.

Key words Polyvinylidene difluoride membrane, Mucins, Alcian blue staining, Succinic acid, Supported molecular matrix electrophoresis, *O*-Glycans

1 Introduction

Gel electrophoresis, such as agarose gel electrophoresis or agarose–polyacrylamide composite gel electrophoresis [1], and membrane electrophoresis employing a cellulose acetate membrane or polyvinylidene difluoride (PVDF) membrane treated with polyvinyl alcohol [2] are frequently used for the analytical separation of mucins. The separated mucin is typically visualized with Alcian blue or periodic acid-Schiff (PAS) staining, regardless of the separation method. Mucopolysaccharides and acidic mucins are commonly visualized with Alcian blue staining on tissue sections, electroblotted membranes from gel electrophoresis, and electrophoresed membranes [2–4]. However, this method cannot be used to stain mucins with a low acidic glycan content, such as porcine gastric mucin (PGM), which consists primarily of neutral glycans [2]. PAS staining can selectively visualize glycoproteins, including mucins, regardless of the acidic residue content; however, this

Akihiko Kameyama (ed.), *Mucins: Methods and Protocols*, Methods in Molecular Biology, vol. 2763,
https://doi.org/10.1007/978-1-0716-3670-1_9,

method is incompatible with glycan analysis because periodic acid oxidation degrades glycans. The succinylation-Alcian blue staining method introduced here is compatible with glycan analysis [5] and can visualize mucins on PVDF membranes independent of acidic residues.

Alcian blue is a cationic dye containing four isothiouronium residues and can sensitively bind negatively charged macromolecules (e.g., mucopolysaccharides) through electrostatic forces [6]. Mucin molecules are characterized by high-density clusters of *O*-linked glycans; each monosaccharide unit of the glycan cluster contains multiple hydroxyl groups and can be used for esterification. In succinylation-Alcian blue staining, mucins with a low acidic glycan content are temporarily converted to mucins with a high acidic glycan content through succinylation of the hydroxyl groups of the glycan cluster. The succinylated mucin derivatives can then bind to Alcian blue (Fig. 1). During *O*-glycan revelations such as reductive *β*-elimination and eliminative oximation (*see* Chap. 13), which are frequently used for glycan release from the mucins, succinyl monoesters are readily removed from the glycans. Therefore, succinylation-Alcian blue staining would not influence glycan analysis of mucins (Fig. 2), and this method can be applied to mucins separated via supported molecular matrix electrophoresis (SMME, *see* Chap. 7) and mucins electroblotted on a PVDF membrane. Here, we describe the procedures for succinylation-Alcian blue staining of mucins separated via SMME.

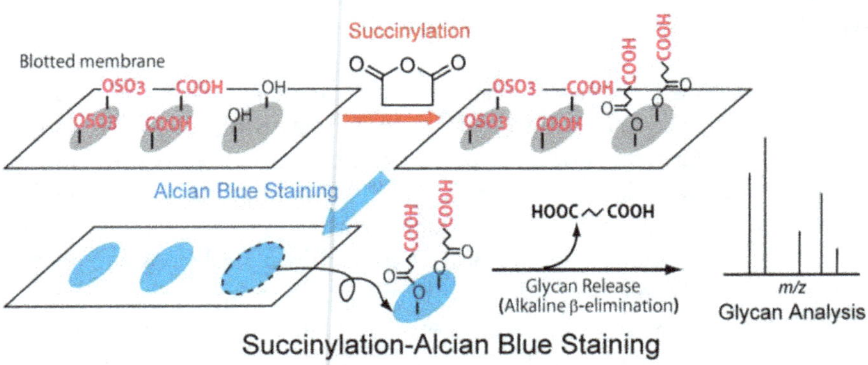

Fig. 1 Schematic diagram of succinylation-Alcian blue staining. Carboxyl groups are introduced to glycan clusters of neutral mucins through succinylation. The succinylated mucins can be stained with Alcian blue. Succinyl groups can be readily removed from glycans during alkaline *β*-elimination, which is a method for releasing *O*-linked glycans. (Reproduced from [5] with permission. Copyright 2012 American Chemical Society)

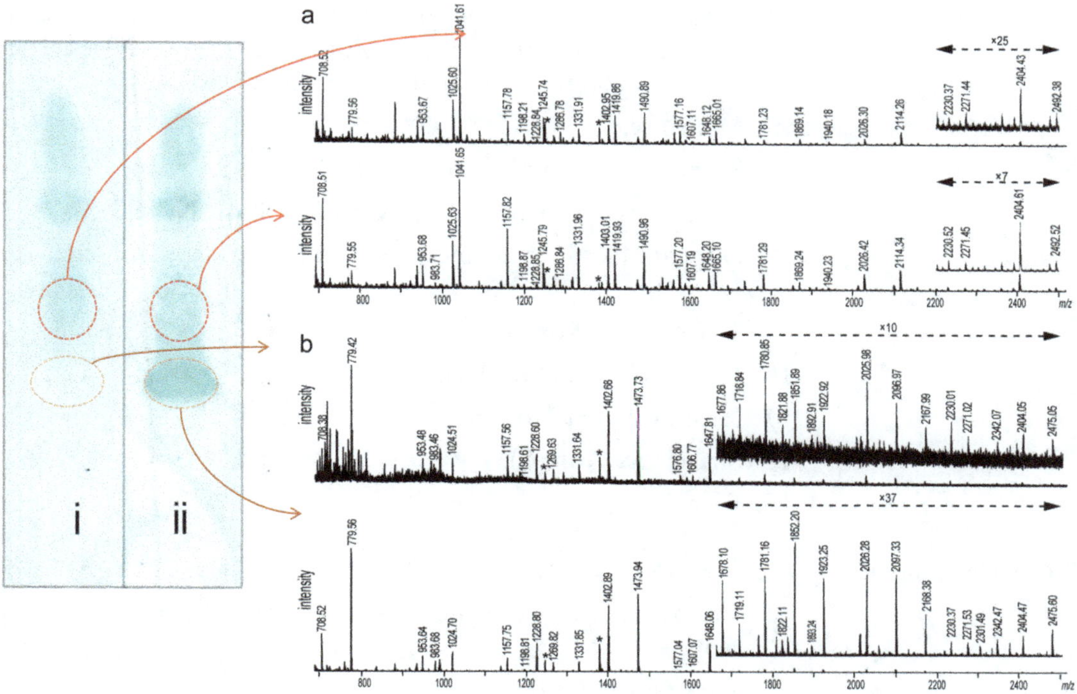

Fig. 2 Matrix-assisted laser desorption ionization time-of-flight mass spectrometer (MALDI-TOF MS) spectra of *O*-linked glycans released from the supported molecular matrix electrophoresis (SMME)-separated mucins of porcine gastric mucin. Left: lane (i) shows Alcian blue staining and lane (ii) shows succinylation-Alcian blue staining. Right: panel "a" displays the MS spectra of acidic glycans and panel "b" shows the MS spectra of neutral glycans. The same signals are observed with both Alcian blue and succinylation-Alcian blue staining

2 Materials

Prepare all solutions using ultrapure water and analytical grade reagents. Ensure that all reagents are prepared and stored at room temperature, unless specified otherwise. For specific instructions on working with PVDF membrane in the presence of mucin, please refer to Chap. 7.

2.1 Pretreatment of Membrane (See Note 1)

1. PVDF membrane with mucin present (*see* **Note 2**).
2. Acetone.
3. Glass Petri dish with lid (diameter: 10 cm, *see* **Note 3**).
4. Tweezer (*see* **Note 4**).
5. Heat block (*see* **Note 5**).
6. Shaker (*see* **Note 6**).

2.2 Succinylation-Alcian Blue Staining

1. Succinic anhydride.
2. Acetonitrile.
3. Pyridine.

4. Alcian blue solution: Weigh 100 mg of Alcian blue 8GX and transfer it to a 200 mL glass bottle. Add 100 mL 0.1% acetic acid to dissolve it. Store at 4 °C (*see* **Note7**).
5. Methanol.
6. Screw capped glass tube (*see* **Note 8**).
7. Glass Petri dish with lid (diameter: 10 cm, *see* **Note 3**).
8. Disposal plastic containers (9 × 12 cm, depth: 2 cm, *see* **Note 9**).
9. Tweezer (*see* **Note 4**).
10. Plastic bag (*see* **Note 10**).
11. Heat block (*see* **Note 5**).
12. Shaker (*see* **Note 6**).
13. Fume hood.

2.3 Image Capture

1. Diethylene glycol.
2. Disposal plastic containers (9 × 12 cm, depth: 2 cm, *see* **Note 9**).
3. Tweezer (*see* **Note 4**).
4. Oriented polypropylene (OPP) film pocket sheet (*see* **Note 11**).
5. ChemDoc XRS plus system (*see* **Note 12**).
6. White light conversion screen (*see* **Note 13**).

3 Methods

After thoroughly drying the membrane, treat it with succinic anhydride and stain it with Alcian blue. Once stained, make the membrane transparent to obtain an image. Perform all procedures at room temperature, unless otherwise specified.

3.1 Pretreatment of Membrane

1. Add approximately 20 mL of acetone to a glass Petri dish. Transfer the electrophoresed SMME membrane into the acetone in the dish, cover the dish with a lid, and gently shake for 30 min to remove water from the membrane (*see* **Note 14**).
2. Remove the membrane from the acetone and keep it at room temperature for 20 min. Then, place it on a heat block at 80 °C for 20 min to dry. Allow the membrane to cool to room temperature before further handling. It is important to note that heating the membrane directly after removal from acetone may cause it to adhere to the heating block.

3.2 Succinylation-Alcian Blue Staining

1. Dissolve 500 mg of succinic anhydride in 4 mL of acetonitrile in a glass tube (*see* **Note 15**).
2. Add 1 mL of pyridine to the solution and mix it using a vortex mixer (*see* **Notes 16** and **17**).
3. Pour the freshly prepared reagent mixture into a glass dish (*see* **Note 3**).
4. Immerse the dried membrane into the reagent mixture and immediately cover the glass dish with the lid (*see* **Notes 16** and **18**).
5. Place the glass dish in a plastic bag and seal it (*see* **Notes 16** and **19**).
6. Keep the plastic bag in the fume hood for 2 h (*see* **Note 20**).
7. Add approximately 20 mL of the Alcian blue solution to a plastic container. Transfer the membrane into the Alcian blue solution and gently shake it for 20 min (*see* **Note 21**).
8. Remove the membrane from the Alcian blue solution and transfer it to another plastic container. Wash and decolorize the membrane with 10 mL of methanol. Discard the blue methanol and repeat the decolorizing procedure with 10 mL of fresh methanol. Repeat this step several times to remove the background (*see* **Note 22**).
9. Take out the membrane and keep it at room temperature for 20 min. Then, dry the membrane on a heat block at 80 °C for 20 min, and allow it to cool to room temperature (*see* **Note 23**).

3.3 Image Capture (Also See Chap. 7)

1. Place the dried membrane on a glass Petri dish.
2. Soak the membrane in 1 mL of diethylene glycol (*see* **Note 24**).
3. Hold the membrane between OPP sheets and remove any air bubbles, if present (*see* **Note 25**).
4. Set the membrane with the OPP sheet on the white light conversion screen (*see* **Note 26**).
5. Capture the image of the membrane using the epi-white mode with the ChemDoc XRS plus system (*see* **Note 27**).

4 Notes

1. The purpose of this treatment is to remove water from the membranes after electrophoresis and blotting, as water can reduce the efficiency of mucin succinylation.
2. PVDF membranes with blotted mucins or SMME membranes. When using SMME membranes, it is crucial to ensure that the membrane is prepared with mixed hydrophilic polymer:

polyvinyl alcohol should be less than one-third of the total hydrophilic polymers. Higher polyvinyl alcohol content can result in a darker background. For the preparation of SMME membranes, refer to Chap. 7.

3. Select a dish that is just large enough to accommodate the membrane. For example, a 10 cm dish can be used for 5 × 6 cm membrane.
4. Flattened tips on forceps make it easier to grip the membrane.
5. The heating block or system used should be capable of holding and heating the membranes to 80 °C.
6. A shaker can be used to shake the Petri dish during the acetone treatment.
7. Store the Alcian blue solution in small portions according to the size of the membranes to be used. For 5 × 6 cm membranes, it is recommended to store them in 20 mL portions.
8. Pyridine emits a strong odor, so ensure that the glass tube is tightly capped.
9. Use a disposable dish just large enough to fit the membrane. For 5 × 6 cm membranes, select a dish with the appropriate size as indicated.
10. Choose a bag that is large enough to fully cover the Petri dish.
11. Use a pocket sheet that can hold the membrane inside. Make a small slit in the pocket sheet to easily insert the membrane.
12. A gel imaging system suitable for visualizing Coomassie blue-stained gels can be used to capture the membrane image.
13. This is used to convert transmitted UV light (transilluminator) to transmitted white light. When using an imager with a white light source, follow the manufacturer's instructions.
14. The membrane may swell slightly when immersed in acetone, but this is not a concern.
15. Adjust the volume of succunic anhydride solution based on the size of the membrane. For example, use 4 mL of reagent solution for 5 × 6 cm membranes. Do not change the concentration.
16. Perform these steps in a fume hood due to the strong odor of pyridine.
17. The amount of pyridine should be 1/5 of the volume of the succinic anhydride solution used.
18. Ensure that the entire surface of the membrane is immersed in the reagent solution. Pouring the solvent directly onto the dried membrane may cause unnecessary scratch-like backgrounds.

19. Cover the Petri dish with a plastic bag to prevent solvent vapor from escaping between the dish and the lid.
20. Shaking is not necessary. If room temperatures are lower, it is advisable to heat the dish to approximately 40 °C.
21. Place the membrane in the staining solution without washing off the reagent solution.
22. The Alcian blue staining solution can be reused approximately five times.
23. Ensure that the membrane is free of water or methanol, as these may affect the transparency of the membrane in the subsequent steps.
24. Adjust the amount of diethylene glycol based on the size of the membrane. As a general guideline, use 40 μL/cm^2.
25. Wear gloves and use your fingers to push out any air bubbles. Without gloves, sebum can adhere to the film and result in a dirty image.
26. Other materials can be used as long as they provide a white background.
27. Acquire the image using the same approach as that for capturing pictures of Coomassie blue-stained gels.

References

1. Holden KG, Yim NCF, Griggs LJ, Weisbach JA (1971) Gel electrophoresis of mucous glycoproteins. I. Effect of gel porosity. Biochemistry 10:3105–3109
2. Matsuno Y-K, Saito T, Gotoh M et al (2009) Supported molecular matrix electrophoresis: a new tool for characterization of glycoproteins. Anal Chem 81:3816–3823
3. Scott JE, Dorling J (1965) Differential staining of acid glycosaminoglycans (mucopolysaccharides) by Alcian blue in salt solutions. Histochemie 5:221–233
4. Schulz BL, Packer NH, Karlsson NG (2002) Small-scale analysis of *O*-linked oligosaccharides from glycoproteins and mucins separated by gel electrophoresis. Anal Chem 74:6088–6097
5. Dong W, Matsuno Y-K, Kameyama A (2012) A procedure for Alcian blue staining of mucins on polyvinylidene difluoride membranes. Anal Chem 84:8461–8466
6. Scott JE, Quintarelli G, Dellovo MC (1964) The chemical and histochemical properties of Alcian blue. Histochemie 4:73–85

Chapter 10

Quantitation of Mucin by Densitometry of an Alcian Blue-Stained Membrane

Mayumi Tamura and Yoichiro Arata

Abstract

It is a challenging task to quantify mucin using conventional protein quantification methods due to the large number of glycans attached to the peptide, which make up approximately 50–90% of its molecular weight. To address this issue, we propose a simple quantification method that involves spotting mucins onto a membrane and staining them with Alcian blue.

Key words Mucin, Quantitation, Alcian blue stain, Polyvinylidene fluoride membranes, Densitometry

1 Introduction

Mucin is a glycoprotein with a high molecular weight (0.5–20 MDa), of which 50–90% is composed of glycans [1, 2]. Current methods for quantifying mucins include separating mucin glycans and analyzing them via liquid chromatography/mass spectrometry (LC/MS), enzyme-linked immunosorbent assay (ELISA), or immune blotting. However, these methods are time-consuming and require expensive equipment and labor; moreover, the Bradford and bicinchoninic acid (BCA) methods, commonly used for protein quantification, are not suitable for mucin quantification. Herein, we present a simple and cost-effective method for quantifying mucins by spotting a mucin solution onto a polyvinylidene fluoride (PVDF) membrane and staining it with Alcian blue, based on the method introduced by Matsuno et al. [3].

Akihiko Kameyama (ed.), *Mucins: Methods and Protocols*, Methods in Molecular Biology, vol. 2763,
https://doi.org/10.1007/978-1-0716-3670-1_10,

2 Materials

All solutions should be prepared using ultrapure water and analytical-grade reagents. All reagents should be stored at 4 °C unless otherwise indicated.

1. PVDF membranes (*see* **Note 1**).
2. Plastic container: The container should be large enough to gently shake the dye solution containing the spotted membrane. For example, a plastic container or Petri dish would be suitable.
3. Shaker.
4. Mucin: commercially available mucin from porcine stomach (*see* **Note 2**).
5. 0.1% Alcian blue stain solution: Dilute 1% Alcian blue 8GX solution tenfold with 3% acetic acid (*see* **Notes 3** and **4**).
6. Methanol.
7. EDTA-PBS: 20 mM NaH_2PO_4, 150 mM NaCl, 2 mM EDTA, pH 7.2 (*see* **Note 5**).
8. Gel imager with analysis software.

3 Methods

3.1 Quantitation of Mucin

1. Prepare the standard mucin solution: Accurately weigh 20 mg of mucin and dissolve it in PBS containing 1 mM EDTA (EDTA-PBS) to make a 2.0 mg/mL solution in a 15- or 50 mL conical tube. Dilute this solution with EDTA-PBS to make 0.2, 0.4, 0.6, 0.8, and 1 mg/mL solutions.
2. Cut the PVDF membrane to an appropriate size (refer to Fig. 1, *see* **Note 6**).
3. Soak the membrane in methanol for 10 s to make it hydrophilic, and then soak the membrane in EDTA-PBS for 10 min to equilibrate (*see* **Note 7**).
4. Drain the buffer well and dry the PVDF membrane until no buffer remains on the membrane (*see* **Note 8**).
5. Add 5 μL of the mucin solution prepared at various concentrations of 0, 0.2, 0.4, 0.6, 0.8, and 1 mg/mL (0, 1, 2, 3, 4, and 5 μg/5 μL) to the pretreated PVDF membrane to plot a standard curve (*see* **Notes 9, 10,** and **11**).
6. Add 5 μL of the sample whose concentration is to be measured.
7. Dry the PVDF membrane on a laboratory table (*see* **Note 12**).

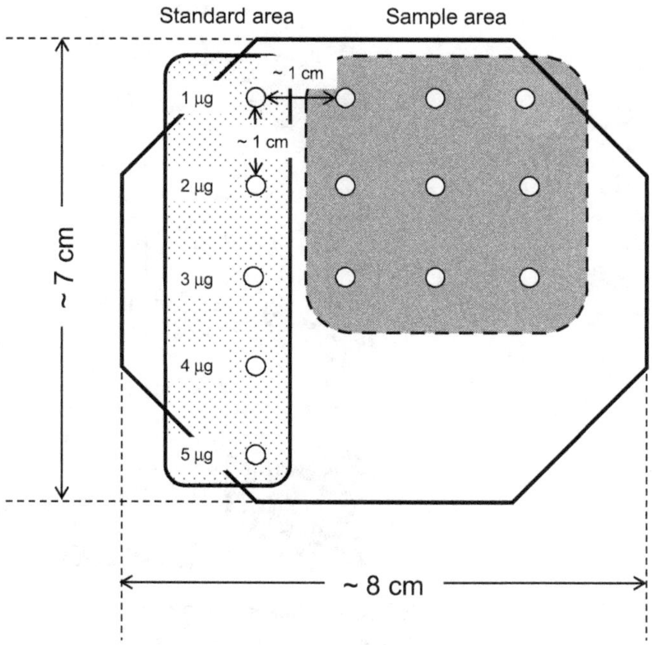

Fig. 1 Polyvinylidene fluoride (PVDF) membrane spotted with mucin. Samples were spotted at approximately 1 cm intervals on plain membranes. The left area of the membrane shows the standard spotting area (coarse dots with five spots for different amounts of mucin), and the right area of the membrane shows the sample spotting area (dark gray with nine sample spots)

8. Add enough 0.1% Alcian blue stain solution (5–10 mL) to soak the membrane and shake it slowly for 1 h for staining (*see* **Note 13**).
9. Discard the Alcian blue stain solution, add methanol, and shake slowly for 5 min. Discard the methanol and repeat this process two more times. If the discarded methanol still contains Alcian blue, repeat the wash process.
10. Dry the PVDF membrane in the same manner as **step 7**.
11. Capture images (*see* Fig. 2) and analyze them with analysis software (*see* **Note 14**).
12. In the case visualized in Fig. 3, samples 1, 2, 3, 4, 5, 6, 7, 8, and 9 were estimated to contain 0.18, 0.82, 0, 0, 0, 0, 1.55, 0.54, and 0 µg of mucin, respectively.

4 Notes

1. We used commercially available PVDF membranes specifically designed for western blotting. Nitrocellulose membranes are not suitable, as they dissolve when immersed in the staining solution. We used Immun-Blot® Low Fluorescence PVDF (Bio-Rad Laboratories, Inc. CA, USA).

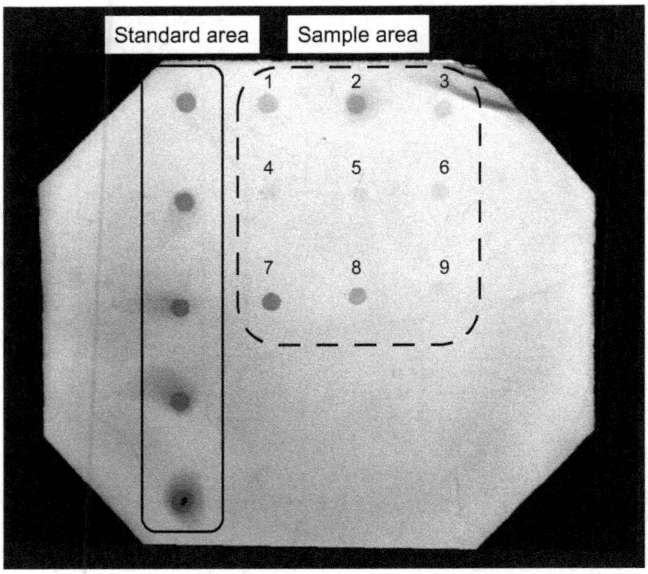

Fig. 2 Alcian blue staining results for PVDF membranes. Standards were spotted in the area enclosed by the solid line, while the samples (No. 1–9) were spotted in the area enclosed by the dotted line

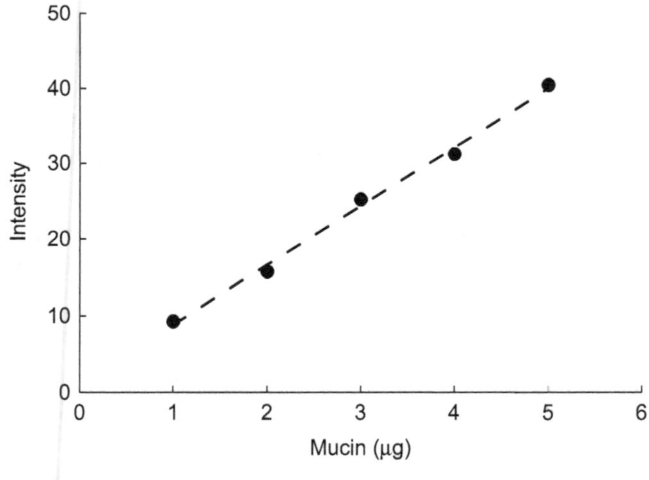

Fig. 3 Standard curve of mucin. The *x*-axis represents the amount of mucin (μg), and the *y*-axis represents the intensity of Alcian blue staining

2. We used Sigma-Aldrich® mucin from porcine stomach (Type III, partially purified powder).
3. For ease of handling, we used 1% Alcian blue 8GX solution (#66011, Sigma-Aldrich®, Merck KGaA, Darmstadt, German). Dilution was done immediately before use.

4. Alcian blue staining is pH dependent: at pH 2.5, acidic mucopolysaccharides with carboxy and sulfate groups are stained blue, while at pH 1 and below, only acidic mucopolysaccharides with sulfate groups are stained blue.
5. We used this method to examine the interaction between Galectin-2 (Gal-2) and mucin. To prevent oxidation of the Gal-2 protein, which loses its sugar-binding activity when oxidized, we used PBS containing 1 mM EDTA.
6. The size of the membrane is approximately 7 × 5 cm for a 5 × 4 spot. We cut off the four corners to fit the membrane into a Petri dish.
7. After immersion in methanol, it is important to completely submerge the membrane in the buffer to ensure even staining.
8. If buffer remains on the membrane, the mucin solution added in the next step may become diluted, resulting in uneven staining.
9. It is not necessary to dry the membrane after each spot is made. Drying the membrane after all spotting is complete is sufficient.
10. Adding too little or too much mucin solution for spotting on the PVDF membrane may result in poor performance. In our case, a spotting volume of 2 μL was difficult to handle and did not result in consistent values; contrarily, a spotting volume of 10 μL was excessive and took too long to dry.
11. The calibration curve for mucin concentration showed linearity in the range of approximately 0.06–5 μg (0.0125–1 mg/mL, 5 μL) for the porcine stomach-derived mucin we used.
12. If the membrane is not dried enough, mucin in the sample solution may run off when immersed in the staining solution.
13. We used a seesaw-type shaker.
14. We photographed the membrane with a ChemiDoc XRS plus system (Bio-Rad Laboratories, Inc. CA, USA) and used Image Lab, dedicated image analysis software, to quantify the spot densities numerically, graph them, and calculate the final quantification results.

Acknowledgments

The authors thank Mr. Shohei Kamei and Ms. Moeko Nakajima for their technical assistance. This work was supported by research grants from the Teikyo University and Josai University.

References

1. Brockhausen I, Wandall HH, Ten Hagen KG et al (2022) O-GalNAc glycans. In: Varki A, Cummings RD, Esko JD et al (eds) Essentials of glycobiology [Internet], 4th edn. Cold Spring Harbor, New York
2. Bansil R, Turner BS (2006) Mucin structure, aggregation, physiological functions and biomedical applications. Curr Opin Colloid Interface Sci 11:164–170
3. Matsuno YK, Saito T, Gotoh M et al (2009) Supported molecular matrix electrophoresis: a new tool for characterization of glycoproteins. Anal Chem 81:3816–3823

Chapter 11

Quantitation of MUC5AC and MUC5B by Stable Isotope Labeling Mass Spectrometry

Giorgia Radicioni and Mehmet Kesimer

Abstract

Mucins MUC5AC and MUC5B are large glycoproteins that play an essential role in the innate defense of epithelial surfaces and their quantitation in biological samples would be informative about the health status of the tissue/samples they are derived from. However, they are difficult to study and quantify with traditional methods such as ELISA and western blot, due to their size, heterogeneity, and high degree of glycosylation. We successfully implemented a stable isotope labeling mass spectrometry approach for absolute quantification of mucin macromolecules. Here, in detail, we describe this accurate and sensitive liquid chromatography and mass spectrometry (LC-MS) method applied for both MUC5AC and MUC5B quantification in diverse and complex biological samples.

Key words MUC5AC, MUC5B, Proteomics, Stable isotope labeling mass spectrometry

1 Introduction

MUC5AC and MUC5B are the major secreted gel-forming mucins of the airways and play a central role in protecting airway surfaces [1–4]. In health the gel-forming mucins are secreted at an optimal concentration to maintain homeostasis and effective mucociliary clearance. In certain lung diseases, however, their increased/decreased concentrations contribute to disease pathogenesis. Thus, the concentrations of MUC5AC and MUC5B in airways secretions make excellent potential biomarkers to understand the initiation and progression of lung diseases. It has been shown, for example, that MUC5AC dramatically increases in hypersecretory diseases, such as asthma [5] and chronic obstructive pulmonary disease (COPD) [6, 7]. Accurately measuring the concentrations of MUC5AC and MUC5B, however, is non-trivial. The unique properties of the gel-forming mucins, their high degree of polymerization and extensive *O*-linked glycosylation, make several of the techniques commonly used to quantify proteins ineffective [8].

Akihiko Kameyama (ed.), *Mucins: Methods and Protocols*, Methods in Molecular Biology, vol. 2763,
https://doi.org/10.1007/978-1-0716-3670-1_11,

Mass spectrometry, specifically stable isotope labeling mass spectrometry, is one of the protein quantification techniques that is effective with mucins [9]. The base principle of stable isotope labeling mass spectrometry is to identify unique enzymatically cut peptides that are not repeated within the protein sequence, and to compare their signal with a stable isotope labeled counterpart injected at the same time at a known concentration. The isotope labeled peptides are identical to the natural peptides (originating from the enzymatic digestion), except for their mass: some atoms have been replaced with heavier, stable isotopes, which results in a higher mass for the labeled peptide. Stable isotopically labeled peptides share the same physiochemical properties and chemical reactivities as their non-labeled counterparts, as well as the same retention time (*see* Fig. 1a, b).

We recently described an MUC5AC and MUC5B shotgun quantification approach based on stable isotope labeling mass spectrometry with targeted selected ion monitoring data-independent acquisition (tSIM-DIA) [6]. In our workflow (*see* Fig. 2), samples are prepared by filter-aided sample preparation (FASP): we use disposable centrifugal ultrafiltration units to efficiently reduce, alkylate, and enzymatically digest our samples. We previously used label-free proteomics to identify tryptic peptides that would be good candidates for absolute quantification and purchased stably labeled isotopes of the chosen sequences (*see* **Note 1**). Stably labeled internal standard peptides are added at a known concentration to the tryptic digested samples, the mixture is then fractionated by HPLC, and analyzed by mass spectrometry. By comparing the area of the two peptides (the naturally occurring light vs. the stably labeled heavy peptides) we can calculate the concentration, which reflects the concentration of the proteins, e.g., MUC5AC and MUC5B (*see* Fig. 3a) [7]. Stable isotope labeling mass spectrometry is an accurate and reproducible way to quantify mucin, and here we provide the step-by-step details of sample preparation, data acquisition, and analysis.

2 Materials

Prepare all solutions using ultrapure water and analytical grade reagents. Do not add sodium azide to the reagents. Prepare and store all reagents at room temperature (unless indicated otherwise).

2.1 Sample Preparation

1. Guanidine hydrochloride (GuHCl), assay percent range ≥ 99.5%, 8 M stock solution: Dissolve 76.42 g in 100 mL of ultrapure water.
2. 100 mM dithiothreitol: Dissolve 15.5 mg in 1 mL of 8 M GuHCl.

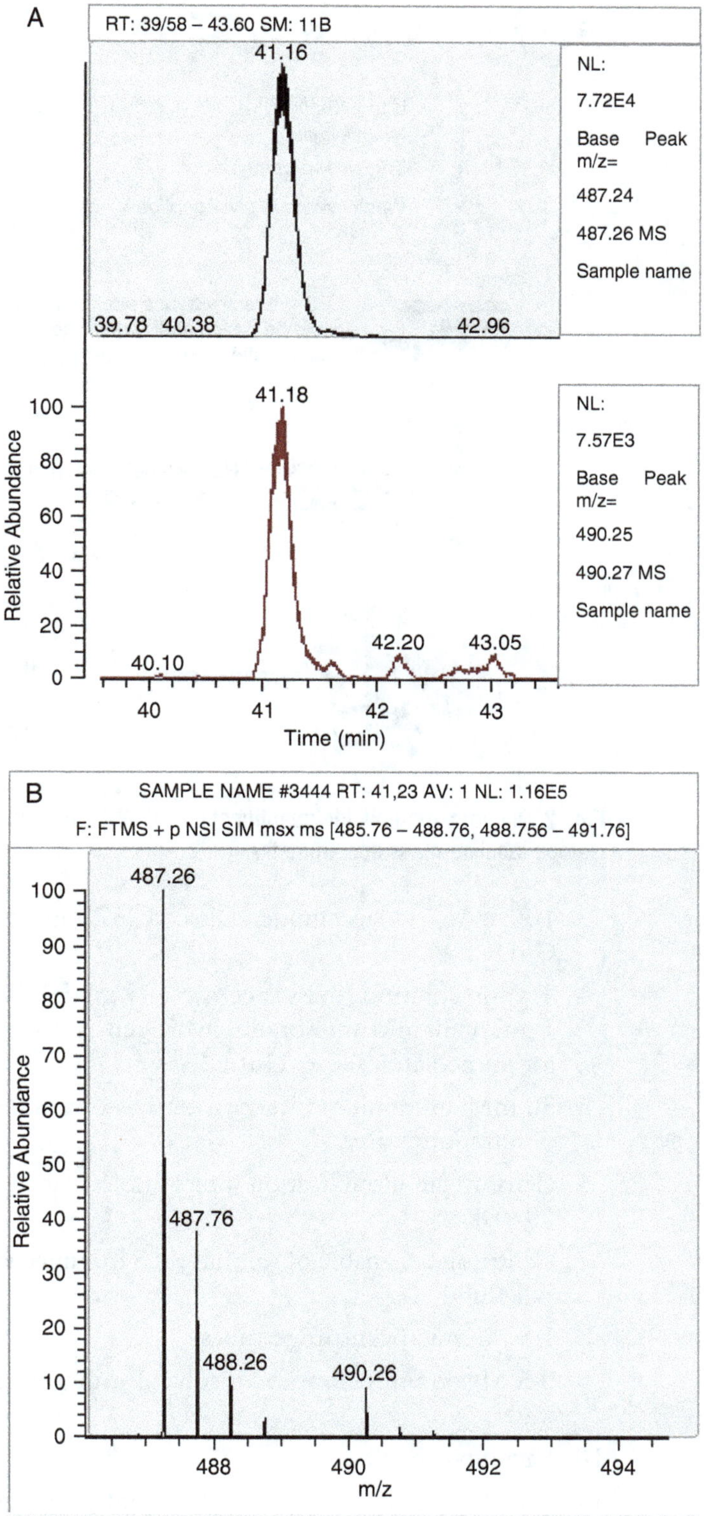

Fig. 1 Example of mass spectrometry detection of MUC5B peptide TWLVPDSR. (**a**) Co-elution of natural peptide (top, in black) and its heavy isotope counterpart (bottom, in red). (**b**) Full spectrum showing both parental ion: *m/z* 487.26 (light, natural peptide from the sample) and 490.26 (heavy isotope from the internal standard)

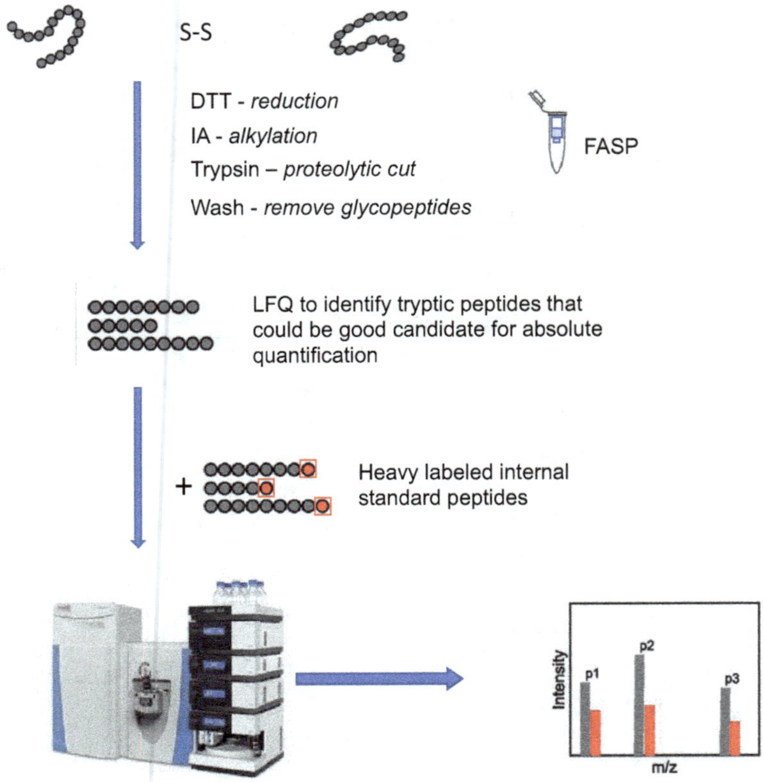

Fig. 2 Diagram method for quantification of MUC5AC and MUC5B by stable isotope labeling mass spectrometry

3. 100 mM iodoacetamide: Dissolve 37 mg in 1 mL of 8 M GuHCl.
4. Trypsin solution (mass spectrometry grade): Dissolve 20 μg in 1 mL of 50 mM ammonium hydrogen carbonate (NH_4HCO_3) for immediate use (*see* **Note 2**).
5. 50 mM ammonium bicarbonate: Dissolve 3.95 g in 1000 mL of ultrapure water.
6. Centrifugal ultrafiltration units molecular weight cutoff of 10 kDa.
7. Centrifuge. Capable of spinning 1.5 mL microcentrifuge tubes at 14,000 × *g*.
8. 1.5 mL microcentrifuge tubes.
9. 0.5 M sodium chloride: Dissolve 3 g in 100 mL of ultrapure water.
10. Vacuum centrifuge.
11. 0.1% formic acid: Dissolve 100 μL of formic acid in 100 mL of ultrapure water.

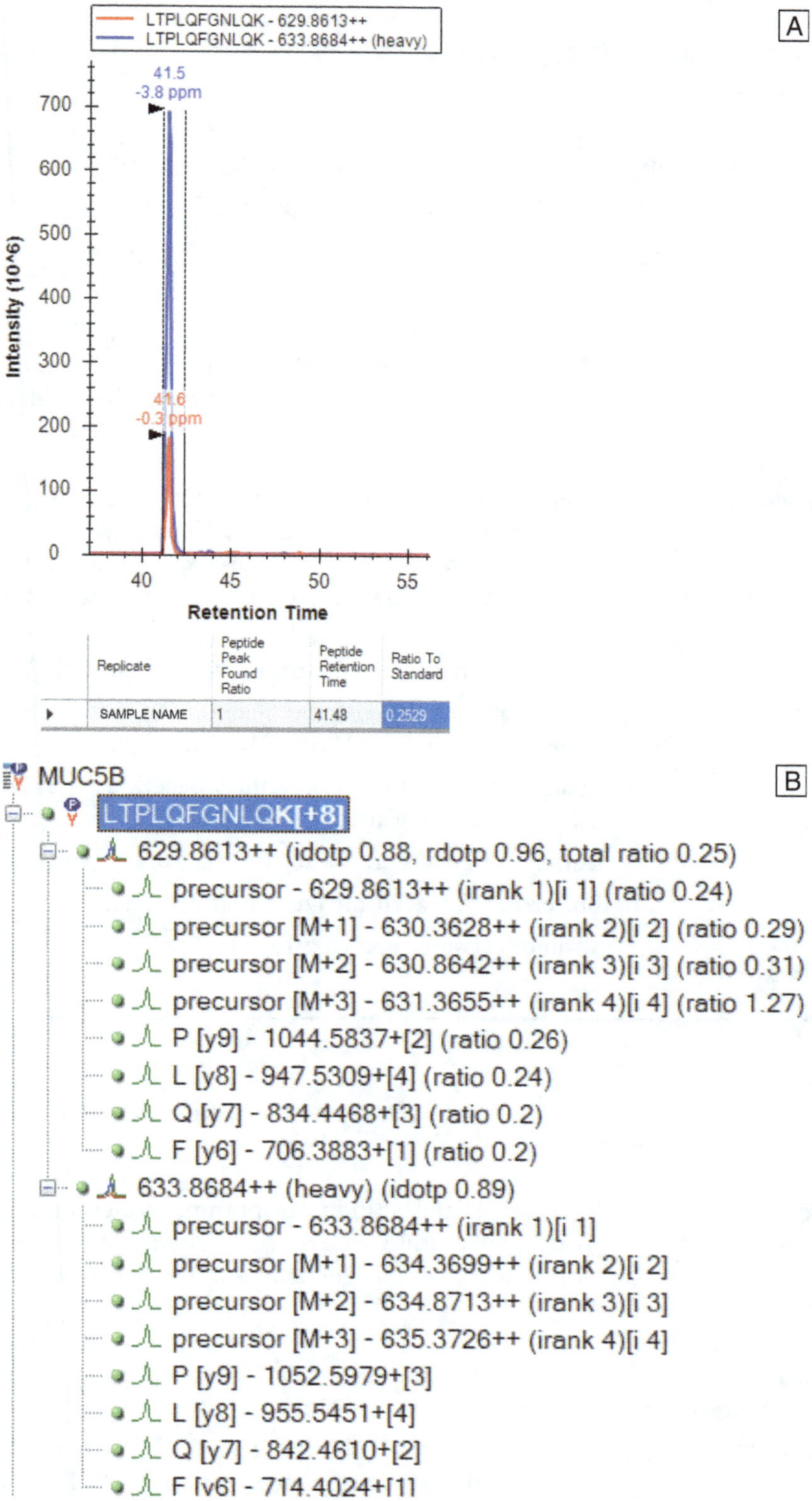

Fig. 3 Example of raw file analysis in Skyline software. (**a**) The table in the software shows the sample name ("replicate"), the peptide retention time, and the H/L ratio calculated by the software ("ratio to standard"). (**b**) Transitions selected for peptide LTPLQFGNLQK quantification

Table 1
Synthetized heavy labeled peptides for MUC5B and MUC5AC

MUC5B	MUC5AC
G[Val($^{13}C_5$; ^{15}N)]VQLSDWR	AEDAPGVPLR [Arg($^{13}C_6$, $^{15}N_4$)]
TWL[Val($^{13}C_5$; ^{15}N)]VPDSR	GTFLDDTGK [Lys($^{13}C_6$; $^{15}N_2$)]
LTPLQFGNLQK [Lys($^{13}C_6$; $^{15}N_2$)]	GTDSGDFDTLENLR [Arg($^{13}C_6$, $^{15}N_4$)]

12. Stable isotope labeling mass spectrometry Thermo Scientific HeavyPeptide AQUA (concentrations of 5 pmol/μL ± 25% with purity >97%). Three peptides for each mucin, as reported in Table 1.

2.2 LC-MS and Data Analysis

1. Liquid chromatography and mass spectrometry (LC-MS) system: Dionex ultimate 3000 RSLCnano system coupled to a hybrid quadrupole orbitrap mass spectrometer with a nano spray source (Q-Exactive, Thermo Fisher, Bremen, Germany).
2. Trap column: Thermo Scientific Acclaim PepMap 100, 100 μm × 2 cm, nanoViper C18 5 μm 100 Å.
3. Analytical column: Thermo Scientific Acclaim PepMap RSLC, 75 μm × 15 cm, nanoViper C18 2 μm 100 Å.
4. Loading buffer: 0.1% (v/v) trifluoroacetic acid and 2% acetonitrile in ultrapure water.
5. Buffer A: 97% water, 3% acetonitrile, 0.1% formic acid.
6. Buffer B: 0.1% acetonitrile, 0.1% formic acid.
7. Skyline software (version v1.4).

3 Methods

Carry out all procedures at room temperature unless specified otherwise.

3.1 Sample Preparation and Storage

1. Biological samples (induced sputum, broncho-alveolar lavages (BAL), cell culture secretions, etc.) are diluted 1:1 with 8 M GuHCl and kept refrigerated (4 °C) until further use (*see* **Note 3**).

3.2 Reduction, Alkylation, and Trypsin Digestion by FASP

1. Load each sample (50–400 μL) into a disposable centrifugal ultrafiltration unit (*see* **Note 4**).
2. Add the prepared dithiothreitol (DTT) solution 1:10 to the sample so that the final concentration of DTT is 10 mM, and incubate for 1 h at 65 °C (*see* **Note 5**).

3. Add the prepared iodoacetamide solution 1:5 to the sample so that the final concentration of iodoacetamide is 20 mM and incubate for 1 h at room temperature in the dark (*see* **Note 5**).
4. Centrifuge samples at 14,000 × *g* for 15–20 min or until all of the liquid has gone through the filter (*see* **Note 6**).
5. Wash twice by adding with 100 μL of 50 mM ammonium hydrogen carbonate (NH_4HCO_3), spinning for 15–20 min at 14,000 × *g*. Discharge the flow-through.
6. Put the filter unit into a clean microcentrifuge tube.
7. Add 2 μg of trypsin (100 μL of the stock solution for each sample) and incubate 18 h at 37 °C.
8. Wash twice with 100 μL 50 mM ammonium hydrogen carbonate (NH_4HCO_3) (*see* **Note 7**).
9. Wash (once) with 50 μL 0.5 M sodium chloride (NaCl) (*see* **Note 8**).
10. Remove and discard the filter inserts (peptides to be analyzed are in the flow through).
11. Concentrate peptides by a vacuum centrifuge system (*see* **Note 9**).
12. Dissolve the sample into a known volume (30–50 μL) of 0.1% formic acid in water (*see* **Note 10**).

3.3 Stable Isotope Labeling Mass Spectrometry

1. Solubilize each internal standard in 0.1% formic acid.
2. Pool three heavy isotope internal standard peptides for both MUC5AC and MUC5B to a final concentration of 500 fmol/μ L for each peptide.
3. Add internal standard mixture to each sample to a final concentration of 100 fmol/μL for each peptide (2 μL of the internal standard and 8 μL of sample).
4. For each sample run, inject 1 μL of the mixture of the tryptic-digested peptides and internal standards into the LC.
5. The LC-MS run is 32 min long, with the gradient reported in Table 2.
6. Acquire data by a target method combining a tSIM scan method with a time-scheduled duplexed DIA method. Settings are reported in Table 3.

3.4 Results Analysis

All raw files obtained from tSIM-DIA analyses are opened in Skyline (version v1.4), which extracts the ratio between the corresponding endogenous and internal standard peak (L/H) for each peptide.

Table 2
Buffer gradient for targeted mass spectrometry

Min	%B
0–2	4
2–20	30
20–28	80
29–32	4

Table 3
Setting for mass spectrometry method (*see* Note 11)

Properties of the method	
Use lock masses	Off
Chromatogram peak width	15 s
Method duration	32 min
General	
Runtime	32 min
Polarity	Positive
Targeted-SIM	
Resolution	70,000
AGC target	5.00E + 05
Maximum IT	200 ms
MSX count	6
Isolation window	3.0 *m/z*
Scan range	400 to 800 *m/z*
DIA	
Resolution	35,000
AGC target	2.00E + 05
Maximum IT	Auto
Loop count	2
MSX count	2
MSX isochronous	On
Isolation window	3.0 *m/z*
Fixed first mass	–
(N)CE/stepped	27

1. Create a peptide list (edit > insert > peptides) with the sequence of the peptides to analyze. The software will list both natural and heavy peptides.
2. For each peptide add the heavy modification (right click on the peptide sequence > modify). Right click on the light one and select "all the transitions," select also "synchronize isotope label types," so the same transitions will be checked also for the heavy peptide.
3. Upload the standard raw file and deselect the transition not present in the spectrum (indicated by a red or yellow dot) (*see* **Note 12**).
4. Upload all the other files. The software will create a table with the results automatically, but it is suggested to manually check the peaks selection (Fig. 3b).
5. Export the results into Microsoft Excel.
6. Average the light isotope/heavy isotope area ratio of the three peptides for MUC5AC and then for MUC5B.
7. Calculate the MUC5B and MUC5AC concentrations (pmol/mL) using the following equation:

$$\text{Protein concentration} = [\text{L/H} \times \text{C} \times \text{a/b}^{*}\ \text{c/d}],$$

where L/H is the average area ratio between light and heavy peptides, C is the concentration of the internal standard injected into the LC, a is the volume used to resuspend the peptides, b is the sample's starting volume, c/d is the dilution factor for mixing the sample and the internal standard (10/8).

4 Notes

1. To achieve a robust, reliable, and reproducible protein absolute quantification protocol, the choice of peptide is crucial. First, it is necessary to run several samples using label free proteomics to generate a list of all peptides produced via tryptic digestion (in silico digestion is insufficient). In general good candidate peptides for absolute quantification should be easily proteolyzed by trypsin, have a robust and reproducible signal in the given mass spectrometer, and they should be detectable in all the samples. Moreover, peptides used for absolute quantification should meet the following criteria:
   - Unique for the protein to be quantified. If the peptide is shared with other proteins, it is not possible to determine from which protein it is generated, therefore that peptide cannot be used to quantify the protein of interest. This is extremely important for MUC5AC and MUC5B, because they share a lot of their sequence.

- Not repeated within the protein sequence. If the peptide is present in repeated regions of the protein, it is not possible to convert its signal to the number of times that the peptide is detected. For this reason and to have an accurate quantification of the protein, the ratio of peptide: protein must be 1:1.
- There are some restrictions in the peptide sequences, regarding the amino acids that can undergo chemical modification. Some chemical modifications could change the mass of a peptide so that that it can no longer be detected at the m/z set in the method. Peptides used for quantification cannot have cysteine residues due to the possibility of disulfide bond formation. Similarly methionine (which can undergo oxidation), glutamic acid at the N-terminal (which can transform into pyroglutamic acid very easily under acidic conditions), asparagine followed by glycine (which is subject to deamination with formation of aspartic acid), and glutamine followed by glycine (which can also undergo the deamination process, but at a slower rate) should also be avoided in candidate peptides.
- The length of the peptide should be between 7 and 25 amino acids, to ensure a possible synthesis. However, it is recommended to contact the company producing the peptides to check their guidelines for peptide length.

2. It is important to use mass spectrometry grade trypsin because its residues have been modified by reductive methylation, resulting in a very active and stable enzyme that is also extremely resistant to autolytic digestion.
3. Adding 8 M GuHCl (chaotropic agent) allows proteins to unfold and makes them more accessible to proteolytic cleavage. In the same way it destroys all the proteolytic activity of proteases that might be present in the samples, leading to sample stability. Ideally the natural proteolytic activity in the samples should be stopped as soon as possible as uncontrolled protein cuts will create semi-tryptic peptides, which will interfere with identification and quantification of peptides.
4. The FASP filter volume capacity is 500 μL, do not overfill it (400 μL is ideal).
5. Discard any unused solution.
6. In general 400 μL take around 15–20 min at 14,000 × *g* to go through the filter. However, after every centrifugation, before proceeding with the next step, make sure to spin until all of the liquid has gone through the filter.

7. This step allows for the separation of tryptic peptides (flow-through) from large glycopeptides (which remain on the filter) that are not suitable for the LC-MS method used in this protocol.
8. This step helps to reduce any possible ionic interaction between peptides and the filter.
9. Ammonium bicarbonate should be evaporated through this procedure.
10. If it is not possible to inject the samples right after freeze-drying, they can be stored in a freezer (−20 °C) before adding formic acid.
11. The parameters in Table 3 are specific for the Q-Exactive hybrid quadrupole orbitrap mass spectrometer (Thermo Fisher, Bremen, Germany). There are equations and values for all the instrument parameters, which depend on the instrument, that are usually provided by companies in reference material (e.g., for Thermos Fisher it can be found at https://assets.thermofisher.com/TFS-Assets/CMD/Reference-Materials/wp-65147-ms-q-exactive-orbitrap-scan-modes-wp65147-en.pdf).
12. It is good practice to run a quality control sample consisting of only a heavy isotope internal standard mixture. This will allow the identification of the best transitions for the heavy peptides and check the retention time and the mass spectrometry signal quality (among the other calibration and check points usually performed for the mass spectrometer). It is also suggested to have an MUC5AC and MUC5B standard solution to prepare with each batch of samples to assess the quality of the whole process (reduction, alkylation, trypsinization, etc.).

Acknowledgments

We thank the Kesimer lab and the Marsico Lung Institute for helping and supporting during the development of this method. The described method was developed as part of the SPIROMICS ancillary study which is mainly supported by National Institutes of Health NIH/NHLBI R01HL110906 (MK) and partially supported by R01HL103940 (MK) and 5U01HL137880-05.

References

1. Kesimer M, Kirkham S, Pickles RJ et al (2009) Tracheobronchial air-liquid interface cell culture: a model for innate mucosal defense of the upper airways? Am J Physiol Lung Cell Mol Physiol 296:L92–L100
2. Buisine MP, Devisme L, Copin MC et al (1999) Developmental mucin gene expression in the human respiratory tract. Am J Respir Cell Mol Biol 20:209–218

3. Kirkham S, Sheehan JK, Knight D et al (2002) Heterogeneity of airways mucus: variations in the amounts and glycoforms of the major oligomeric mucins MUC5AC and MUC5B. Biochem J 361:537–546
4. Kesimer M, Ehre C, Burns KA et al (2013) Molecular organization of the mucins and glycocalyx underlying mucus transport over mucosal surfaces of the airways. Mucosal Immunol 6: 379–392
5. Cho JL, Ling MF, Adams DC et al (2016) Allergic asthma is distinguished by sensitivity of allergen-specific CD4+ T cells and airway structural cells to type 2 inflammation. Sci Transl Med 8(359):359ra132. https://doi.org/10.1126/scitranslmed.aag1370
6. Kesimer M, Ford AA, Ceppe A et al (2017) Airway mucin concentration as a marker of chronic bronchitis. N Engl J Med 377(10): 911–922. https://doi.org/10.1056/NEJMoa1701632
7. Radicioni G, Ceppe A, Ford AA et al (2021) Airway mucin MUC5AC and MUC5B concentrations and the initiation and progression of chronic obstructive pulmonary disease: an analysis of the SPIROMICS cohort. Lancet Respir Med 9(11):1241–1254. https://doi.org/10.1016/S2213-2600(21)00079-5
8. Thornton DJ, Rousseau K, McGuckin MA (2008) Structure and function of the polymeric mucins in airways mucus. Annu Rev Physiol 70: 459–486
9. Kuzyk MA, Smith D, Yang J et al (2009) Multiple reaction monitoring-based, multiplexed, absolute quantitation of 45 proteins in human plasma. Mol Cell Proteomics 8(8):1860–1877. https://doi.org/10.1074/mcp.M800540-MCP200

Part III

Preparation and Analysis of Mucin Glycans

Chapter 12

Preparation of *O*-Glycans from Mucins Using Hydrazine Treatment

Yukinobu Goso and Makoto Kurihara

Abstract

Mucin glycomic analysis is crucial owing to the participation of mucin *O*-glycans in several biological functions. Liquid chromatographic analysis of fluorescently labeled glycans is an effective tool for glycomic analysis. The first step of this analysis involves the release of *O*-glycans from mucins. As no enzyme is known to release all glycans, chemical methods are required for the process; therefore, hydrazine treatment is a commonly used chemical method. It enables the release of *O*-glycans from mucin while preserving the aldehyde group at the reducing end. This ensures that the reducing end can be modified using fluorescent reagents. However, it is also accompanied by the degradation of the glycans through a process called "peeling." Here, we describe a method for releasing glycans from mucins using hydrazine treatment with minimal "peeling."

Key words Mucin, *O*-glycan, Hydrazine, Peeling, Mass spectrometry, Glycomics

1 Introduction

Mucins are high-molecular weight (>500 kDa) glycoproteins produced by epithelial cells and highly glycosylated with mucin-type (GalNAc-type) *O*-glycans [1]. Their functions include protecting the mucosal surface and regulating host-microbiota interaction, and they are also involved in carcinogenesis [2]. As *O*-glycans drive the function of mucins, glycomic analysis is required to comprehensively elucidate its role in biological processes [3, 4]. Liquid chromatographic analysis of fluorescently labeled glycans is an effective method to achieve high sensitivity and high throughput. Thus, releasing *O*-glycan from mucins is crucial to evaluate its function. Currently, no enzyme is known to release glycans entirely. Therefore, chemical methods are required for the release of these biomolecules.

Reductive alkaline β-elimination is commonly used to release *O*-glycans from glycoproteins; however, it causes loss of the

Akihiko Kameyama (ed.), *Mucins: Methods and Protocols*, Methods in Molecular Biology, vol. 2763,
https://doi.org/10.1007/978-1-0716-3670-1_12,

reducing-end aldehyde group required for fluorescent tagging. Therefore, hydrazine [5–7], ammonia [8], or simple alkali [9, 10] treatments are used to obtain reducing-end aldehyde group-bearing glycans. However, the 3-*O*-substitution of reducing-end *N*-acetylgalactosamine (GalNAc) with a saccharide causes glycan degradation. This degradation process is called "peeling." In ammonia and simple alkali treatments, late-stage degradation occurs against one of the "peeled" products containing a Morgan–Elson chromogen [11]. This results in the loss of the Morgan–Elson chromogen and the formation of a glycan with a GlcNAc molecule at the reducing end. Moreover, the resulting glycan cannot be readily distinguished from non-degraded *O*-glycans using mass spectrometry (MS). Therefore, we used hydrazine treatment to release glycans and examined the optimal reaction conditions with minimal "peeling" [11–13].

We have proposed a mechanism for the release of glycans using hydrazine treatment (*see* Fig. 1a) and the subsequent re-*N*-acetylation process (*see* Fig. 1b) [11, 12]. The process of release of glycan **II**, GlcNAcβ1–6 (Galβ1–3)GalNAc, from the corresponding mucin moiety (**I**) is shown here. Usually, saccharides may be attached to the terminal galactose (Gal) and *N*-acetylglucosamine (GlcNAc) of Glycan **II**. Glycan **II** is released through β-elimination caused by hydroxide ion (reaction ***a***), which is produced through the reaction of hydrazine with water (reaction ***b***). Glycan **II** reacts with hydrazine to produce hydrazone **III** (reaction ***c***). It also reacts with the hydroxide ion to produce Gal (**IV**) and saccharide **V**, comprising GlcNAc and a Morgan–Elson chromogen (reaction ***d***); in other words, it undergoes "peeling." Next, Gal (**IV**) and the saccharide **V** react with hydrazine to produce their hydrazone derivatives, **VI** and **VII**, respectively (reactions ***f*** and ***g***). Hydrazone **III** is considered stable for a hydroxide ion. Therefore, no further "peeling" occurs. During the subsequent re-*N*-acetylation process, hydrazones **III** and **VI** reverts to the original glycans **II** and **IV** via their acetylhydrazones **VIII** and **IX**, respectively [11, 14]. However, **X** produced from **VII** is not converted to **V** [11]. Moreover, **II** and **IV** can be fluorescently labeled using the aldehyde group at the reducing end [15]. However, **X** cannot occur because there is no reducing-end aldehyde group. Thus, the release of glycans through hydrazine treatment leads to a simplified interpretation.

The critical process for the release of the glycan from mucin using hydrazine treatment is the suppression of the "peeling" process. We found that the weak acids, including malonic acid, suppressed the "peeling" under hydrazine gas treatment (*see* Figs. 2 and 3) [11, 13]. This may be attributed to the oxonium ions formed by the reaction of malonic acid with water (reaction ***e***), which may regulate the reactions ***d*** (peeling) and reaction ***a*** (-β-elimination) (*see* Fig. 1a) [11, 13]. As water is produced during the formation process of hydrazone (reactions ***c***, ***f***, and ***g*** in Fig. 1a),

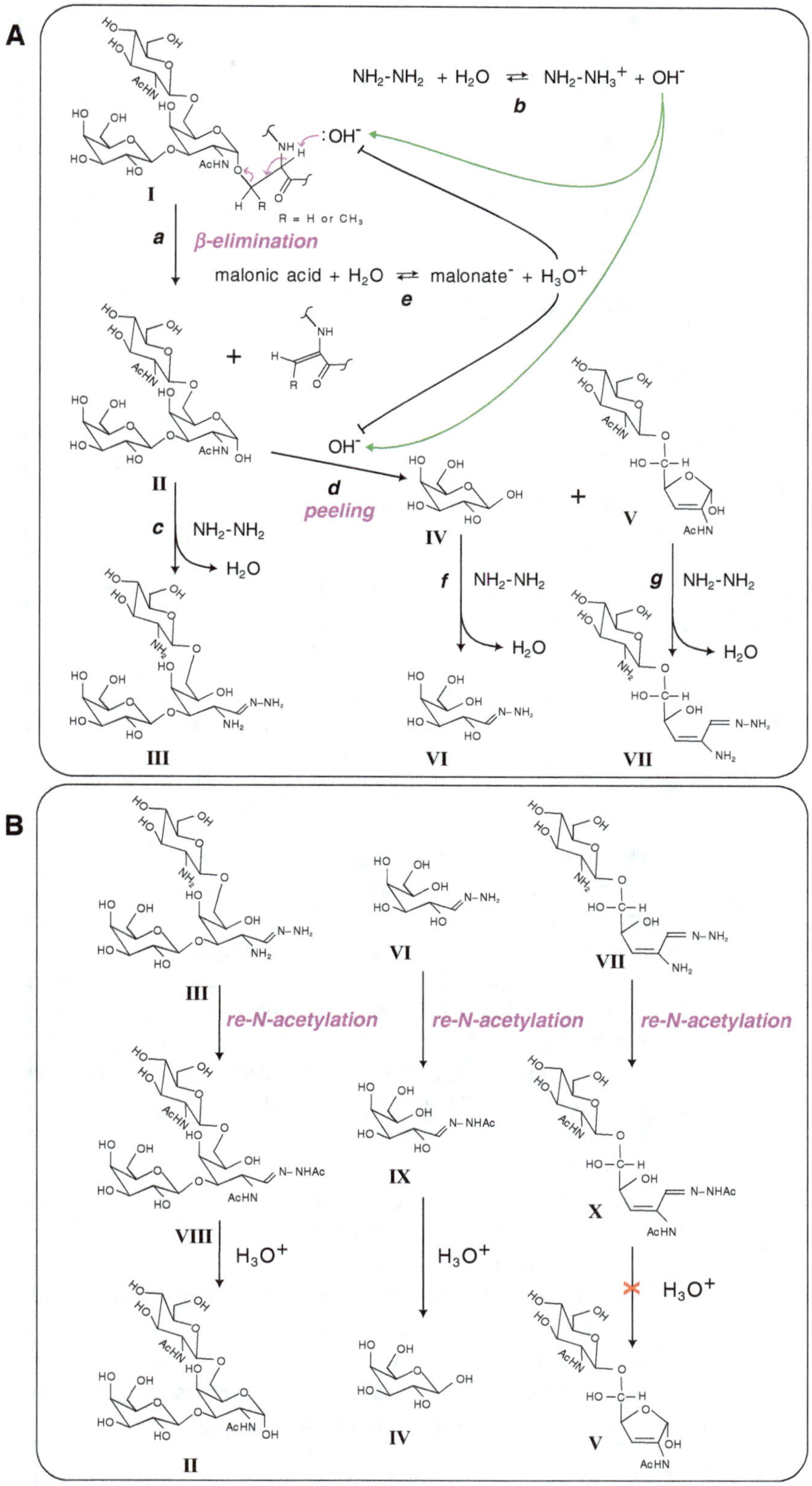

Fig. 1 Hypothesized mechanism of hydrazine treatment (**a**) and the subsequent re-*N*-acetylation (**b**). Ac: acetyl group

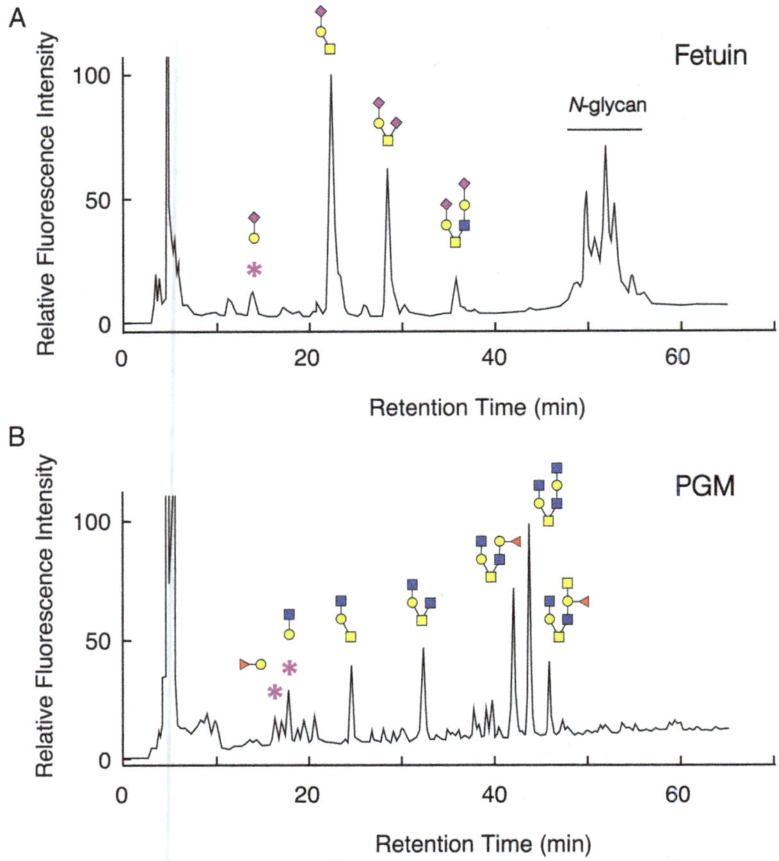

Fig. 2 Released *O*-glycans through hydrazine treatment from the model glycoproteins. Fetuin (**a**) and porcine gastric mucin (PGM) (**b**) were treated with anhydrous hydrazine gas at 60 °C for 6 h in the presence of malonic acid, and analyzed with TSKgel Amide-80 chromatography after being fluorescently tagged with anthranilic acid. The proposed glycan structures, as determined by tandem mass spectrometry are illustrated. Yellow circles, blue squares, yellow squares, red triangles, and purple diamonds denote Gal, GlcNAc, GalNAc, Fuc, and Neu5Ac, respectively. Pink asterisks indicate the "peeled" glycans. (Reproduced from [11] with permission with some modifications. Copyright 2017 American Chemical Society)

and a hydroxide ion is produced from the reaction of hydrazine with water, the absence of malonic acid in the reaction system would promote "peeling."

This chapter describes methods for preparing *O*-glycans using hydrazine gas treatment with minimal "peeling." Here, bovine fetuin and partially purified porcine gastric mucin were used as model glycoproteins (*see* Fig. 2), whereas rat gastric mucins were utilized as research subjects (*see* Fig. 3).

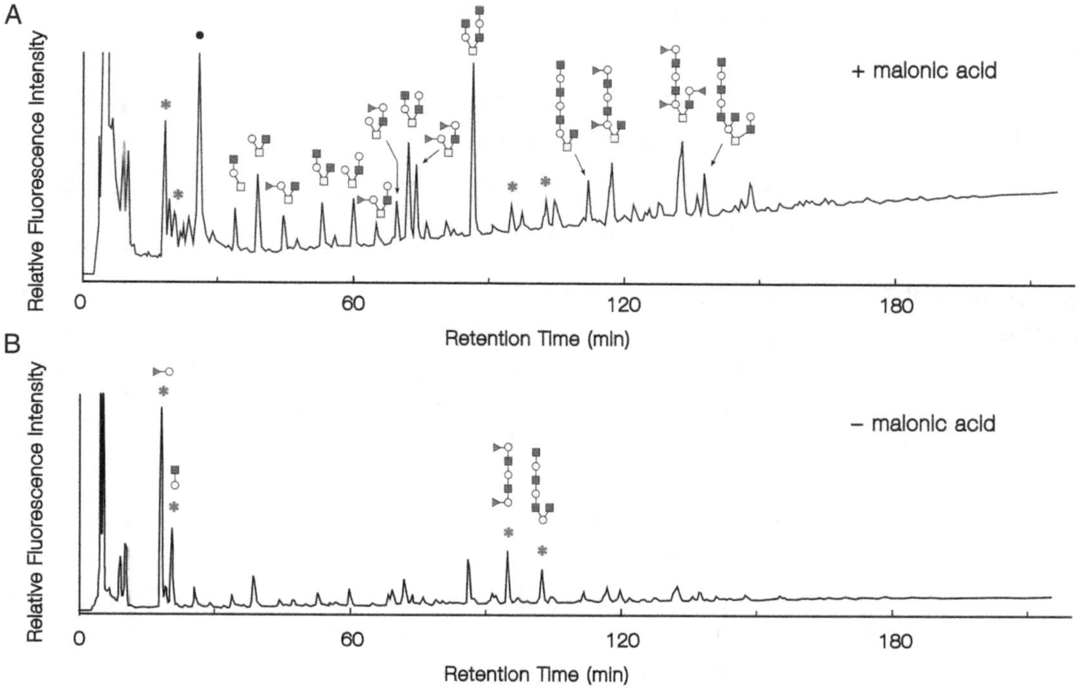

Fig. 3 TSKgel Amide-80 chromatography of anthranilic acid-labeled glycans obtained from rat gastric mucin treated using hydrazine gas at 60 °C for 6 h in the presence (**a**) or absence (**b**) of malonic acid. Proposed glycan structures, as determined by tandem mass spectrometry are illustrated according to Fig. 2. Pink asterisks indicate the "peeled" glycans. The nonglycan peak is marked with a closed dot. (Reproduced from [11] with permission with some modifications. Copyright 2017 American Chemical Society)

2 Materials

2.1 Mucins and Model Glycoproteins

1. Partially purified porcine gastric mucin (*see* **Note 1**).
2. Fetuin (fetal calf serum).
3. Rat gastric mucin (prepared from rat stomach, *see* **Note 2**).

2.2 Hydrazine Treatment

1. Glass test tubes (10 × 75 mm).
2. 200 mM Malonic acid.
3. Lyophilizer.
4. Waters Pico-Tag reaction vials (*see* **Note 3**) and Waters Pico-Tag Work Station (*see* **Note 4**).
5. Vacuum desiccator connected to vacuum pump, P_2O_5.
6. Anhydrous hydrazine (*see* **Note 5**).
7. Solid carbon dioxide.
8. Plastic glove bag.
9. Hybridization oven.
10. Sulfuric acid trap.

11. Saturated aqueous $NaHCO_3$ solution.
12. Acetic anhydride.
13. Cation exchange resin (H^+-form) (*see* **Note 6**).
14. Anion exchange resin (acetate-form) (*see* **Note 7**).
15. Empty column cartridges (6 mL).
16. 0.4 M pyridinium acetate, pH 5.0.

2.3 Analysis of O-Glycans

1. Screw cap 1.5 mL tube.
2. Anthranilic acid (AA, 2-aminobenzoic acid) labeling solution: Anthranilic acid (30 mg) and $NaBH_3CN$ (30 mg) are dissolved in 1 mL of methanol containing 4% (w/v) NaOAc-$3H_2O$ and 2% (w/v) boric acid. The solution can be prepared at the time of use.
3. Dry block heater.
4. Acetonitrile.
5. HILIC SPE (hydrophilic interaction liquid chromatography solid phase extraction) column (3 mL) (*see* **Note 8**).
6. 97% (v/v) aqueous acetonitrile solution.
7. Centrifugal evaporator.
8. PD-MiniTrap G-10 column (Merck) (*see* **Note 9**).
9. 0.1 M pyridinium acetate, pH 5.0.
10. Reversed phase SPE column (1 mL) (*see* **Note 10**).
11. 5% (v/v) acetonitrile aqueous solution.
12. HILIC HPLC column (*see* **Note 11**).
13. High performance liquid chromatography (HPLC) system consisting of a binary pump, an auto sampler, a fluorescent detector, and a column oven.
14. Mobile phase A: 0.2% (v/v) triethylamine and 0.2% (v/v) acetic acid in water.
15. Mobile phase B: 0.1% (v/v) acetic acid in acetonitrile.
16. The Matrix-assisted laser desorption/ionization time-of-flight mass spectrometry (MALDI-TOF MS) instrument (*see* **Note 12**).
17. Matrix solution: 50 mg/mL 2,5-dihydrobenzoic acid (DHB) in methanol/water (1:1; v/v).
18. Component identification: GlycoMod online tool (https://web.expasy.org/glycomod/).
19. Structure verification: GlycoWorkBench, downloadable Java tool (https://glycoworkbench.software.informer.com/).

3 Methods

All procedures are performed at room temperature unless specified otherwise.

3.1 Release of O-glycan Using Hydrazine Treatment

1. 1 mL of solution containing 150 μg mucins or model glycoproteins in a test tube (10 × 75 mm) is added to 10 μmol malonic acid (50 μL of 200 mM malonic acid), and then lyophilized (*see* **Note 13**).
2. The tube is placed into a Waters Pico-Tag reaction vial, and the contents are dried in vacuo over P_2O_5 in the vacuum desiccator for over 16 h (*see* **Note 14**).
3. Under a dry nitrogen atmosphere in a plastic glove bag, anhydrous hydrazine (150 μL) is added into the reaction vials but outside the test tubes (*see* **Note 15**).
4. Vials are cooled using solid carbon dioxide, immediately evacuated to <200 Pa and sealed, using the Waters Pico-Tag Work Station with a vacuum pump and a cooling trap.
5. Vials are heated at 60 °C for 6 h using a hybridization oven.
6. After cooling the vials, hydrazine is removed in vacuo using a Waters Pico-Tag Work Station with a vacuum pump and a cooling trap. The remaining hydrazine is then further removed over sulfuric acid in the vacuum desiccator (*see* **Note 16**).
7. The released *O*-glycans are dissolved in 200 μL of ice-cold saturated $NaHCO_3$. Next, re-*N*-acetylation is performed by adding 5 × 2 μL of acetic anhydride at 3 min intervals in an ice bath, and then adding 10 μL for 15 min at room temperature.
8. The solution is added to 1.2 mL of water.
9. A cation exchange resin is packed to empty column cartridge by adding a 4 mL slurry of Dowex 50Wx4 (H^+-form) /water (1:1 (w/v)). The columns are washed with 20 mL of water.
10. An anion exchange resin is packed to empty column cartridge by adding a 3.3 mL slurry of Toyopearl QAE (acetate-form) / water (1:2.5 (w/v)). The columns are washed with 16 mL of water.
11. A cation exchange resin column is set over an anion exchange resin column.
12. Solutions from the re-*N*-acetylated samples are pipetted to the upper cation exchange column. The column is subsequently washed with 25 mL of water, and the eluent is collected.
13. The cation exchange resin column is removed.

14. The anion exchange resin column is washed with 15 mL water, and the eluent from the column is collected.
15. Eluates obtained in **steps 12** and **14** are combined. Glycan in the solution is designated as the "neutral glycan."
16. Acidic glycans are eluted from the anion exchange column with 10 mL of 0.4 M pyridinium acetate, pH 5.0.
17. The neutral and acidic glycan solutions are lyophilized and re-dissolved in appropriate amount of water.

3.2 Analysis of O-Glycans

AA-derivatization of glycans and HPLC of AA-labeled glycans generally follow the method of Anumula [15].

1. AA-labeling solution (100 μL) is added to the glycan samples (20 μL) in 1.5 mL screw cap tubes. The mixture is heated at 80 °C for 1 h using a dry block heater.
2. After cooling, 30 μL of water and 850 μL of acetonitrile are added (final water content of approximately 5% (v/v)) to the solution.
3. A DPA-6S SPE tube is washed with 10 mL of acetonitrile.
4. The sample is applied to the column (*see* **Note 17**).
5. The columns are washed with 8 mL of 97% (v/v) aqueous acetonitrile solution, and the eluting solution is discarded.
6. AA-labeled glycans are eluted with 2.5 mL of water.
7. The collected solutions are evaporated using a centrifugal evaporator.
8. A PD-MiniTrap G-10 column is washed with 6 mL of 0.1 M pyridinium acetate, pH 5.0.
9. AA-labeled glycans are dissolved in 100 μL of 0.1 M pyridinium acetate, pH 5.0. They are subsequently applied to the column.
10. The column is washed with 300 μL of 0.1 M pyridinium acetate, pH 5.0, and the eluting solutions are discarded.
11. AA-labeled glycans are eluted with 2 mL 0.1 M pyridinium acetate, pH 5.0 (*see* **Note 18**).
12. The collected solution is evaporated using a centrifugal evaporator and re-dissolved in 500 μL of water.
13. OASIS HLB (30 mg/cc) Cartridge (Waters) is washed with 2 mL of acetonitrile and then with 2.5 mL of water.
14. The sample is loaded into the column. Next, AA-labeled glycans are eluted with 2 mL of 5% (v/v) acetonitrile aqueous solution.
15. The collected solution is evaporated using a centrifugal evaporator.

16. A TSKgel Amide-80 column is equilibrated with 13% mobile phase A and 87% mobile phase B, with a flow rate of 0.7 mL/min at either 30 °C or 70 °C.
17. After the sample is injected into the column, it elutes over 87–57% mobile phase B in 60 min for analytical purpose and 87–52% mobile phase B in 210 min for preparative purposes. Fluorescence is detected at 360 nm/425 nm (emission/excitation wavelength).
18. During preparative chromatography, peak fractions are collected using a fraction collector or manually. They are subsequently concentrated using a centrifugal evaporator.
19. An AA-glycan sample is spotted on a stainless steel MALDI plate.
20. DHB matrix solution (0.5 μL) is added on top of the sample.
21. The spot is dried before MS analysis.
22. Mass spectra are acquired using a MALDI mass spectrometer operating in positive mode at an accelerating voltage 20 kV.
23. Tandem mass spectrometry is performed in LIFT mode.
24. The GlycoMod tool is accessed.
25. *m/z* values of peaks are inputted. The relevant options are filled as follows: monoisotopic, positive, Na^+, *O*-linked oligosaccharide/derivatized oligosaccharide, monosaccharide/underivatized, reducing terminal derivative: AA, mass: 137.13.
26. GlycoMod is started. The output is a list of potential components.
27. Manual interpretation is performed based on *O*-glycan databases.
28. GlycoWorkBench is launched.
29. Putative structures for the *m/z* value are drawn. Mass options are set as [MONO, Und, Na, 0, AA].
30. The peak list is inputted into GlycoWorkBench.
31. Using the ANNOTATE tool, the fragments to the *MS/MS* peak list for all possible structures is matched.
32. Using the FRAGMENT tool, a theoretical fragmentation of the putative structures is performed.
33. The most likely structure from both **steps 31** and **32** is identified.

4 Notes

1. Partially purified porcine gastric mucin was prepared as previously described [16]. Briefly, porcine gastric mucin (Type II,

crude, Merck) was digested using pepsin (3× incubation with 0.5% (w/w) pepsin (Merck) overnight at 37 °C in a solution adjusted to pH 2.5 using concentrated hydrochloric acid). Mucin becomes a glycopeptide with a molecular weight of approximately 500 kDa [17], whereas other (glyco)proteins are digested. After the pH was added to 6.5 using 5 M NaOH, 5% (w/w) cetylpyridinium chloride was added to a supernatant obtained through centrifugation (15,000 × *g* for 30 min at 4 °C). The solution was stirred overnight at 4 °C and centrifuged (21,500 × *g* for 60 min at 4 °C) to remove glycosaminoglycans. The supernatant was dialyzed using water at 4 °C overnight, after 2 h running water dialysis. The intradialysis fluid was then subjected to ethanol precipitation (final ethanol concentration = 33% (v/v)). The supernatant obtained through centrifugation (8,000 × *g* for 60 min at 4 °C) was subjected to further ethanol precipitation (final ethanol concentration = 50% (v/v)). Finally, the precipitate obtained through centrifugation was dissolved in water and lyophilized.

2. Rat gastric mucins were prepared as previously described [18]. The mucosa was scraped off from the stomach and homogenized in 6 M guanidinium chloride, pH 7.4, containing 2% (v/v) Triton X-100, 50 mM Tris–HCl, 100 mM dithiothreitol, and protease inhibitors. Then the mucin was extracted. After alkylation using iodoacetamide, the sample solution was applied to a Sepharose CL-6B column equilibrated with 4 M guanidinium chloride, pH 7.4, containing 0.5% (v/v) Triton X-100, 50 mM Tris–HCl, and protease inhibitors. The excluded fraction from the column was further purified using CsCl density gradient centrifugation. Subsequently, the mucin fraction was collected and desalted using ethanol precipitation.
3. Similar vials, e.g., Borosilicate Glass Hydrolysis Derivatization Vials (Thermo), can be used.
4. Only a vacuum pump with a cooling trap and vacuum gauge can be used.
5. Hydrazine with lower moisture content should be used, because water enhances the "peeling" reaction [12]. Testing several commercially available reagents with the model glycoproteins is recommended.
6. Dowex 50Wx4 resin (200-400 mesh, 454 g, Dow Chemical) is prewashed sequentially with 3 L of 2 M NaOH, 4 L of water, 6 L of 3 M HCl, and 2 L of water. Another cation exchange resin may be available.
7. Toyopearl QAE-550C resin (500 mL, TOSOH) is prewashed with 1 L of 0.5 M HCl, 1.5 L of water, 5 L of 0.5 M NaOH, 2 L of 0.5 M NaOAc, 0.5 L of 0.5 M acetic acid and 2.5 L of water,

sequentially, and stored at 4 °C. Another anion exchange resin may also be available.

8. Discovery DPA-6S SPE tube (250 mg/3 mL, Merck) or equivalent products.
9. Equivalent gel permeation mini columns may be available.
10. OASIS HLB (30 mg/cc) Cartridge (Waters) or equivalent products.
11. TSKgel Amide-80 column (4.6 × 250 mm, TOSOH) or equivalent products.
12. A Bruker Autoflex III mass spectrometer or an equivalent equipment.
13. The absence of malonic acid causes further "peeling" as shown in Fig. 3.
14. In the next procedure, the addition of hydrazine into the reaction vials is made in a plastic bag. Therefore, a desiccator that is able to pass through the opening of the plastic bag should be used.
15. Hydrazine is highly toxic. Therefore, all protocols should be performed in a hood with a ventilation system.
16. In the later removal stage, a sulfuric acid trap is installed between the desiccator and the vacuum pump.
17. As the AA-labeled glycans are insoluble in 95% (v/v) aqueous acetonitrile solution, the sample should be stirred vigorously and then applied to the column in suspension.
18. The remaining AA reagent and others are removed.

Acknowledgments

This work was supported by the Japan Society for the Promotion of Science KAKENHI grant 25460934.

References

1. Dekker J, Rossen JWA, Büller HA et al (2002) The MUC family: an obituary. Trends Biochem Sci 27:126–131
2. Corfield AP (2015) Mucins: a biologically relevant glycan barrier in mucosal protection. Biochim Biophys Acta 1850:236–252
3. Karasawa F, Shiota A, Goso Y et al (2012) Essential role of gastric gland mucin in preventing gastric cancer in mice. J Clin Invest 122: 923–934
4. Bergstrom K, Liu X, Zhao Y et al (2016) Defective intestinal mucin-type *O*-glycosylation causes spontaneous colitis-associated cancer in mice. Gastroenterology 151:152–164
5. Kuraya N, Hase S (1992) Release of *O*-linked sugar chains from glycoproteins with anhydrous hydrazine and pyridylamination of the sugar chains with improved reaction conditions. J Biochem 112:122–126
6. Iwase H, Ishii-Karakasa I, Fujii E et al (1992) Analysis of glycoform of *O*-glycan from human myeloma immunoglobulin A1 by gas-phase hydrazinolysis following pyridylamination of oligosaccharides. Anal Biochem 206:202–205

7. Patel T, Bruce J, Merry A et al (1993) Use of hydrazine to release in intact and unreduced form both *N*- and *O*-linked oligosaccharides from glycoproteins. Biochemistry 32:679–693
8. Huang Y, Mechref Y, Novotny MV (2001) Microscale nonreductive release of *O*-linked glycans for subsequent analysis through MALDI mass spectrometry and capillary electrophoresis. Anal Chem 73:6063–6069
9. Karlsson NG, Packer NH (2002) Analysis of *O*-linked reducing oligosaccharides released by an in-line flow system. Anal Biochem 305:173–185
10. Yamada K, Hyodo S, Matsuno Y et al (2007) Rapid and sensitive analysis of mucin-type glycans using an in-line flow glycan-releasing apparatus. Anal Biochem 371:52–61
11. Goso Y, Sugaya T, Ishihara K et al (2017) Comparison of methods to release mucin-type *O*-glycans for glycomic analysis. Anal Chem 89: 8870–8876
12. Goso Y, Tsubokawa D, Ishihara K (2009) Evaluation of conditions for release of mucin-type oligosaccharides from glycoproteins by hydrazine gas treatment. J Biochem 145:739–749
13. Goso Y (2016) Malonic acid suppresses mucin-type *O*-glycan degradation during hydrazine treatment of glycoproteins. Anal Biochem 496:35–42
14. Merry AH, Neville DCA, Royle L et al (2002) Recovery of intact 2-aminobenzamide-labeled *O*-glycans released from glycoproteins by hydrazinolysis. Anal Biochem 304:91–99
15. Anumula KR (2008) Unique anthranilic acid chemistry facilitates profiling and characterization of Ser/Thr-linked sugar chains following hydrazinolysis. Anal Biochem 373:104–111
16. Iwase H, Ishii I, Ishihara K et al (1988) Release of oligosaccharides possessing reducing-end *N*-acetylgalactosamine from mucus glycoprotein in Streptomyces sp. OH-11242 culture medium through action of endo-type glycosidase. Biochem Biophys Res Commun 151: 422–428
17. Carlstedt I, Sheehan JK, Corfield AP et al (1985) Mucous glycoproteins: a gel of a problem. Essays Biochem 20:40–76
18. Goso Y, Ishihara K, Kurihara M et al (1999) Rat gastric mucins recognized by monoclonal antibodies RGM21 and HIK1083: isolation of mucin species characteristic of the surface and glandular mucosa. J Biochem 126:375–381

Chapter 13

Eliminative Oximation of *O*-Glycans from Mucins

Akihiko Kameyama

Abstract

The large variety and high concentration of *O*-glycans are characteristic properties of mucins and play a crucial role in their unique functions. Analyzing the *O*-glycans of mucins is essential for investigating the functions of mucins. Eliminative oximation is an aqueous reaction that can be used to obtain *O*-glycan oximes from mucins. Using diazabicyclo undec-7ene (DBU) as a base, an organic superbase that can be removed with an organic solvent during solid-phase extraction, and adding hydroxylamine to the reaction mixture in advance, the *O*-glycans released from the mucin are immediately converted to the corresponding glycan oximes. The glycan oxime can then be fluorescently labeled with a fluorescent labeling reagent and 2-picoline borane via reductive amination. *O*-glycans that have been fluorescently labeled can be analyzed using conventional HPLC techniques.

Key words Mucin, *O*-linked glycan, Glycan oxime, Hydroxylamine, Reductive amination

1 Introduction

One of the characteristic properties of mucins is their extensive glycosylation, which plays important roles in mucin functions, including mucosal protection, infection prevention, and inflammation regulation. Alterations in glycosylation are often observed in tumors. Analyzing the glycans of mucins is essential for studying their functions. Mucins typically contain both *N*- and *O*-glycans, and *N*-glycans can be easily released from mucins by *N*-glycanases or hydrazinolysis [1–3]. On the other hand, *O*-glycans can be liberated through β-elimination, which is initiated via a proton abstraction with a base from the α-position of glycosylated serine or threonine. However, the base used in this reaction also abstracts the proton from the 2-position of the reducing end of the released glycans, triggering the glycan degradation process known as "peeling." To prevent peeling, reductive β-elimination has been developed and is widely used as a standard method for releasing *O*-glycans from mucins [4, 5]. During reductive elimination, the reducing ends of the released *O*-glycans are reduced to alditols prior

Akihiko Kameyama (ed.), *Mucins: Methods and Protocols*, Methods in Molecular Biology, vol. 2763,
https://doi.org/10.1007/978-1-0716-3670-1_13,

to the peeling reaction. While the alditols remain stable under basic conditions, they cannot be labeled with a fluorescent tag for highly sensitive analysis using HPLC and capillary electrophoresis (*see* Chap. 15). To release *O*-glycans while preserving the integrity of the reducing end, hydrazinolysis under mild conditions has been employed thus far.

The *O*-glycans of mucins can be released via β-elimination using anhydrous hydrazine as a base in hydrazinolysis. The released glycans immediately react with hydrazine to form their respective hydrazones. Since hydrazones decompose more slowly than unbound glycans in the presence of bases, they can be recovered by halting the reaction at a predetermined time. Hydrazine is a strong nucleophile that attacks the amide residues of *N*-acetyl hexosamines and sialic acids of glycans. Re-acetylation is necessary to obtain the original glycans. Notably, this procedure converts Neu5Gc to Neu5Ac. In addition, hydrazinolysis employs anhydrous hydrazine, a hazardous chemical, and requires completely anhydrous conditions, necessitating extreme caution and specialized equipment.

Eliminative oximation, as described here, is a novel method for releasing *O*-glycans from mucins and obtaining the corresponding oximes that can be labeled with fluorescent tags [6]. Despite the fact that the reaction occurs under aqueous conditions, the outcomes are identical to those of hydrazinolysis. Various bases, such as sodium hydroxide and lithium hydroxide, can be used in the reaction, but we chose the organic superbase diazabicyclo undec-7ene (DBU), which can be removed with an organic solvent during solid-phase extraction. By adding hydroxylamine to the reaction mixture beforehand, the *O*-glycans released from the mucin are immediately converted into their corresponding oximes. Glycan oximes are more stable than glycan hydrazones under basic conditions; however, "peeling" occurs when the reaction is prolonged under heated conditions. In addition, hydroxylamines are highly nucleophilic and may degrade *N*-glycolyl groups, which are more susceptible than *N*-acetyl groups. Therefore, reactions at 50 °C for more than 2 h should be avoided for samples that may contain NeuGc [7]. Eliminative oximation yields glycan oximes that can be fluorescently labeled with an aniline-based fluorescent labeling reagent, such as 2-amidebenzamide and anthranilic acid, using 2-picoline borane as a reducing reagent [8]. Conventionally, labeled *O*-glycans are analyzed via HPLC with a fluorescent detector (*see* Fig. 1). Although not presented here, eliminative oximation can also be applied to mucin bands separated via supported molecular matrix electrophoresis (SMME) [7, 9] and *O*-glycan oximes prepared from mucins can also be immobilized on hydrazide beads [10]. The same condition as the eliminative oximation also slowly cleaves *N*-glycans from glycoproteins [7]. To avoid confusion,

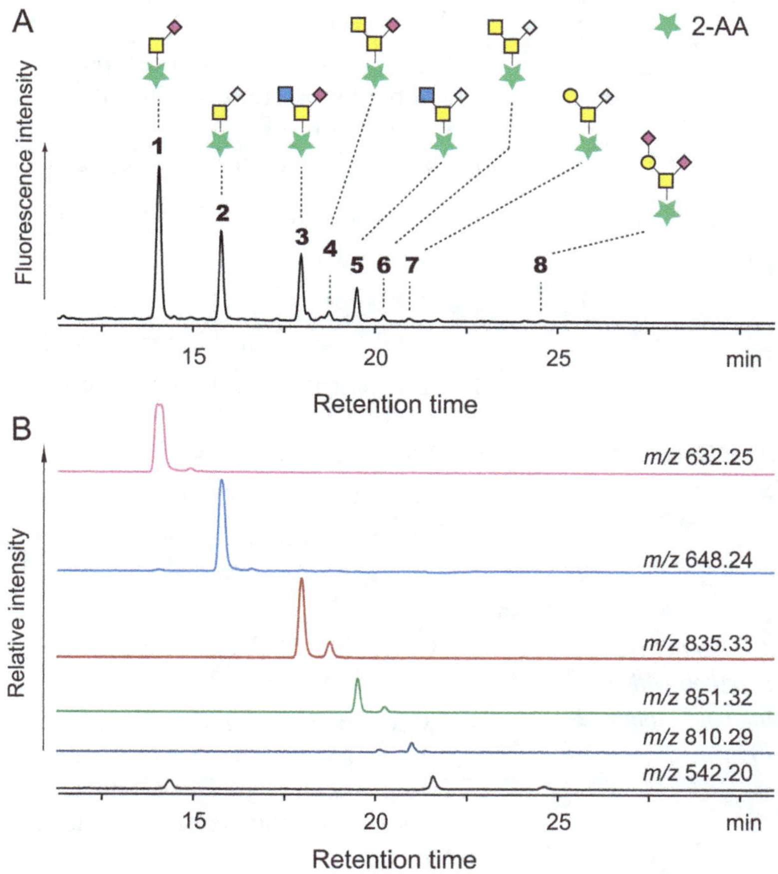

Fig. 1 Analysis of 2-AA labeled *O*-glycans in bovine submaxillary mucin (BSM) prepared via eliminative oximation. (**a**) HPLC chromatogram detected with a fluorescence detector (excitation: 330 nM, emission: 420 nM). (**b**) Selected ion chromatograms of the indicated *m/z* values via LC-MS. Glycan structures are depicted using CFG graphical notation for glycans. (Reproduced from [6] with permission. Copyright 2019 Elsevier)

removal of *N*-glycans before the eliminative oximation is recommended.

2 Materials

Prepare all solutions with ultrapure water and analytical grade reagents. Unless otherwise specified, prepare and store all reagents at room temperature. The materials typically used in Subheadings 3.1 and 3.2 are detailed below.

1. Acetonitrile.
2. Cleanup spin column (a component of BlotGlyco™ *O*-Glycan Sample Preparation Kit, Sumitomo Bakelite Co., Ltd., Tokyo, Japan) (*see* **Note 1**).
3. Gravity flow manifold (*see* **Note 2**).
4. Microcentrifuge (*see* **Note 3**).
5. Vortex mixer.
6. Heat block.
7. 1.5 mL microcentrifuge tube (*see* **Note 4**).
8. 5 mL centrifuge tube (*see* **Note 4**).

2.1 Eliminative Oximation

1. Mucin solution (*see* **Note 5**).
2. 50% hydroxylamine solution (*see* **Notes 6** and 7).
3. Diazabicyclo undec-7ene (DBU) (*see* **Note 6**).
4. Acetone.

2.2 Reductive Amination with 2-Aminobenzamide (2-AB)

1. 10% acetic acid: Add 180 μL of water and 20 μL of acetic acid in a 1.5 mL microcentrifuge tube and mix well (*see* **Note 8**).
2. 1.0 M labeling reagent solution: Weigh 27 mg of 2-aminobenzamide and transfer it to a 1.5 mL microcentrifuge tube. Add 200 μL of 0.24 M 2-picoline borane solution (solution "b") into the tube and dissolve 2-aminobenzamide (*see* **Notes 8** and **9**).
   (a) 10% acetic acid in methanol: Add 180 μL of methanol and 20 μL of acetic acid in a 1.5 mL microcentrifuge tube and mix well (*see* **Notes 10** and **11**).
   (b) 0.24 M 2-picoline borane solution: Weigh 5 mg of 2-picoline borane and transfer it to a 1.5 mL microcentrifuge tube. Add 200 μL of 10% acetic acid in methanol (solution "a") into the tube and dissolve 2-picoline borane (*see* **Note 11**).
3. HPLC system equipped with a fluorescent detector (ex. 330 nM, em. 420 nM).

3 Methods

Both 50% hydroxylamine and DBU are basic reagents. Their contact with eyes may cause severe damage. Always use the reagents while wearing appropriate protective equipment, such as gloves and goggles. In the event of skin contact, cleanse immediately under flowing water. In this method, *O*-glycans released from mucins are collected as oxime derivatives and subsequently labeled with fluorescent labeling reagents typically used for glycan analysis (*see*

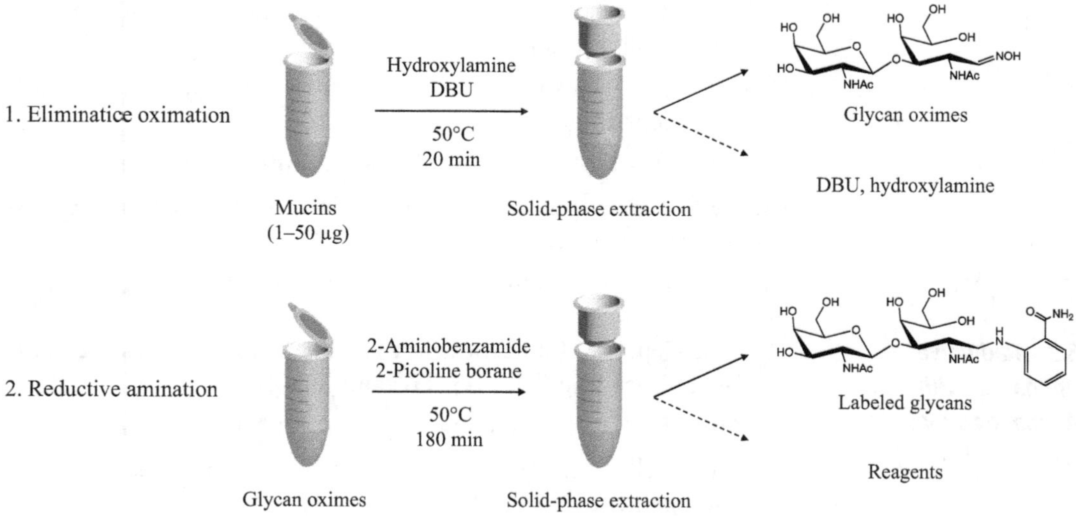

Fig. 2 Schematic diagram of the procedure. Gal β1-3GalNAc is shown as an example of *O*-glycans

Fig. 2). We have confirmed that both 2-AB and anthranilic acid (2-AA) can be fluorescently labeled without issue. Note that the 8-aminopyrene-1,3,6-trisulfonic acid (APTS) used in capillary electrophoresis cannot be used to recover labeled glycans using the method presented here.

3.1 Eliminative Oximation

1. Add 50% hydroxylamine solution (8 μL) and mucin solution (16 μL) in a 1.5 mL microcentrifuge tube and mix well with a vortex mixer (*see* **Note 6**).
2. Add 16 μL of DBU to the mixture (*see* **Note 6**).
3. Mix well with the vortex mixer and centrifuge with a microcentrifuge.
4. Incubate at 50 °C for 20 min in a heat block. **Steps 5, 6,** and 7 are done while waiting 20 min (*see* **Note 12**).
5. Set the cleanup spin column onto a gravity flow manifold.
6. Wash the cleanup spin column with 200 μL of water and then 600 μL of acetonitrile twice.
7. Add 200 μL of cold acetone to a new 5 mL centrifuge tube.
8. Transfer the heated mixture into the cold acetone in the 5 mL centrifuge tube (*see* **Note 13**).
9. Dilute the mixture with acetonitrile up to 4 mL and mix by pipetting.
10. Pass the entire volume of the diluted solution through the cleanup column (prepared in **step 6**) in several batches. Discard the pass-through fraction.
11. Wash the cleanup column with 600 μL of acetonitrile twice. Discard the washings.

12. Centrifuge the cleanup column using the microcentrifuge to remove all liquid from the column.
13. Set the cleanup column onto a new 1.5 mL microcentrifuge tube.
14. Add 100 μL of double distilled water onto the cleanup column.
15. Centrifuge the cleanup column using the microcentrifuge and collect the eluted solution.
16. Lyophilize the solution using a centrifuge evaporator.

3.2 Reductive Amination with 2-Aminobenzamide (2-AB) (See Note 14)

1. Add 25 μL of 10% acetic acid to a tube containing dried glycans (*see* Subheading 3.1, **step 16**) and dissolve it.
2. Add 25 μL of the labeling reagent solution into the tube.
3. Mix well with a vortex mixer and centrifuge using a microcentrifuge.
4. Incubate at 50 °C for 3 h in a heat block. Perform **steps 5, 6,** and 7 while waiting for 3 h.
5. Set the cleanup spin column onto a gravity flow manifold.
6. Wash the cleanup spin column with 200 μL of water and then 600 μL of acetonitrile twice.
7. Add 4 mL of acetonitrile in a new 5 mL centrifuge tube.
8. Transfer the heated mixture **(step 4)** into the 5 mL centrifuge tube.
9. Dilute the mixture with acetonitrile up to 4 mL and mix by pipetting.
10. Pass the entire volume of the diluted solution through the cleanup column (prepared in **step 6**) in several batches. Discard the pass-through fraction.
11. Wash the cleanup column with 600 μL of acetonitrile twice. Discard the washings.
12. Centrifuge the cleanup column with the microcentrifuge to remove all liquid from the column.
13. Set the cleanup column onto a new 1.5 mL microcentrifuge tube.
14. Add 100 μL of double distilled water to the cleanup column.
15. Centrifuge the cleanup column with the microcentrifuge and collect the eluted solution.
16. Analyze the solution via HPLC equipped with a fluorescent detector (*see* **Note 15**).

4 Notes

1. Cleanup spin column is the only item used in this kit. The others are not used. Another similar kit is the EZGlyco *O*-glycan prep kit (Sumitomo Bakelite Co., Ltd.). If you use the latter kit, please follow the instructions of the manufacturer. The cleanup device included in the other kits used for fluorescent labeling of glycans may also be used, but in most cases, it is specialized for fluorescent labeling of *N*-linked glycans, so the cleaning conditions should be considered when using it.
2. Vacuum manifolds may also be used, but vacuum pump should not be employed. GL SPE mini-manifold and GL SPE gravity flow manifolds (5010–50,150, 5010–50,430, GL Science Inc., Tokyo, Japan) are utilized.
3. We use CHIBITAN-II (Merck KGaA, Germany).
4. Use only tubes that are not coated with a low adsorption polymer.
5. Mucin solutions may contain the following denaturants, detergents, and buffers: dithiothreitol (DTT), iodoacetamide (IAA), NP-40, urea, 1% Triton X-100, 0.1% sodium dodecyl sulfate (SDS), PBS, and 0.1 M Tris–HCl. However, guanidine hydrochloride is not acceptable.
6. Goggles and gloves should be used for handling.
7. Any surplus chemicals should be diluted with 10% acetone and disposed of according to local laws.
8. Use 25 μL per sample. When processing a large number of samples, increase the volume accordingly.
9. Since it is challenging to accurately measure 27 mg, rapidly measure out 2-picoline borane and calculate the amount of solution required based on this quantity. Calculate the required quantity as 3.4 mg/sample when processing a large number of samples.
10. Use immediately after preparation. Ethyl acetate will be generated.
11. Since it is challenging to precisely measure 5 mg, rapidly measure out 2-picoline borane and calculate the quantity of solution required based on this amount. Calculate the required quantity as 0.6 mg/sample when processing a large number of samples.
12. Maintain precise temperature control. Higher temperatures increase peeling and lower temperatures decrease yield.
13. Hydroxylamine is quenched by acetone.

14. Other labeling reagents such as anthranilic acid (2-AA), can be used instead of 2-AB.
15. Regarding HPLC analysis of fluorescent labeled *O*-glycans, *see* Chap. 15.

References

1. Pallesen LT, Pedersen LRL, Petersen TE et al (2007) Characterization of carbohydrate structures of bovine MUC15 and distribution of the mucin in bovine milk. J Dairy Sci 90:3143–3152
2. Taniguchi T, Woodward AM, Magnelli P et al (2017) N-glycosylation affects the stability and barrier function of the MUC16 mucin. J Biol Chem 292:11079–11090
3. Patel T, Bruce J, Merry A, Bigge C et al (1993) Use of hydrazine to release in intact and unreduced form both *N*- and *O*-linked oligosaccharides from glycoproteins. Biochemistry 32: 679–693
4. Carlson DM (1968) Structures and immunochemical properties of oligosaccharides isolated from pig submaxillary mucins. J Biol Chem 243:616–626
5. Kotsias M, Kozak RP, Gardner RA et al (2019) Improved and semi-automated reductive β-elimination workflow for higher throughput protein *O*-glycosylation analysis. PLoS One 14: e0210759
6. Kameyama A, Thet Tin WW, Toyoda M et al (2019) Sakaguchi M. A practical method of liberating *O*-linked glycans from glycoproteins using hydroxylamine and an organic superbase. Biochem Biophys Res Commun 513:186–192
7. Kameyama A, Dissanayake SK, Thet Tin WW (2018) Rapid chemical de-*N*-glycosylation and derivatization for liquid chromatography of immunoglobulin N-linked glycans. PLoS One 13:e0196800
8. Sato S, Sakamoto T, Miyazawa E et al (2004) One-pot reductive amination of aldehydes and ketones with α-picoline-borane in methanol, in water, and in neat conditions. Tetrahedron 60: 7899–7906
9. Matsuno Y-K, Saito T, Gotoh M et al (2009) Supported molecular matrix electrophoresis: a new tool for characterization of glycoproteins. Anal Chem 81:3816–3823
10. Kameyama A, Nishijima R, Yamakoshi K (2021) Bmi-1 regulates mucin levels and mucin *O*-glycosylation in the submandibular gland of mice. PLoS One 16:e0245607

Chapter 14

9-Fluorenylmethyl Chloroformate Labeling for *O*-Glycan Analysis

Keita Yamada

Abstract

Structural analysis of *O*-glycans from mucins and characterization of the interaction of these glycans with other biomolecules are essential for a full understanding of mucins. Various techniques have been developed for the structural and functional analysis of glycans. While 9-fluorenylmethyl chloroformate (Fmoc-Cl) is generally used to protect amino groups in peptide synthesis, it can also be used as a glycan-labeling reagent for structural analysis. Fmoc-labeled glycans are strongly fluorescent and can be analyzed with high sensitivity using liquid chromatography-fluorescence detection (LC-FD) analysis as well as being analyzed with high sensitivity by matrix-assisted laser desorption/ionization-time of flight mass spectrometry (MALDI-TOF MS). Fmoc-labeled glycans can be easily delabeled and converted to glycosylamine-form or free (hemiacetal or aldehyde)-form glycans that can be used to fabricate glycan arrays or synthesize glycosyl dendrimers. This derivatization allows for the isolation from biological samples of glycans that are difficult to synthesize chemically, as well as the fabrication of immobilized-glycan devices. The Fmoc labeling method promises to be a tool for accelerating *O*-glycan structural analysis and an understanding of molecular interactions. In this chapter, we introduce the Fmoc labeling method for analysis of *O*-glycans and fabrication of *O*-glycan arrays.

Key words Glycosylamine, 9-Fluorenylmethyl chloroformate (Fmoc-Cl), *O*-Glycan, Liquid chromatography fluorescence detection (LC-FD), Hydrophilic interaction liquid chromatography (HILIC), Glycan array, Polylactosamine

1 Introduction

Serine (Ser)/threonine (Thr)-linked glycans (*O*-glycans) play roles in the immune function of mucins by adsorbing bacteria and viruses [1, 2]. *O*-Glycans have also been proposed to modulate signal transduction by interaction with a variety of biomolecules [3]. A comprehensive characterization of mucins requires analysis of the structures of the *O*-glycans contained in them and interactions of the *O*-glycans with other biomolecules. *O*-glycosylation occurs as a number of types including *O*-*N*-acetyl glucosamine (GlcNAc)-, *O*-fucose (Fuc)-, *O*-mannose (Man)-, *O*-glucose (Glc)-, and *O*-*N*-

Akihiko Kameyama (ed.), *Mucins: Methods and Protocols*, Methods in Molecular Biology, vol. 2763,
https://doi.org/10.1007/978-1-0716-3670-1_14,

acetyl galactosamine (GalNAc)-types, differentiated by the sugar attached to the oxygen atom of the Ser or Thr side chain [4]. Of these *O*-glycans, *O*-GalNAc-type glycans are the most widely expressed in vertebrates and have very diverse structures. As *O*-GalNAc-type glycans are found in large quantities in mucins, they are sometimes referred to as mucin-type *O*-glycans [5]. To analyze their complex structures, glycans should first be released from core proteins and separated by liquid chromatography (LC) or capillary electrophoresis (CE). Mass spectrometry (MS) analysis is essential for obtaining information on glycan structures. Various derivatization methods have been developed to assist in LC, CE, and MS analysis of glycan [6]. The most common derivatization methods involve fluorescent labeling combined with reductive amination or permethylation [7]. Beyond these well-established derivatization methods, derivatization of glycans with amine-labeling reagents has recently come into use. Here, reduced hemiacetal-form glycans are converted to glycosylamine-form glycans, which are then reacted with amine-labeling reagents. As the amine-labeling reaction proceeds under milder conditions than those of reductive amination, the risk of sialic acid desorption during labeling is minimized. The most popular labeling reagent for glycosylamine-form glycans is RapiFluor-MS, which is based on *N*-hydroxysuccinimidyl carbamate groups. Kits for glycan analysis using this reagent are now available [8]. Another useful glycosylamine-form-glycan-labeling reagent is 9-fluorenylmethyl chloroformate (Fmoc-Cl), which is commonly used to protect amine groups in peptide synthesis [9]. As Fmoc-labeled glycans have strong fluorescence, they can be analyzed with high sensitivity using liquid chromatography fluorescence detection (LC-FD). The detection sensitivity for Fmoc-labeled glycans in LC-FD is three times higher than for glycans labeled using common reductive amination reagents [9]. Further, Fmoc-labeled glycans can be easily delabeled and converted to glycosylamine-form or free (hemiacetal or aldehyde)-form glycans suitable for preparing glycan interaction analysis tools such as glycan arrays and glycan dendrimers. The Fmoc labeling method enables isolation from biological samples of glycans with complex structures that are difficult to chemically synthesize and the creation of arrays and other tools for interaction analysis. In this chapter, we introduce methods for analysis of *O*-glycans using Fmoc labeling.

The Fmoc labeling method is easily applied to the analysis of asparagine (Asn)-linked glycans (*N*-glycans), as glycosylamine-form *N*-glycans can be recovered quantitatively and easily from core proteins using endoglycosidases such as peptide *N*-glycanase F [9, 10]. However, the Fmoc labeling method has not commonly been applied in the analysis of *O*-glycans because it has been difficult to recover glycosylamine-form *O*-glycans. We developed a method for *O*-glycan analysis using Fmoc labeling (*see* Fig. 1)

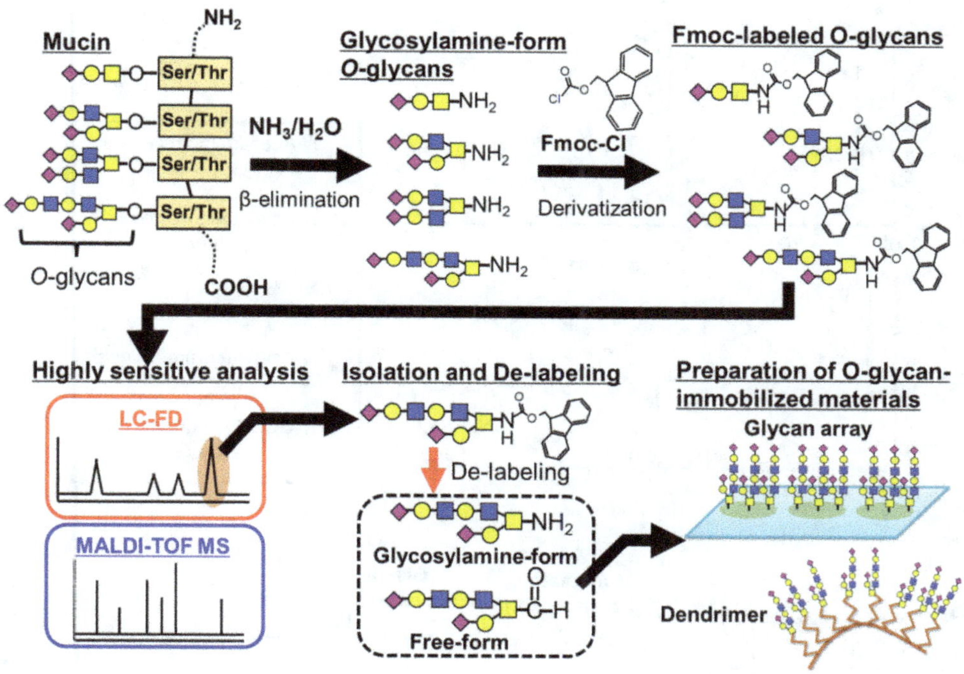

Fig. 1 Scheme for *O*-glycan analysis using Fmoc derivatization

[11]. When *O*-glycans are released from proteins by ammonia-based β-elimination, the released *O*-glycans are present in a glycosylamine form. Glycosylamine-form *O*-glycans are typically converted to free-form glycans in aqueous solutions below pH 7.0 but are stable in solution above pH 8.0. Glycosylamine-form *O*-glycans in aqueous solution at pH 8.0 can easily be labeled with Fmoc-Cl, and the resulting Fmoc-labeled *O*-glycans can be analyzed with high sensitivity by LC-FD (*see* Fig. 2). The major *O*-glycans in bovine submaxillary gland mucin (BSM) and bovine fetuin were detected using FD-LC analysis, as were complex *O*-glycans from cultured cancer cells. Fmoc-labeled *O*-glycans were detected with high sensitivity using matrix-assisted laser desorption/ionization-time of flight mass spectrometry (MALDI-TOF MS) [11]. It was also possible to isolate Fmoc-labeled *O*-glycans using hydrophilic interaction liquid chromatography (HILIC) and obtain glycosylamine-form or free-form *O*-glycans by delabeling. We have successfully immobilized glycosylamine-form *O*-glycans obtained by delabeling on epoxy resin-coated slides. The resulting immobilized-*O*-glycan device can be used to detect interactions with lectins (*see* Fig. 3). We believe the Fmoc labeling method is capable of accelerating structural analysis of *O*-glycans as well as analysis of interactions of *O*-glycans with other biomolecules.

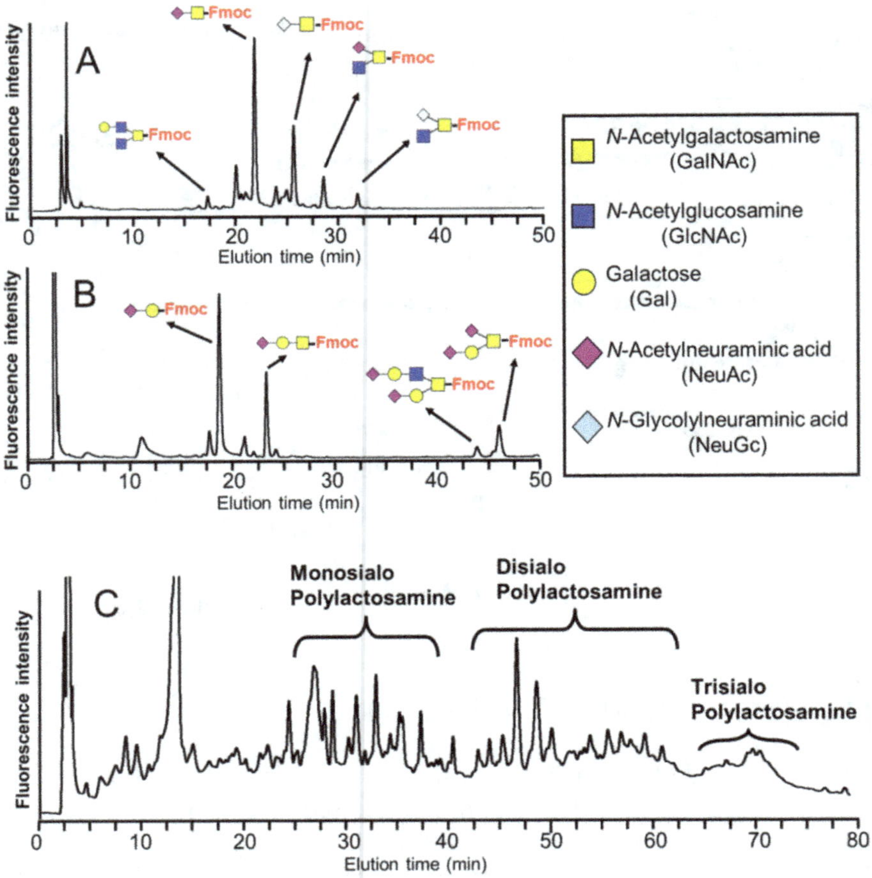

Fig. 2 FD-LC analysis of Fmoc-labeled *O*-glycans. Fmoc-labeled *O*-glycans prepared from (**a**) BSM, (**b**) bovine fetuin, and (**c**) mucin of MKN45 cells were analyzed using HILIC LC-FD. Sialyl-Tn (NeuAcα2–6GalNAc)- and core 3 (GlcNAcβ1–3GalNAc)-type *O*-glycans, incorporating N-acetylneuraminic acid (NeuAc) or *N*-glycolylneuraminic acid (NeuGc) groups, were identified in BSM. Sialyl-T (NeuAcα2-3Galβ1–3GalNAc)-, disialyl-T (NeuAcα2-3Galβ1–3(NeuAcα2–6) GalNAc)-, and disialylated Core 2-type *O*-glycans (NeuAcα2-3Galβ1–3 (NeuAcα2-3Galβ1-4GlcNAcβ1–6) GalNAc), incorporating galactose (Gal) groups, were identified in bovine fetuin. Multiple polylactosamine-type *O*-glycans containing one to three sialic acid residues were identified in MKN45. When Fmoc-labeled *O*-glycans are analyzed under these conditions, they are separated by differences in molecular weight and the number of acidic groups

2 Materials

Prepare all solutions using ultrapure water and analytical grade reagents. Do not add sodium azide. Prepare and store reagents at room temperature unless otherwise indicated.

2.1 Recovery of Glycosylamine-Form O-Glycans from Mucin Using Ammonia-Based β-Elimination

1. A centrifugal concentrator with a vacuum pump and low-temperature trap. Alkali-resistant filters should be connected.
2. Glycoprotein (1 ~ 1000 μg) or mucin samples (*see* **Note 1**).
3. Ammonium carbonate.

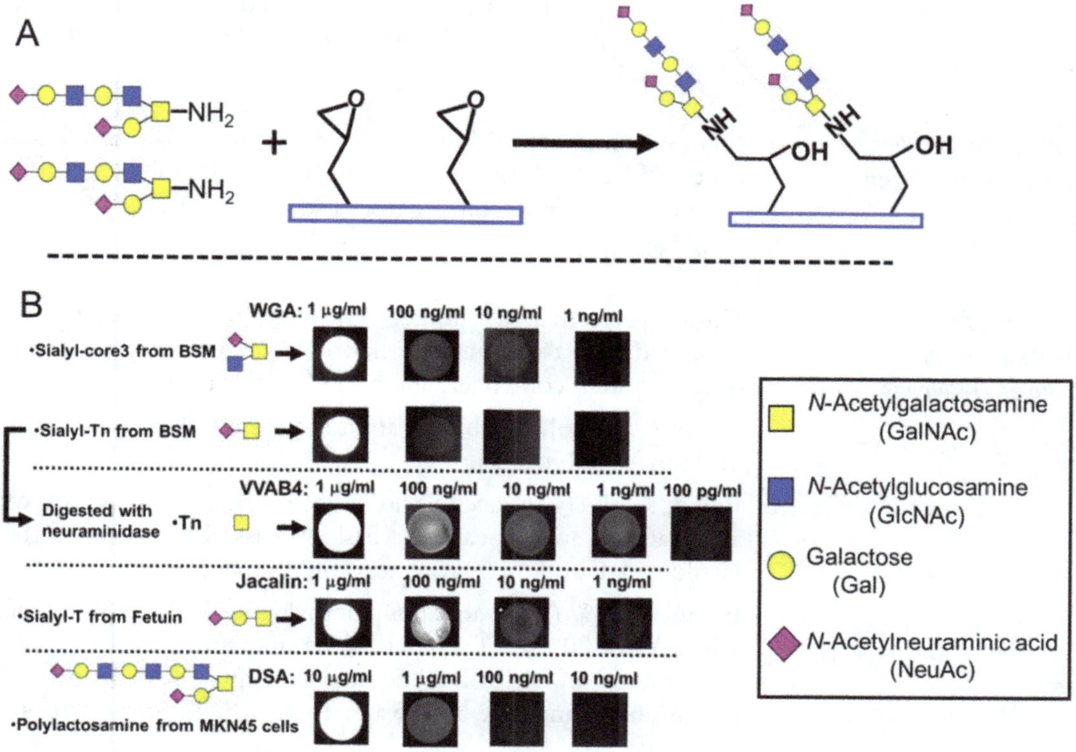

Fig. 3 Fabrication of a glycan array with glycosylamine-form *O*-glycans. (**a**) Immobilization reaction of glycosylamine-form *O*-glycans on epoxy-coated slides. (**b**) Analysis of interactions between glycosylamine *O*-glycan arrays and lectins. Fmoc-labeled *O*-glycans prepared from BSM and bovine fetuin and MKN45 cells were converted to glycosylamine-form *O*-glycans by delabeling and immobilized on the epoxy-coated slide. We successfully detected interactions between *O*-glycans and lectins using this glycan array with biotinylated lectin. Cy3-labeled streptavidin was added and fluorescence on the array surface was observed using an evanescent-field fluorescence-assisted scanner. Wheat germ agglutinin (WGA) recognizes the *N*-acetyl group in GlcNAc and NeuAc residues. *Vicia villosa* lectin (VVAB4) recognizes Tn antigen (i.e., GalNAc residues). *Datura stramonium* agglutinin (DSA) recognizes repeating units of *N*-acetylglucosamine and Gal-GlcNAc, referred to as polylactosamine. Jacalin recognizes T- and sialyl-T antigens. (Reproduced from [11] with permission. Copyright 2013 American Chemical Society))

4. Ammonium-carbonate-saturated aqueous ammonia solution: Weigh 3.0 g of ammonium carbonate into a 50 mL of centrifuge tube. Add 10 mL of 28% aqueous ammonia and mix. After allowing the solution to stand at room temperature for 5 min, draw off the supernatant. The solution should be prepared freshly before use.
5. 10× sodium phosphate buffer (300 mM sodium phosphate pH 8.4): Weigh 4.046 g of disodium hydrogen phosphate (Na_2HPO_4) and 0.216 g of sodium dihydrogen phosphate (NaH_2PO_4) into a glass beaker. Add 90 mL of water and dissolve the phosphate salts. Transfer the solution to a

100 mL measuring flask, add water to 100 mL and mix. Store at 4 °C. This solution should be diluted tenfold before being added to the sample.

2.2 Fmoc Labeling of Glycosylamine-Form O-Glycans

1. Fmoc-Cl solution: Weigh 8.0 mg of Fmoc-Cl into a 1.5 mL tube. Add 1.0 mL of acetone and dissolve. Fmoc-Cl solution should be prepared freshly before use.
2. Chloroform.

2.3 HILIC LC-FD Analysis of Fmoc-Labeled O-Glycans

1. High performance liquid chromatography instrument equipped with two pumps, a fluorescence detector, a column oven, a system controller, and an LC workstation.
2. Column: A polymer-based amino column (4.6 mm i.d. × 250 mm) (*see* **Note 2**).
3. Eluent A (2% (v/v) acetic acid in acetonitrile): Add 20 mL of acetic acid to a 1-L measuring flask or glass beaker. Add acetonitrile to 1 L and mix. Store at room temperature or 4 °C.
4. Eluent B (3% (v/v) aqueous triethylamine, 5% (v/v) acetic acid): Add 500 mL of water to a 1-L measuring flask or glass beaker. Add 50 mL of acetic acid and mix. Add 30 mL of triethylamine and mix. Add water to make 1 L. Store at room temperature or 4 °C.

2.4 Recovery of Glycosylamine-Form and Free-Form O-glycans by Delabeling

1. *N, N*-Dimethylformamide (DMF) (*see* **Note 3**).
2. Morpholine.
3. Diethyl ether (*see* **Note 3**).
4. 0.1% (v/v) aqueous acetic acid: Add 9.99 mL of water and 0.01 mL of acetic acid to 15 mL centrifuge tubes and mix. Store at room temperature or 4 °C.

2.5 Fabrication of Glycan Array with Glycosylamine-Form O-Glycans

1. Microarray-grade epoxy-coated glass slide (NEXTERION Slide E, SCHOTT, Mainz, Germany) (*see* **Note 4**).
2. Incubation chamber and chamber sheet.
3. Silicon rubber sheet.
4. Spotting solution: Dilute the spotting solution for DNA microarrays with an equal volume of water (*see* **Note 5**).
5. Tris-buffered saline (TBS): Use commercial products.
6. Washing solution (TBS containing 1% (v/v) Triton-X 100): Add 495 mL of TBS and 5 mL of Triton-X 100 to a glass beaker and mix. Store at 4 °C.
7. Blocking solution (TBS containing 2% (w/v) bovine serum albumin (BSA)): Weigh 800 mg of BSA into a 50 mL centrifuge tube. Add TBS to the centrifuge tube to 40 mL. Store at 4 °C.

8. Binding solution (TBS containing 1% (v/v) Triton-X 100, 1 mM manganese chloride ($MnCl_2$), 1 mM calcium chloride ($CaCl_2$), and 1% (w/v) BSA): Weigh 6.0 mg of $MnCl_2$ tetrahydrate and 3.4 mg of $CaCl_2$ into a 50 mL centrifuge tube. Weigh 300 mg of BSA into the centrifuge tube. Add 30 mL of washing solution to the centrifuge tube and mix. Store at 4 °C.

3 Methods

Carry procedures out at room temperature unless specified otherwise.

3.1 Preparation of Glycosylamine-Form O-Glycans by Ammonia-Based β-Elimination

1. Add 100 mg of ammonium carbonate and 1 mL of ammonium carbonate-saturated aqueous ammonia solution to dried glycoprotein or mucin samples (1–1000 μg) and mix using a vortex mixer.
2. Seal the sample tube and incubate at 37 °C for 24 h (*see* **Note 6**).
3. Evaporate the sample solution to remove ammonia using a centrifugal concentrator.
4. Add 200 μL of 30 mM sodium phosphate pH 8.4 to the dried sample and evaporate to remove ammonium carbonate (*see* **Note 7**).
5. Add 200 μL of water to the dried sample and evaporate to remove ammonium carbonate. Repeat this step twice (*see* **Note 7**).
6. Collect the dried materials containing glycosylamine-form *O*-glycans (*see* **Note 8**).

3.2 Fmoc Labeling of Glycosylamine-Form O-Glycans

1. Add 300 μL of water to the dried glycosylamine-form *O*-glycans from Subheading 3.1 (*see* **Note 9**).
2. Add 200 μL of Fmoc-Cl solution to the glycosylamine-form *O*-glycan aqueous solution and mix using a vortex mixer.
3. Incubate the mixture at 37 °C for 80 min.
4. Add 300 μL of chloroform to the mixture and mix with a vortex mixer.
5. Centrifuge the mixture at 18,000 × *g* for 2 min and collect the aqueous upper layer.
6. Add 150 μL of water and 300 μL of chloroform to the aqueous phase and mix using a vortex mixer.
7. Centrifuge the mixture at 18,000 × *g* for 2 min and collect the aqueous upper layer.

8. Evaporate the aqueous phase. The recovered dried sample contains Fmoc-labeled *O*-glycans. This sample can be stored at −20 °C for several months (*see* **Note 10**).

3.3 Analysis of Fmoc-Labeled O-Glycans by HILIC LC-FD

1. Set the excitation wavelength of the fluorescence detector to 266 nM and the emission wavelength to 310 nM.
2. Set the column temperature to 50 °C.
3. Set the combined flow rate of the two pumps to 1.0 mL/min and maintain this flow rate during the analysis.
4. Set up a linear gradient program for eluent A and eluent B. From 0 to 2 min after the start of the analysis, hold eluent B at 15%, then increase eluent B from 15 to 80% from 2 to 82 min and hold at this composition for a further 10 min.
5. Before starting the analysis, equilibrate the column with 15% eluent B for 15 min.
6. Dissolve Fmoc-labeled *O*-glycans in water and inject onto the column (*see* **Note 11**).
7. Typical analytical results for Fmoc-labeled *O*-glycans are shown in Fig. 2.

3.4 Recovery of Glycosylamine-Form and Free-Form O-Glycans Using Fmoc Elimination

1. Dissolve Fmoc labeled glycan fractions from LC-FD in 20 μL of water.
2. Add 30 μL of DMF and 20 μL of morpholine to the Fmoc labeled glycans solution and mix using a vortex mixer.
3. Incubate the mixture at 37 °C for 30 min (*see* **Note 12**).
4. Add 500 μL of diethyl ether to the mixture and mix vigorously using a vortex mixer.
5. Centrifuge the mixture at 18,000 × g for 2 min and collect the aqueous lower phase.
6. Evaporate the aqueous phase using a centrifugal concentrator. The glycosylamine-type *O*-glycans are obtained at this point (*see* **Note 13**).
7. Add 30 μL of 0.1% (v/v) aqueous acetic acid to the dried samples and incubate at room temperature for 1 h.
8. Evaporate the mixture using a centrifugal concentrator. The label-free-form *O*-glycans are obtained at this point (*see* **Note 14**).

3.5 Fabrication of O-Glycan Arrays Using Glycosylamine-Form O-Glycans

The mechanism of immobilization of glycosylamine-form *O*-glycans onto slides is shown in Fig. 3a. The primary amine of the glycosylamine undergoes a nucleophilic reaction with the epoxy group on the slide, leading to immobilization of the glycosylamine-form *O*-glycans on the slide.

1. Perform **steps 1–6** of Subheading 3.4 to recover glycosylamine-form *O*-glycans.
2. Add 50% (v/v) aqueous spotting solution to the glycosylamine-form *O*-glycans (*see* **Note 15**).
3. Spot 1 μL of glycosylamine-form *O*-glycan solution onto an epoxy-coated glass slide attached to a silicon rubber sheet to print glycan spots with a diameter of approximately 1 mM (*see* **Note 16**).
4. Incubate the glass slide in the incubation chamber containing wet paper at 25–30 °C for 3 h to complete immobilization.
5. Rinse the immobilized-*O*-glycan glass slide three times with the washing solution to remove excess non-immobilized materials.
6. Add the blocking solution to the immobilized-*O*-glycan glass slide and incubate for 1.5 h in the incubation chamber.
7. Remove the blocking solution and rinse the glass slide three times with the washing solution.
8. Rinse the immobilized-*O*-glycan glass slide three times with water and air-dry it. This slide can be used as a glycan array (*see* **Note 17**).
9. Glycan-binding molecules are dissolved in binding solution and added to the prepared array (*see* **Note 18**).

Typical results from adding lectin to immobilized-*O*-glycan slides are shown in Fig. 3b.

4 Notes

1. *N*-glycans are also recovered by amine-based β-elimination. When amine treatment is applied directly to crude protein fractions from cells or tissues, it is difficult to distinguish *N*-glycans from *O*-glycans. Use of this reaction with mucin fractions purified from cells or tissues is thus preferable. *See* [12] for purification of mucin fractions.
2. A commercial amino column (e.g., Shodex Asahipak NH2P-50 4E (Resonac Inc., Tokyo, Japan)) can be used.
3. It is preferable to add molecular sieves to the reagent.
4. Epoxy-coated glass slides deteriorate when exposed to humidity and should be stored in a desiccator.
5. A commercial spotting solution (e.g., Matsunami spotting solution (Matsunami Glass Ind., Ltd., Kishiwada City, Japan) and NEXTERION Spot (SCHOTT)) can be used.

6. Ammonia-based β-elimination may be carried out at 60 °C for 48 h. At a reaction temperature of 60 °C, the amount of *O*-glycans released from the protein is higher. However, the peeling reaction degrades 80% of the released *O*-glycans.
7. It is advisable to heat the product to at least 60 °C during evaporation to facilitate ammonium carbonate removal.
8. After glycosylamine-form *O*-glycans are recovered, Fmoc labeling should be carried out immediately.
9. To label glycosylamine-form *O*-glycans obtained by a method other than in Subheading 3.1 with Fmoc-Cl, dissolve the glycans in 300 μL of 20 mM sodium phosphate pH 8.4.
10. If excess Fmoc-Cl remains, the sample should be purified using a HILIC cartridge. The HILIC cartridge purification method shown in Chap. 15 can be applied (*see* Chap. 15). However, some small O-glycans are lost when this method is used.
11. The volume of sample injected onto the column should be less than 10 μL. While the sensitivity depends on the detector, in general several fmol of glycans can be detected.
12. If delabeling does not proceed to completion, increase the reaction time.
13. Glycosylamine-form *O*-glycans cannot be stored for a long time. When glycosylamines are recovered, they should be used immediately for preparation of arrays or other immobilized-glycan materials.
14. Free-form *O*-glycans can be stored for a long time.
15. Adjust the concentration of *O*-glycan solution to 10 pmol/μL or higher. The higher the glycan concentration, the better the array quality.
16. Spotting can be performed with micropipettes, but a spotter for inkjet-type microarrays can also be used.
17. The prepared glycan arrays can be stored at 4 °C for 2–3 months.
18. The glycan-binding molecule should be labeled with a fluorescent reagent. If the molecule cannot readily be labeled with a fluorescent reagent, a fluorescent-labeled antibody can be used for detection. When measuring biotinylated molecules, fluorescent-labeled streptavidin is used for detection. While there are a range of fluorescent reagents available, Cy3 (excitation wavelength 550 nM, emission wavelength 570 nM) or Alexa Fluor 555 (excitation 555 nM, emission 568 nM) can be used for highly sensitive detection given a low background signal. Fluorescent detection on arrays is possible without the use of a scanner. Particularly if the glycan spot size is large, a general fluorescent image analyzer can be used.

Acknowledgments

This work was supported by the Japan Society for the Promotion of Science KAKENHI (Grant Number 23 K06075). We would like to thank Editage (www.editage.com) for English language editing.

References

1. Sun X, Li D, Duan Z (2021) Structural basis of glycan recognition of rotavirus. Front Mol Biosci 8:658029
2. Bergstrom K, Xia L (2022) The barrier and beyond: roles of intestinal mucus and mucin-type O-glycosylation in resistance and tolerance defense strategies guiding host-microbe symbiosis. Gut Microbes 14(1):2052699
3. Tran DT, Ten Hagen KG (2013) Mucin-type O-glycosylation during development. J Biol Chem 288(10):6921–6929
4. Zhang L, Ten Hagen KG (2019) O-linked glycosylation in Drosophila melanogaster. Curr Opin Struct Biol 56:139–145
5. Brockhausen I, Wandall HH, Hagen KGT et al (2022) O-GalNAc glycans. In: Varki A, Cummings RD, Esko JD et al (eds) Essentials of glycobiology, 4th edn. Cold Spring Harbor Laboratory Press, New York
6. Kinoshita M, Yamada K (2022) Recent advances and trends in sample preparation and chemical modification for glycan analysis. J Pharm Biomed Anal 207:114424
7. Kang P, Mechref Y, Kyselova Z et al (2007) Comparative glycomic mapping through quantitative permethylation and stable-isotope labeling. Anal Chem 79(16):6064–6073
8. Lauber MA, Yu YQ, Brousmiche DW et al (2015) Rapid preparation of released N-Glycans for HILIC analysis using a labeling reagent that facilitates sensitive fluorescence and ESI-MS detection. Anal Chem 87(10): 5401–5409
9. Kamoda S, Nakano M, Ishikawa R et al (2005) Rapid and sensitive screening of N-glycans as 9-fluorenylmethyl derivatives by high-performance liquid chromatography: a method which can recover free oligosaccharides after analysis. J Proteome Res 4(1):146–152
10. Nakano M, Higo D, Arai E et al (2009) Capillary electrophoresis-electrospray ionization mass spectrometry for rapid and sensitive N-glycan analysis of glycoproteins as 9-fluorenylmethyl derivatives. Glycobiology 19(2):135–143
11. Yamada K, Hirabayashi J, Kakehi K (2013) Analysis of O-glycans as 9-fluorenylmethyl derivatives and its application to the studies on glycan array. Anal Chem 85(6):3325–3333
12. Matsuno YK, Saito T, Gotoh M et al (2009) Supported molecular matrix electrophoresis: a new tool for characterization of glycoproteins. Anal Chem 81(10):3816–3823

Chapter 15

Liquid Chromatography and Capillary Electrophoresis Analysis of 2AA-Labeled *O*-Glycans

Keita Yamada

Abstract

To reveal *O*-glycan structures in mucins, it is necessary to release covalently bound *O*-glycans from the polypeptide backbone and derivatize to a form suitable for structural analysis. Various derivatization methods can now be applied in the analysis of *O*-glycans following the development of *O*-glycan release methods. Among the many derivatization methods available, we prefer to use fluorescent labeling with 2-aminobenzoic acid (anthranilic acid, 2AA). 2AA-labeled *O*-glycans can be detected with high sensitivity using liquid chromatography fluorescence detection (LC-FD) analysis because of the strong fluorescence. In addition, as 2AA has a carboxyl group that carries a negative charge, 2AA-labeled *O*-glycans can be analyzed with high sensitivity in negative ion mode mass spectrometry. Furthermore, because the negative charge of 2AA provides a driving force for electrophoresis, 2AA-labeled *O*-glycans can be analyzed using capillary electrophoresis (CE) and capillary affinity electrophoresis. High detection sensitivity and versatility are key advantages of the 2AA-labeling method. Here we present our preferred LC-FD and CE analytical methods for 2AA-labeled *O*-glycans.

Key words 2-Aminobenzoic acid (2AA), *O*-Glycans, Liquid chromatography fluorescence detection (LC-FD), Hydrophilic interaction liquid chromatography (HILIC), Capillary gel electrophoresis (CGE), Capillary affinity electrophoresis (CAE), Lectin, Polylactosamine, Online concentration

1 Introduction

Mucins contain a large number of serine (Ser)/threonine (Thr)-linked glycans (*O*-glycans), which represent one of the important components determining the characteristics of mucins [1]. *O*-glycans have diverse structures and are often clustered in particular regions of mucins. It is difficult to analyze the detailed structures of *O*-glycans while they remain bound in mucins. To analyze *O*-glycans, it is necessary to release them from the core protein and derivatize them into a suitable form for analysis. However, it has long been difficult to recover *O*-glycans that retain reducing ends suitable for derivatization, as there was no effective method of release from proteins comparable to that available for asparagine

Akihiko Kameyama (ed.), *Mucins: Methods and Protocols*, Methods in Molecular Biology, vol. 2763,
https://doi.org/10.1007/978-1-0716-3670-1_15,

(Asn)-linked glycans (*N*-glycans). To address this problem, a number of methods for *O*-glycan release have been developed, including in-line flow [2, 3], hydrazinolysis [4], ammonia-based β-elimination [5], and eliminative oximation [6] (*see* Chaps. 12, 13, and 14). The availability of these techniques has made it possible to apply various derivatization methods in *O*-glycan analysis.

Fluorescent labeling using reductive amination is the most common derivatization method used for sensitive and quantitative analysis of glycans. While there are many fluorescent reagents available for reductive amination [7], such as 2-pyridylamine, 2-aminobenzamide (AB), and 8-aminopyrene-1,3,6-trisulfonic acid (APTS), we prefer to use 2-aminobenzoic acid (2AA) for *O*-glycan analysis (*see* Fig. 1). 2AA-labeled glycans show strong fluorescence, with an intensity ten times higher than that of AB-labeled glycans [8, 9]. In addition, glycans labeled with 2AA are very stable and their fluorescence persists for a long time. The 2AA labeling method thus allows highly sensitive detection of glycans using liquid chromatography-fluorescence detection analysis (LC-FD). Further, 2AA-labeled glycans can be detected with high sensitivity in negative-ion-mode mass spectrometry (MS) because 2AA has a carboxyl group that carries a negative charge. This property is highly effective in MS analysis of minor acidic glycans including sulfated, phosphorylated, and glucuronidated glycans [10, 11]. 2AA-labeled glycans are often analyzed using a combination of LC-FD and MS analyses. The negative charge of 2AA can be used as a driving force for electrophoresis, so 2AA-labeled glycans can be easily analyzed using capillary electrophoresis (CE). The fluorescence intensity of 2AA is less than that of APTS, the standard labeling reagent for CE analysis. However, 2AA-labeled glycans can be detected with high sensitivity using capillary electrophoresis-laser-induced fluorescence (CE-LIF) using a He-Cd laser [12]. As described above, any combination of LC-FD, MS, and CE-LIF analysis can be applied to analysis of 2AA-labeled glycans. This versatility is an advantage of the 2AA labeling method. In this chapter, we present the LC-FD and CE-LIF techniques that we prefer to use for the analysis of 2AA-labeled *O*-glycans prepared from biological samples.

Various modes have been developed for LC-FD separation of 2AA-labeled glycans, including reversed-phase, hydrophilic interaction liquid chromatography (HILIC), and anion exchange chromatography modes. Among these modes, we most often analyze 2AA-labeled *O*-glycans in HILIC mode using polymer-based amino columns (NH_2 columns) [13]. Under these conditions, acidic *O*-glycans including sialylated and sulfated *O*-glycans tend to be retained more strongly on the column than neutral *O*-glycans (*see* Fig. 2). The NH_2 column thus enables separation of neutral from acidic *O*-glycans and is useful for comparing amounts of these structural classes in biological samples. In addition, *O*-glycans are

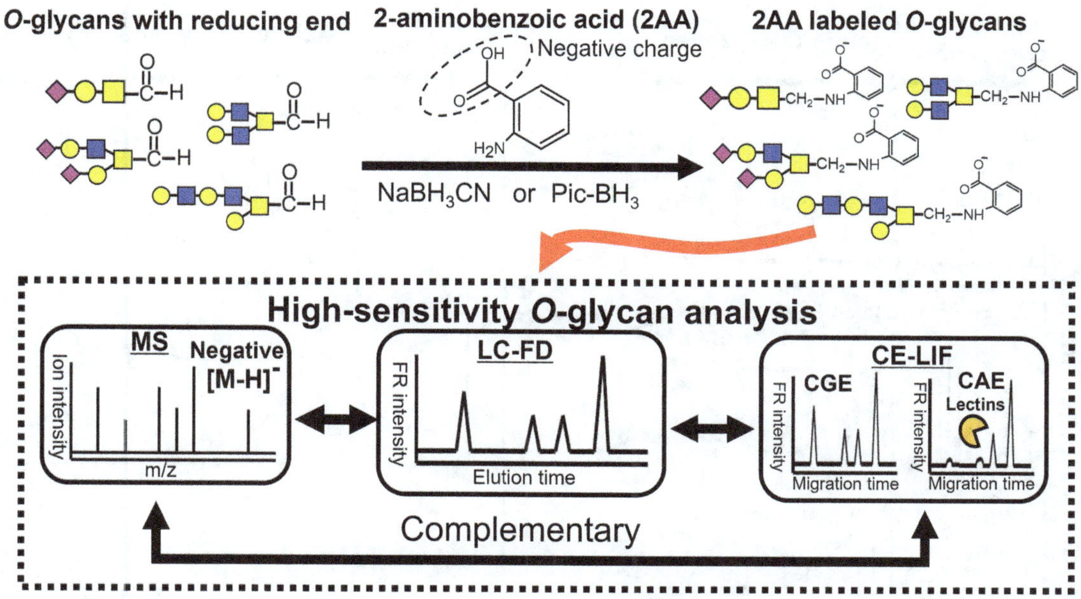

Fig. 1 Scheme of 2AA-labeled *O*-glycan analysis
2AA-labeled glycans can be measured with high sensitivity using liquid chromatography fluorescence detection (LC-FD) and negative ion mode MS. Further, as 2AA has a negative charge, it can also be used for capillary electrophoresis (CE) and Capillary affinity electrophoresis (CAE) analysis. 2AA-labeled glycans can be analyzed by integrating data obtained from each analytical method

separated according to differences in molecular weight in this context. When a sample containing polylactosamine (a repeating structure comprising galactose and *N*-acetylglucosamine)-type *O*-glycans is analyzed, multiple *O*-glycan peaks with different numbers of *N*-acetyllactosamine units are detected, and complex chromatograms are obtained. HILIC LC-FD is useful in obtaining a complete picture of the *O*-glycan profile in biological samples. Further, 2AA-labeled glycans isolated by HILIC can be analyzed using MS to obtain detailed structural information.

CE separation modes used in the analysis of 2AA-labeled glycans include capillary zone electrophoresis, micellar electrokinetic chromatography, and capillary gel electrophoresis (CGE). Among these modes, we often use the CGE method for analysis of 2AA-labeled *O*-glycans (*see* Fig. 3). An electroosmotic flow-suppressed capillary and a Tris-borate separation buffer (pH 8.3) containing polyethylene glycol are used in the CGE method. 2AA-labeled glycans are separated based on the mass-to-charge ratio and the weak molecular sieving effect of polyethylene glycol. The CGE method can rapidly separate complex 2AA-labeled *O*-glycan mixtures obtained from biological sources such as cultured cancer cells (*see* Fig. 3). Isomeric separation is also possible due to the interaction of boric acid in the separation buffer with glycans.

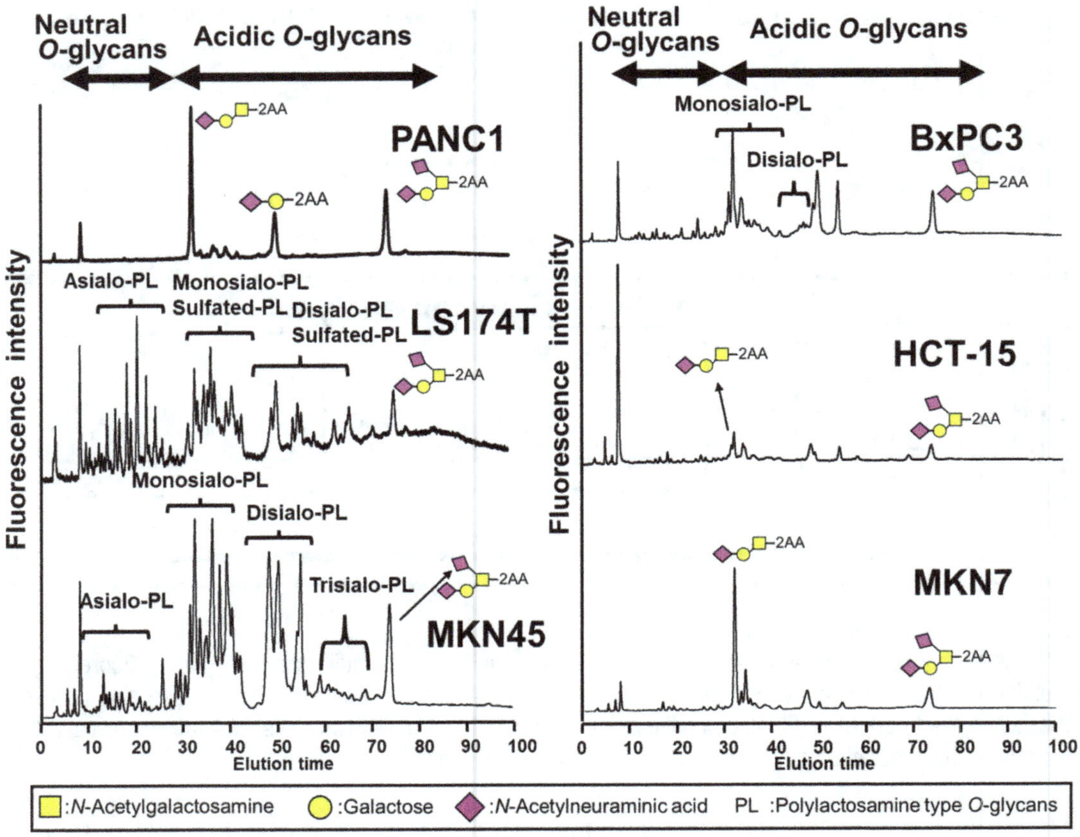

Fig. 2 Hydrophilic interaction liquid chromatography (HILIC) LC-FD analysis of 2AA-labeled *O*-glycans derived from epithelial cancer cell lines

2AA-labeled *O*-glycans prepared from colon adenocarcinoma cell lines (LS174T and HCT-15), gastric adenocarcinoma cell lines (MKN45 and MKN7), and pancreatic cancer cell lines (PANC1 and BxPC3) were analyzed using HILIC LC-FD. Neutral *O*-glycans are observed around 10–30 min, and acidic *O*-glycans including sialylated and sulfated glycans are observed after 30 min. Glycans with higher contents of sialic acid and sulfate groups tend to be retained more strongly on the column. Glycan preparations from LS174T and MKN45 cells, which contain large amounts of polylactosamine-type *O*-glycans, show complex chromatograms. Glycans from PANC1 and MKN7 cells, in which sialyl-T (NeuAcα2-3Galβ1–3GalNAc) and disialyl-T (NeuAcα2-3Galβ1–3(NeuAcα2–6)GalNAc) are the major components, show simple chromatograms. Details of the *O*-glycans produced by the cells are given in [13]. LC-FD analysis is useful to get a full picture of the *O*-glycan profile in biological samples

This effect is useful for detection of Tn antigen (*N*-acetylgalactosamine (GalNAc)-Ser/Thr) [14].

One of the advantages of CE analysis is that it is easy to analyze interaction between glycans and glycan-binding molecules in capillary affinity electrophoresis (CAE) mode (*see* Fig. 4) [15]. When glycan-binding molecules such as lectins or antibodies are added to the separation buffer, glycan-binding molecules bind to the migrating glycans, and the migration velocity of glycans decreases (*see* Fig. 4a). The interaction between glycans and glycan-binding

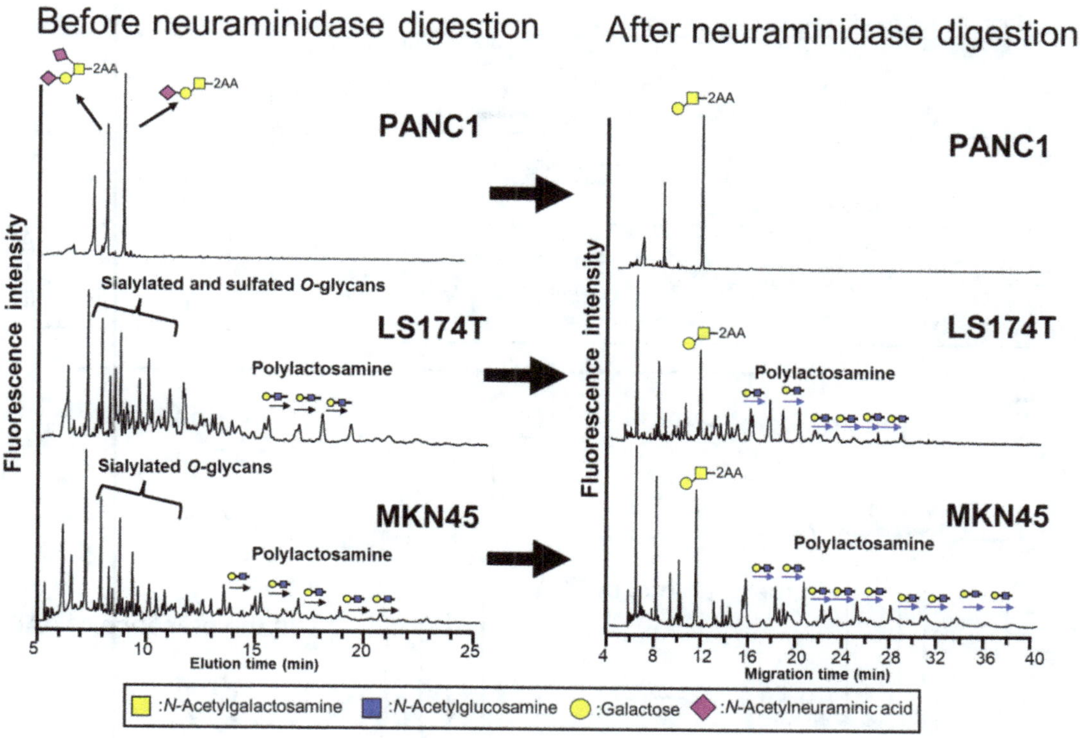

Fig. 3 CGE analysis of 2AA-labeled *O*-glycans derived from epithelial cancer cell lines 2AA-labeled *O*-glycans prepared from PANC-1, LS174T, and MKN45 cells were analyzed using the CGE method. Acidic *O*-glycans such as sialylated and sulfated glycans have a high rate of migration, and most are detected within 15 min. In contrast, neutral *O*-glycans are observed after 12 min (left). Sialylated polylactosamine-type *O*-glycans with high molecular weights from MKN45 cells are detected after 15 min. When analyzing *O*-glycan mixtures obtained following neuraminidase digestion, most *O*-glycans other than sulfated glycans are separated by differences in molecular weight (right), making this method suitable for evaluating elongation of *O*-glycans in biological samples. Affinity electrophoresis can be performed by adding lectin to the separation buffer (Fig. 4)

molecules can be analyzed by evaluating the change in the migration behavior. CAE is useful not only for analyzing the binding specificity of glycan-binding molecules, but also for obtaining structural information on glycans. It is possible to identify partial structures for unknown glycans in biological samples by analyzing interactions between unknown glycans and lectins with known binding properties. CAE is very useful as a tool to complement analytical data obtained from mass spectrometry because it provides structural information including glycosidic bond and isomer information, which are difficult to obtain using mass spectrometry.

CE analysis is useful for some glycan samples, but is not suitable for analysis of low concentration solutions due to the limited amount of sample that can be loaded into a narrow capillary and the short path length for detection. Several online concentration methods have been developed to increase the amount of sample

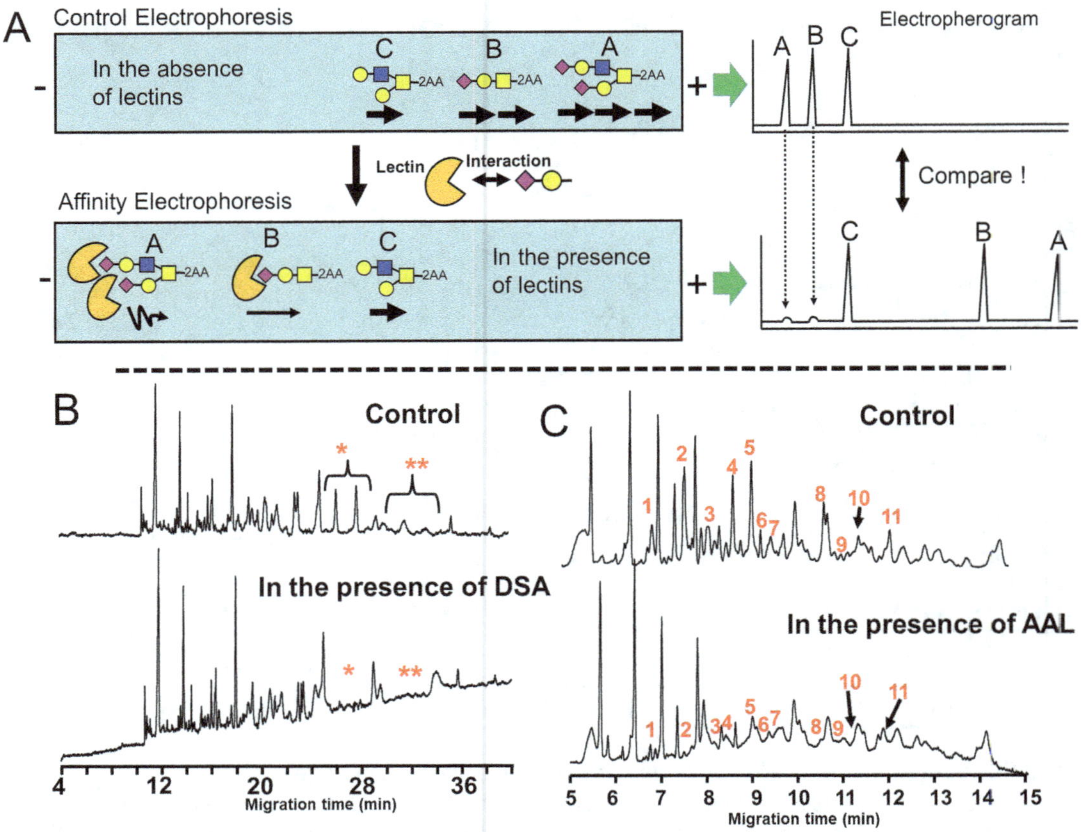

Fig. 4 CAE analysis of 2AA-labeled *O*-glycans
(**a**) Overview of the CAE method: Electrophoresis is first performed in the absence of lectin to obtain control data, then performed again in the presence of lectin. By comparing the electropherograms, glycans interacting with lectins can be identified. The migration of *O*-glycans interacting with lectins is retarded, thus delaying the appearance of the *O*-glycan peaks. (**b**) Detection of the interaction of *Datura stramonium* agglutinin (DSA) with asialo-*O*-glycans from LS174T cells using CAE. DSA binds strongly to the repeating structure of N-acetylglucosamine, making it suitable for detection of polylactosamine-type *O*-glycans [17]. Multiple peaks observed appearing after 20 min have disappeared from the electropherogram apparently due to retarded migration (bottom). The results show that LS174T cells express multiple polylactosamine-type *O*-glycans. (**c**) Detection of the interaction of *Aleuria aurantia* lectin (AAL) with *O*-glycans from LS174T cells using CAE. AAL lectins bind strongly to *O*-glycans including Fucα1–3-(N-acetylglucosamine (GlcNAc)) residues. A delay in the migration of several peaks is observed (bottom). The results show that the major *O*-Glycan of LS174T contains Fucα1–3GlcNAc residues. By using lectins with known high specificity, detailed structural information can be obtained

that can be introduced into the capillary [16]. We increased the detection sensitivity for 2AA-labeled *O*-linked glycans in CE-LIF 36-fold using head-column field-amplified sample stacking (HC-FASS) (*see* Fig. 5), enabling highly sensitive CE analysis of 2AA-labeled *O*-glycans.

As outlined above, various analytical tools including LC-FD, CGE, and CAE can be used to analyze 2AA-labeled *O*-glycans. We

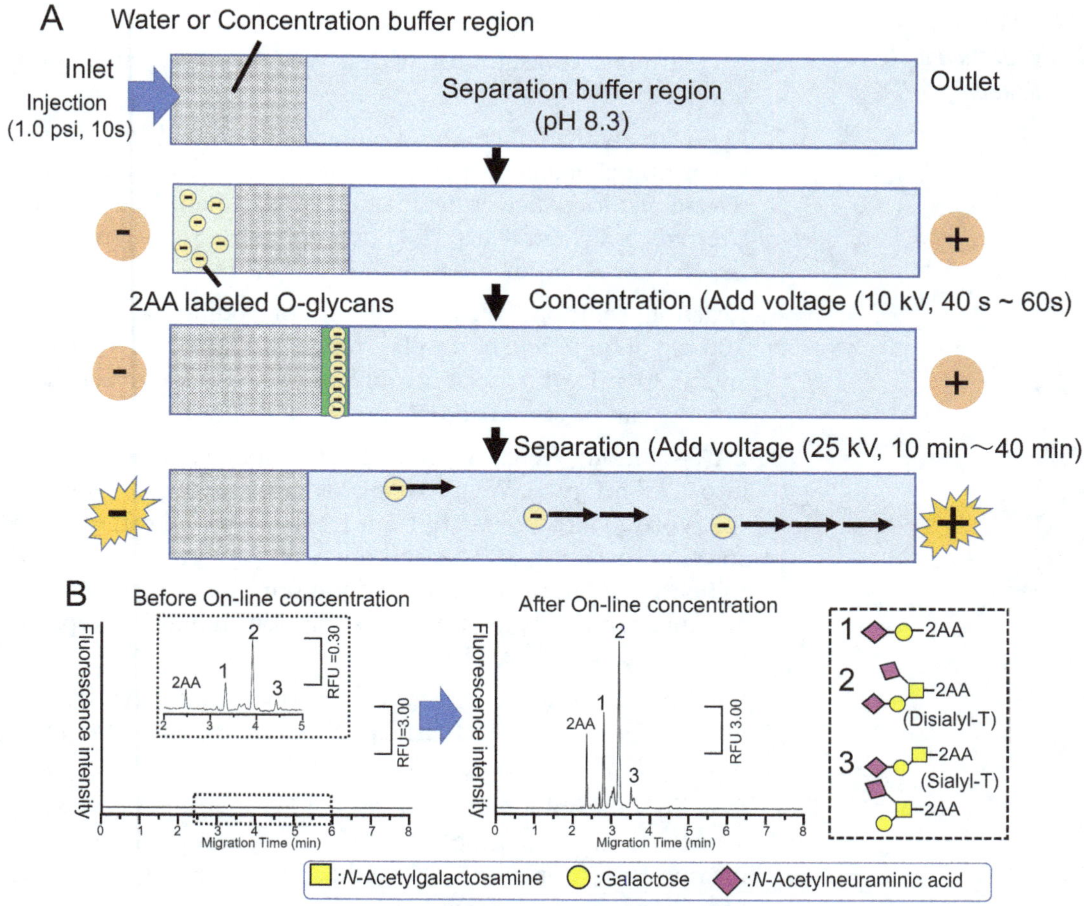

Fig. 5 Online concentration for CGE analysis of 2AA-labeled *O*-glycans
(**a**) An overview of the head-column field-amplified sample stacking (HC-FASS) online concentration method: A small amount of water or concentration buffer is introduced into a capillary filled with separation buffer using pressure injection. 2AA-labeled *O*-glycans are introduced using electrical injection. As the transfer rate of 2AA-labeled *O*-glycans is faster in water or concentration buffer than in separation buffer, 2AA-labeled *O*-glycans concentrate at the interface between the two buffers. This method allows a large amount of glycan sample to be introduced into the capillary while minimizing sample diffusion. (**b**) Analysis of 2AA-labeled *O*-glycans from casein glycomacropeptide using online concentration. When analyzed without online enrichment, the signal of *O*-glycans is weak (left). When online enrichment is used, the O-glycan signals are enhanced by more than 36-fold and are clearly detected (right)

believe these techniques can be effective tools for the analysis of *O*-glycans present in trace amounts in mucin samples.

2 Materials

Prepare all solutions using ultrapure water and analytical grade reagents. Do not add sodium azide. Prepare and store reagents at room temperature unless otherwise indicated.

2.1 2AA Derivatization and Purification

1. Extraction manifold.
2. A centrifugal concentrator with a vacuum pump, and a low temperature trap.
3. *O*-glycans with reducing ends: *O*-glycans should be released from glycoproteins (1–500 μg) using non-reducing chemical release methods such as hydrazinolysis (*see* Chap. 12), eliminative oximation (*see* Chap. 13), ammonia-based β-elimination (*see* [5]), or an in-line flow method (*see* **Note 1**).
4. Reaction solution: Weigh 400 mg of sodium acetate and 200 mg of boric acid into a glass tube. Add 10 mL of methanol and 0.30 mL of water and mix using a vortex mixer. The mixture can be stored at 4 °C for about 1 month.
5. 2AA solution: Weigh 15 mg of 2-aminobenzoic acid (2AA) into a 1.5 mL tube. Weigh 15 mg sodium cyanotrihydroborate (*see* **Note 2**) into a separate 1.5 mL tube. Add 0.5 mL reaction solution to 15 mg of 2AA and mix using a vortex mixer. Add the entire solution to 15 mg of sodium cyanotrihydroborate and mix gently by pipetting. This solution should be prepared freshly before use (*see* **Note 3**).
6. Solid-phase extraction cartridge for purification of 2AA-labeled glycans: For example, InertSep NH2 cartridges (200 mg/3 mL, GL Sciences Inc., Tokyo, Japan).
7. Conditioning solution for InertSep NH2 cartridges (2% (v/v) acetic acid in acetonitrile): Add 20 mL of acetic acid to a 1-L measuring flask or glass beaker. Add acetonitrile to 1 L and mix. Store at room temperature or 4 °C.
8. Recovery solution for InertSep NH2 cartridges (3% (v/v) aqueous triethylamine, 5% (v/v) acetic acid.): Add 500 mL of water to a 1-L measuring flask or glass beaker. Add 50 mL of acetic acid and mix. Add 30 mL of triethylamine and mix. Add water to make 1 L. Store at room temperature or 4 °C.
9. Washing solution for InertSep NH2 cartridges (mixture of conditioning solution and recovery solution at a ratio of 4:1 (v/v)): Add 400 mL conditioning solution and 100 mL of recovery solution to a 500 mL graduated cylinder or glass beaker and mix. Store at room temperature or 4 °C.

2.2 HILIC LC-FD Analysis

1. High performance liquid chromatography instrument equipped with two pumps, a fluorescence detector, a column oven, a system controller, and an LC workstation.
2. Column: A polymer-based amino column (4.6 mm i.d.× 250 mm) (*see* **Note 4**).
3. Eluent A (2% (v/v) acetic acid in acetonitrile): The method of preparation is the same as for the conditioning solution in Subheading 2.1, **item 4**.

4. Eluent B (3% (v/v) aqueous triethylamine, 5% (v/v) acetic acid): The method of preparation is the same as for the recovery solution in Subheading 2.1, **item 5.**

2.3 Capillary Gel Electrophoresis

1. CE systems equipped with a laser-induced fluorescence detection system (P/ACE MDQ or PA800 series; AB Sciex LLC, Framingham, MA) with a He-Cd laser (325 nM) should be used.
2. DB-1capillary; J&W DB-1 GC column (0.1 mm i.d., 40 cm effective length, 50 cm total length, Agilent Technologies, Santa Clara, CA) (*see* **Note 5**). Insert this capillary into the capillary cartridge. Remove a small area (about 0.3–0.5 cm) of the outer coating of the capillary to detect sample fluorescence inside the capillary (*see* **Note 6**).
3. 100 mM Tris-borate buffer pH 8.3: Add 200–300 mL water to a glass beaker. Weigh out 6.057 g of Tris base and add to the glass beaker. Mix with a magnetic stirrer. Slowly add saturated boric acid solution (2.50 g of boric acid in 50 mL of water) to the Tris solution until the pH reaches 8.3, monitoring using a pH meter. Transfer the entire solution to a 500 mL measuring flask, add water to make 500 mL and mix. Pass the buffer through a 0.22 μm cellulose acetate filter into a glass bottle. Store at 4 °C (*see* **Note** 7).
4. Separation buffer (100 mM Tris-borate buffer containing 5% (w/v) polyethylene glycol 35,000): Weigh 5.0 mg of polyethylene glycol 35,000 into a 1.5 mL tube. Add 1 mL of 100 mM Tris-borate pH 8.3 and mix well using a vortex mixer. Store at 4 °C.

2.4 Capillary Affinity Electrophoresis and Online Concentration for CGE

CE system, capillary, and separation buffer as for CGE analysis (*see* Subheading 2.3).

1. Lectins; *Aleuria aurantia* lectin (AAL; *O*-glycans including Fucα1–3-(N-acetylglucosamine (GlcNAc)), *Datura stramonium* agglutinin (DSA; polylactosamine-type *O*-glycans) [17], *Ulex europaeus* agglutinin-1 (UEA-1, (fucose (Fuc)α1–2-galactose (Gal)-R), *Vicia villosa* lectin (VVL; Tn antigen), *Maackia amrensis* agglutinin (MAA; *N*-acetylneuraminic acid (NeuAc)α 2-3Gal), *Sambucus nigra* agglutinin (SNA; NeuAcα2–6Gal) (*see* **Note 8**).
2. Separation buffer containing lectins; lectin samples dissolved in separation buffer (*see* Subheading 2.3, **items 3** and **4**). Lectins should be used in a concentration range of 1–20 μM (*see* **Note 9**).
3. 4 × Concentration buffer (0.2 M phosphate-citrate pH 4.0): Weigh 2.84 g of disodium hydrogen phosphate (Na_2HPO_4) into a 100 mL grass beaker. Add 60–70 mL of water to the

beaker and mix with a magnetic stirrer. Slowly add saturated citric acid solution (27.0 g of citric acid in 50 mL of water) until the pH reaches 4.0, monitoring with a pH meter. Bring the volume to 100 mL and mix. Pass the buffer through a 0.22 μm cellulose acetate filter into a glass bottle. Store at 4 °C. Dilute fourfold at the time of use (*see* **Note 10**).

3 Methods

Carry procedures out at room temperature unless otherwise specified.

3.1 2AA-Labeling and Purification (See Note 11)

1. Add 20 μL of water to *O*-glycans with a reducing end (approximately 50 μg) in a tightly closed microtube and mix using a vortex mixer.
2. Add 60 μL of 2AA solution to the glycan solution and mix gently.
3. Incubate the reaction mixture at 80 °C for 70 min.
4. After the reaction mixture has cooled to room temperature, add 3 mL of conditioning solution and mix using a vortex mixer. 2AA-labeled *O*-glycans are purified from this solution using InertSep NH2 cartridges.
5. Connect InertSep NH2 cartridges to the manifold and pass 3 × 3 mL of water through the cartridge.
6. Pass 3 × 3 mL of conditioning solution through the cartridge.
7. Pass 2AA reaction mixture from **step 4** through the cartridge and collect the eluate (*see* **Note 12**).
8. Pass the eluate collected in **step 7** through the cartridge again.
9. Pass 3 × 3 mL of washing solution through the cartridge.
10. Pass 3 × 1 mL of recovery solution through the cartridge and collect the eluate.
11. Dry the eluate collected in **step 10** in a centrifugal concentrator (*see* **Note 13**).
12. The dried sample represents 2AA-labeled *O*-glycans. The sample is redissolved in water and analyzed using LC-FD and CE. This sample can be stored at −20 °C for several years.

3.2 Analysis of 2AA-Labeled O-glycans Using HILIC LC-FD

1. Set the excitation wavelength in the fluorescence detector to 350 nM and the emission wavelength to 425 nM.
2. Set the column temperature to 50 °C.
3. Set the combined flow rate of the two pumps to 1.0 mL/min and maintain this flow rate during the analysis.

4. Set up a linear gradient program for eluent A and eluent B. From 0 to 2 min after the start of analysis, hold eluent B at 30%, then increase eluent B from 30 to 95% from 2 to 82 min and hold at this composition for another 10 min.
5. Before starting the analysis, equilibrate the column with 30% B eluent for 15 min.
6. Dissolve 2AA-labeled *O*-glycans in water and inject onto the column (*see* **Note 14**).
7. Typical analytical results for 2AA-labeled *O*-glycans from cultured cancer cells are shown in Fig. 2.

3.3 Analysis of 2AA-Labeled O-glycans Using CGE

1. Attach a DB-1 capillary cartridge to a CE system.
2. Add 1.9 mL of 100 mM Tris-borate pH 8.3 and 0.2 mL of separation buffer to a vial. Add sample solution (10 μL) containing 2AA-labeled *O*-glycans to a separate vial (*see* **Note 15**). Place these vials in the sample tray of the CE system.
3. Pass water through the DB-1 capillary for 2.0 min at 20 psi.
4. Pass 100 mM Tris-borate pH 8.3 through the DB-1 capillary for 1.5 min at 20 psi.
5. Pass separation buffer through the DB-1 capillary for 2.0 min at 20 psi.
6. 6 Inject the sample solution containing 2AA-labeled *O*-glycans into the DB-1 capillary at 1 psi for 10 s (pressure injection).
7. Run electrophoresis at 25 kV for 40 min with the inlet and outlet of the capillary in a vial containing 100 mM Tris-borate pH 8.3. When applying the voltage, the inlet side of the capillary should be set as the cathode and the outlet side as the anode (*see* **Note 16**).
8. Repeat **steps 3–6** for the next sample to be analyzed. After completing the measurements, pass water through the capillary at 20 psi for 5 min.

Typical analytical results for 2AA-labeled *O*-glycans prepared from cultured cancer cells are shown in Fig. 3.

3.4 Analysis of 2AA-Labeled O-glycans Using CAE

An overview of the procedure is given in Fig. 4a.

1. Perform CGE measurements as described in Subheading 3.3 to obtain control data for 2AA-labeled *O*-glycans.
2. Add 0.2 mL of separation buffer containing lectin to a small vial and place it in the sample tray of the CE system (*see* **Note 17**). Do not change the capillary or other buffer solutions.
3. Pass water through the DB-1 capillary for 2.0 min at 20 psi.
4. Pass 100 mM Tris-borate pH 8.3 through the DB-1 capillary for 1.5 min at 20 psi.

5. Pass separation buffer containing lectin through the DB-1 capillary for 1.5 min at 20 psi.
6. Inject the sample solution containing the 2AA-labeled *O*-glycans into the DB-1 capillary at 1 psi for 10 s (pressure injection).
7. Run electrophoresis at 25 kV for 40 min with the inlet and outlet of the capillary in a vial containing 100 mM Tris-borate pH 8.3. When applying the voltage, the inlet side of the capillary should be set as the cathode and the outlet side as the anode.
8. After completing the measurements, pass water through the capillary at 20 psi for 2 min (*see* **Note 18**).
9. To evaluate interactions between glycans and lectins, compare the control electropherogram with that obtained after lectin addition.
10. To analyze additional samples, repeat **steps 3–8**.

Typical analytical results for 2AA-labeled *O*-glycans are shown in Fig. 4b.

3.5 Online Concentration for CGE

An overview of the procedure is given in Fig. 5a.

1. Perform CGE **steps 1–4** as described in Subheading 3.3.
2. Inject concentration buffer or water into the DB-1 capillary at 1 psi for 10 s (*see* **Note 19**).
3. Inject the sample solution containing 2AA-labeled *O*-glycans into the DB-1 capillary using an electrical injection method. The inlet side of the DB-1 capillary is dipped in the sample solution and the outlet side is dipped in 100 mM Tris-borate pH 8.3, and voltage is applied at 10 kV for 50 s (*see* **Note 20**). When applying the voltage, the inlet side of the capillary is set as the cathode and the outlet side as the anode.
4. Run electrophoresis at 25 kV for 40 min with the inlet and outlet of the capillary in a vial containing 100 mM Tris-borate pH 8.3. When applying voltage, the inlet side of the capillary should be set as the cathode and the outlet side as the anode (*see* **Note 16**).

Typical analytical results for 2AA-labeled *O*-glycans are shown in Fig. 5b.

4 Notes

1. When releasing *O*-glycans from insoluble proteins extracted from cells or tissues, it is better to pre-digest the proteins with protease from *Streptomyces griseus* (PRONASE,

MilliporeSigma, Burlington, MA). The data presented here were obtained by releasing *O*-glycans from proteins using our glycan releasing in-line flow system [2, 3]. Similar data can be obtained using other *O*-glycan release methods. However, when hydrazinolysis or ammonia-based β-elimination methods are used, *N*-glycans are more likely to be present as contaminants. Therefore, care must be taken when analyzing *O*-glycans obtained using these release methods.

2. Sodium cyanotrihydroborate deteriorates when moist and should be stored in a container with a desiccant. 2-Picoline borane can be substituted for sodium cyanotrihydroborate. To prepare 0.5 mL 2AA solution, 214 mg 2-picoline borane should be used.
3. The 2AA solution should be added to the glycan sample immediately after preparation as it is easily degraded.
4. A commercial amino column (e.g., Shodex Asahipak NH2P-50 4E (Resonac Inc., Tokyo, Japan)) can be used.
5. An eCAP N-CHO coated capillary (0.05 mm i.d., AB Sciex LLC) can also be used, but the detection sensitivity will be lower due to a smaller inner diameter and lower sample volume.
6. The exterior of the capillary can be peeled off by burning it with a low-flame burner for 1–3 s.
7. Warm to room temperature when used for CE analysis.
8. Biotinylated lectins can also be used. Lyophilization of samples are optimal. If samples are dissolved in PBS or another buffer, this should be replaced with separation buffer using ultrafiltration tubes (molecular weight cutoff 10 kDa).
9. The lectin concentration should be less than 10 μM as lectin adsorbs on the capillary wall and makes migration unstable. If binding is not readily observed, change the separation buffer to 100 mM Tris-acetate (pH 7.4) containing 5% (w/v) polyethylene glycol.
10. Commercially available Mcilvaine Buffer (0.2 M citrate-phosphate pH 4.0), diluted fourfold, can also be used.
11. 2AA Glycan Labelling Kit (Ludger Ltd., Abingdon, UK, or Agilent Technologies, Santa Clara, CA) may be used.
12. If precipitates form in the 2AA reaction solution, pass the precipitates through the cartridge too.
13. If the sample is difficult to dry, add 500 μL of acetonitrile and concentrate again in a centrifugal concentrator.
14. The volume of sample solution introduced into the column should be less than 10 μL. While the detection sensitivity

depends on the performance of the detector, several fmol of glycans can be detected.

15. When the volume of sample solution is less than 10 μL, it is difficult to reliably inject the sample.
16. When measuring only *O*-glycans with molecular weights of less than 1000 or acidic *O*-glycans, it is sufficient to apply voltage for 20 min.
17. Measurements can be run 3–4 times in 0.2 mL separation buffer containing lectin.
18. If the capillary is allowed to stand for a long time, lectin adheres to the inner wall and the capillary deteriorates.
19. Injection of water is more effective than injection of stacking buffer when concentrating only neutral *O*-glycans.
20. The longer the voltage is applied, the stronger the concentration effect. However, if the voltage is applied for too long, the peak becomes broad and separation is poor. If concentration is not successful, adjust the length of time that the voltage is applied for.

Acknowledgments

This work was supported by the Japan Society for the Promotion of Science KAKENHI (Grant Number 23 K06075). We would like to thank Editage (www.editage.com) for English language editing.

References

1. Patton S, Gendler SJ, Spicer AP (1995) The epithelial mucin, MUC1, of milk, mammary gland and other tissues. Biochim Biophys Acta 1241(3):407–423
2. Yamada K, Hyodo S, Matsuno YK et al (2007) Rapid and sensitive analysis of mucin-type glycans using an in-line flow glycan-releasing apparatus. Anal Biochem 371(1):52–61
3. Yamada K, Hyodo S, Kinoshita M et al (2010) Hyphenated technique for releasing and MALDI MS analysis of O-glycans in mucin-type glycoprotein samples. Anal Chem 82(17):7436–7443
4. Kozak RP, Royle L, Gardner RA et al (2014) Improved nonreductive O-glycan release by hydrazinolysis with ethylenediaminetetraacetic acid addition. Anal Biochem 453:29–37
5. Huang Y, Mechref Y, Novotny MV (2001) Microscale nonreductive release of O-linked glycans for subsequent analysis through MALDI mass spectrometry and capillary electrophoresis. Anal Chem 73(24):6063–6069
6. Kameyama A, Thet Tin WW, Toyoda M et al (2019) A practical method of liberating O-linked glycans from glycoproteins using hydroxylamine and an organic superbase. Biochem Biophys Res Commun 513(1):186–192
7. Kinoshita M, Yamada K (2022) Recent advances and trends in sample preparation and chemical modification for glycan analysis. J Pharm Biomed Anal 207:114424
8. Kim W, Kim J, You S et al (2019) Qualitative and quantitative characterization of sialylated N-glycans using three fluorophores, two columns, and two instrumentations. Anal Biochem 571:40–48
9. Higel F, Demelbauer U, Seidl A et al (2013) Reversed-phase liquid-chromatographic mass spectrometric N-glycan analysis of biopharmaceuticals. Anal Bioanal Chem 405(8): 2481–2493

10. Yamada K, Kayahara H, Kinoshita M et al (2018) Simultaneous analysis of sulfated and phosphorylated glycans by serotonin-immobilized column enrichment and hydrophilic interaction chromatography. Anal Chem 90(14):8387–8395
11. Yamada K, Suzuki K, Hirohata Y et al (2020) Analysis of minor acidic N-glycans in human serum. J Proteome Res 19(8):3033–3043
12. Maeda E, Kita S, Kinoshita M et al (2012) Analysis of nonhuman N-glycans as the minor constituents in recombinant monoclonal antibody pharmaceuticals. Anal Chem 84(5): 2373–2379
13. Yamada K, Kinoshita M, Hayakawa T et al (2009) Comparative studies on the structural features of O-glycans between leukemia and epithelial cell lines. J Proteome Res 8(2): 521–537
14. Yamada K, Watanabe S, Kita S et al (2010) Determination of Tn antigen released from cultured cancer cells by capillary electrophoresis. Anal Biochem 396(1):161–163
15. Kinoshita M, Kakehi K (2014) Capillary-based lectin affinity electrophoresis for interaction analysis between lectins and glycans. Methods Mol Biol 1200:131–146
16. Kamoda S, Nakanishi Y, Kinoshita M et al (2006) Analysis of glycoprotein-derived oligosaccharides in glycoproteins detected on two-dimensional gel by capillary electrophoresis using on-line concentration method. J Chromatogr A 1106(1–2):67–74
17. Mitsui Y, Yamada K, Hara S et al (2012) Comparative studies on glycoproteins expressing polylactosamine-type N-glycans in cancer cells. J Pharm Biomed Anal 70:718–726

Chapter 16

Preparation of Mucin Glycopeptides by Organic Synthesis

Izuru Nagashima and Hiroki Shimizu

Abstract

Mucins are sugar-rich glycoproteins. Glycoprotein sugar moieties are structurally diverse, making it difficult to obtain naturally pure glycoproteins. Chemical synthesis is a powerful tool for obtaining target or designed compounds. Automated peptide synthesizers are commercially available, and many use the solid-phase peptide synthesis (SPPS) method. In addition, some of these synthesizers apply microwave irradiation to obtain higher reaction yields, thereby enabling the synthesis of 40 to 50 amino acid residual glycopeptides. Theoretically, glycopeptides can be synthesized using methods similar to those used for peptide synthesis, but glycosylated amino acid synthons are less stable than amino acid synthons and are also very expensive. Therefore, they are not suitable for use in large excess amounts. Many of oligosaccharide-linked amino acid synthons are not commercially available, so they must be specially prepared, and they also require careful handling that demands specific organic synthesis experience and techniques. However, monosaccharide-linked amino acid synthons are commercially available and are relatively easy to handle. Here, as an entry into glycopeptide synthesis, we describe a typical glycopeptide synthesis procedure for a 27 amino acid residual MUC1 repeating unit with monosaccharides.

Key words Synthesis, Glycopeptides, *O*-Glycans, MUC1, Mucin

1 Introduction

Most functional proteins secreted on cell surfaces and outside cells in animals are glycosylated and are therefore known as glycoproteins. Protein biosynthesis occurs in the cell cytoplasm in a process called translation. During translation, messenger RNA (mRNA), a template molecule converted from a section of protein-encoding DNA, is read by cell ribosomes to determine the sequence of amino acids in a process called transcription. Thus, the protein sequence is determined by the genes. On the other hand, carbohydrates are introduced during the co-translational or posttranslational modification stages. *N*-Glycans and *O*-glycans are the two most common types of glycoproteins. *N*-Glycans are typically a type of anomeric position of *N*-acetylglucosamine (GlcNAc) linked in the β-conformation to nitrogen on a side chain of asparagine (Asn)

Akihiko Kameyama (ed.), *Mucins: Methods and Protocols*, Methods in Molecular Biology, vol. 2763,
https://doi.org/10.1007/978-1-0716-3670-1_16,

via an amide bond, while *O*-glycans are typically a type of anomeric position of *N*-acetylgalactosamine (GalNAc) linked in the α-conformation to oxygen on side chains of serine (Ser) or threonine (Thr) via ether (acetal) bonds. Mucin is a glycoprotein and one of the main components of mucus released from the mucosal epithelium of the gastrointestinal tract, airways, and salivary glands. It consists of a large molecular mass protein backbone (more than 2 MDa) and a high proportion of oligosaccharide side chains (more than 50% in molecular weight).

Recently, it was revealed that carbohydrates play important roles in cell adhesion, antigen-antibody reactions, viral infection, and other types of cellular communication and that certain specific carbohydrates are found on the surface of cancer cells. Recognition of these carbohydrates would help identify diseases; therefore, obtaining structurally unique carbohydrates or glycoproteins is required for medical and biochemical research. However, the natural diversity of carbohydrates makes this difficult to achieve.

Chemical synthesis is a powerful tool for obtaining target compounds. Although fine organic synthesis of polymers remains challenging, it is possible to synthesize almost any designed compound. Oligosaccharide synthesis has been widely studied in recent decades. Carbohydrates are present as several isomers. For example, 20 isomers of two linked glucose molecules exist because of the different combinations of glycosyl bond conformations (α or β) and variations in regioselectivity (1-1, 1-2, 1-3, etc.) and monosaccharide structures (furanoside or pyranoside). The most important consideration in the chemical synthesis of carbohydrates is controlling these specificities. We achieved this control through glycosylation and selective protection and deprotection of the sugar hydroxy groups (Fig. 1).

Although protein and polymer synthesis are difficult, a variety of peptides have been chemically synthesized. For example, solid-phase peptide synthesis (SPPS, developed by Merrifield, winner of the Nobel Prize in Chemistry, 1984) has become a common method, and automated synthesizers employing SPPS are commercially available. Typical SPPS methods are as follows. Polystyrene polymer gel beads of approximately 0.1 mM diameter attached to suitable linkers were used as the solid-phase resin. Amino acids are elongated individually by cycling of a coupling of 9-fluorenylmethyloxycarbonyl (Fmoc)-protected amino acid synthon to the amino acid on the resin through an amide bond and a deprotection of the Fmoc group of the amino acid on the resin. Once the desired peptide sequence is formed, it is cleaved from the resin, and the protecting groups on the side chains are simultaneously deprotected. Typically, it is designed that they are worked by trifluoroacetic acid (TFA). The resulting crude material was purified using high-performance liquid chromatography (HPLC). In SPPS, purification is carried out only once; therefore, it is

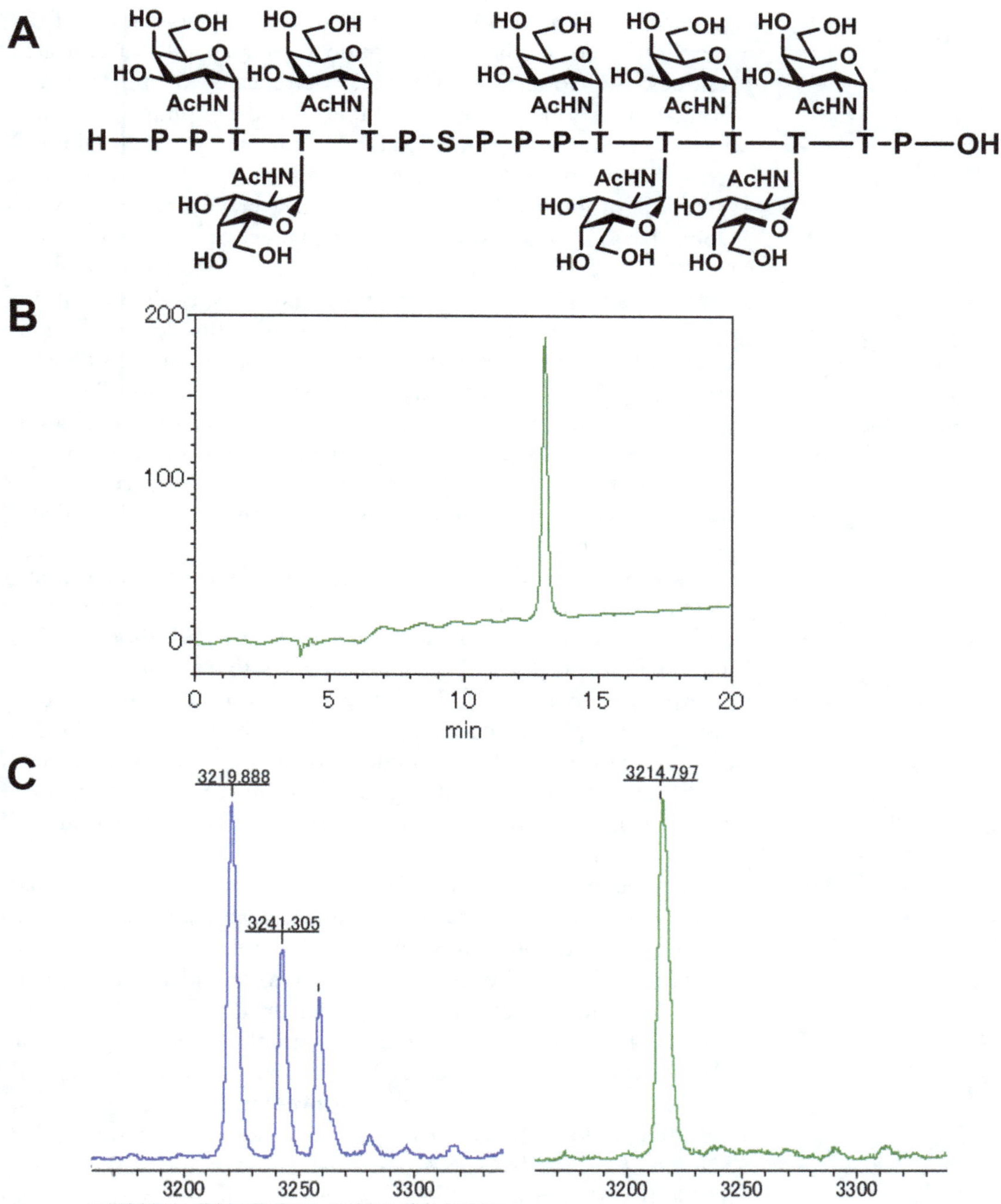

Fig. 1 (**a**) The structure of tandem repeat glycopeptide of MUC2, an example of mucin. (**b**) HPLC data of synthesized MUC2 glycopeptide after purification. Conditions: column, InertSustain C18 (4.6 mM × 250 mM); elution A/B = 95/5 to 85/15 (v/v) over 15 min where eluent A is H_2O (contain 0.1% TFA) and eluent B is CH_3CN (contain 0.1% TFA); flow rate, 1 ml/min, detect, UV 220 nM; column temperature, 30 °C. (**c**) MALDI-TOF MS data by positive (left) and negative (right) modes. Calculated exact mass weight is 3217.4283 ($C_{134}H_{216}N_{24}O_{66}$)

important to achieve a high yield in amino acid elongation. If the intention is to synthesize a peptide comprising 40 amino acid residues, the elongation steps must be repeated 39 times. If the elongation of one amino acid (amino acid coupling and Fmoc group deprotection) proceeded by 95% on average, the total theoretical yield was only 14%. If the amino acid elongation yield increased by only 3% (i.e., to 98% on average), the total yield would be 45%. Therefore, achieving higher reaction yields is key to SPPS, especially for the synthesis of longer peptides. One of the advantages of SPPS is that excess reagents can be easily washed off, so it is common to use excess, more than five equivalents, of amino acid synthons to obtain higher reaction yields. Because SPPS is an artificial synthesis method, it is possible to obtain nonribosomal artificial peptides, such as containing nonnatural amino acids, D-amino acids, and heavy atom substituents. Furthermore, it is possible to synthesize isomers precisely in which the attached GalNAc varies in position and number. Some of isomers of MUC1 glycopeptides are shown in Fig. 2.

Microwave irradiation is commonly used in chemical reactions, and SPPS synthesizers employing microwave radiation are commercially available from suppliers, such as CEM and Biotage. The underlying concept for these synthesizers is that the reaction temperature is increased quickly, and then the reaction time shortens, resulting in higher reaction yields and fewer byproducts. Although these synthesizers are widely used for the synthesis of approximately 40–50 amino acid residual peptides, the microwave heating may lead to side reactions, especially aspartic acid aspartimidation and peptide epimerization.

Glycopeptides are synthesized using a method similar to that used for peptide synthesis. For example, glycosylated amino acid synthons may be used in SPPS (Fig. 3), but they require careful treatment. Glycosyl bonds tend to be weaker under acidic conditions. Moreover, glycosylated amino acid synthons are also expensive. For example, 100 mg of simple Fmoc-Ser[GalNAc $(Ac)_3$-α-D]-OH costs about $1000, meaning that if a synthesis is carried out at 0.1 mmol scale using five equivalents of a reagent (i.e., 0.5 mmol), the cost (for just one reagent) of the required 328 mg of Fmoc-Ser[GalNAc$(Ac)_3$-α-D]-OH would be more than $3000.00. If two of GalNAc moieties are introduced, the cost should be double. We have investigated efficient glycopeptide synthesis methods and have already reported that occasional microwave irradiation in the coupling of glycol amino acids at 50 °C can reduce the amount of reagent required by 1.5 equivalents [1–7]. We also successfully demonstrated higher peptide synthesis yields with microwave irradiation than without microwave irradiation at the same temperature [8]. The SPPS synthesizer being developed by EYELA, Japan, also applies microwave radiation, but its concept is different from other SPPS synthesizers. By the

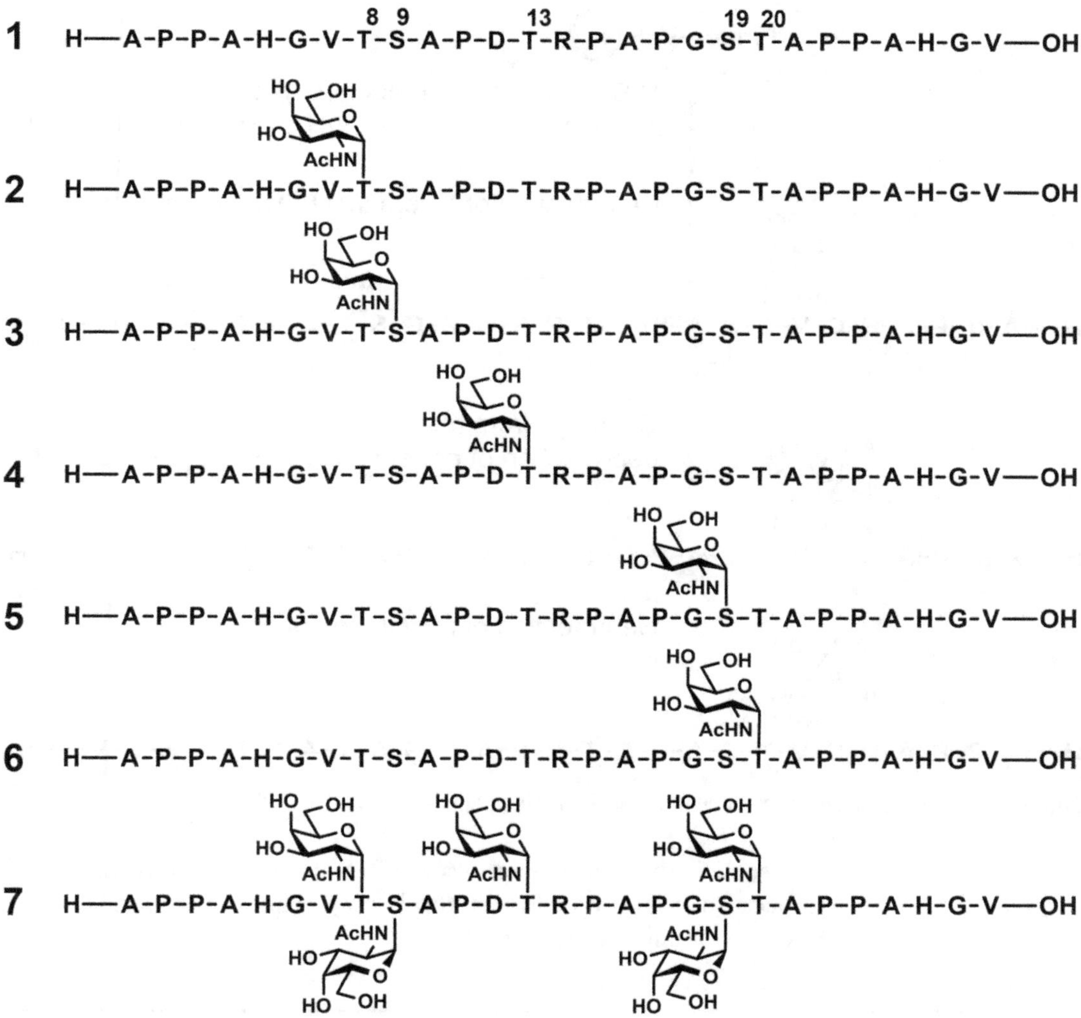

Fig. 2 Structures of the 27 amino acid residual MUC1 peptide and glycopeptides; (**1**) non-glycosylated peptide, (**2**) mono-GalNAc peptide at Thr_8, (**3**) mono-GalNAc peptide at Ser_9, (**4**) mono-GalNAc peptide at Thr_{13}, (**5**) mono-GalNAc peptide at Ser_{19}, (**6**) mono-GalNAc peptide at Thr_{20}, and (**7**) penta-GalNAc peptide at Thr_8, Ser_9, Thr_{13}, Ser_{19}, and Thr_{20}. MUC1 contains 20 amino acid residual repeating units. In chemical synthesis, isomers in which the attached GalNAc varies in position and number can be accurately synthesized

incorporated effective cooling unit, it can irradiate microwaves continuously while maintaining lower reaction temperatures, even at 25 °C. By using this synthesizer, we have achieved to synthesize a peptide in high yields even with only 1.2 equivalents of reagent and the reaction temperature was at 25 °C [8].

We can practically synthesize similar glycopeptides using SPPS. For example, we can synthesize any position of glycosylated MUC1 repeating units or a different number of GalNAc glycosylated MUC1 repeating units (Fig. 2). Herein, we describe typical glycopeptide synthesis procedures by SPPS, commenting on a specific example of MUC1 isomer synthesis (Fig. 3). Although not

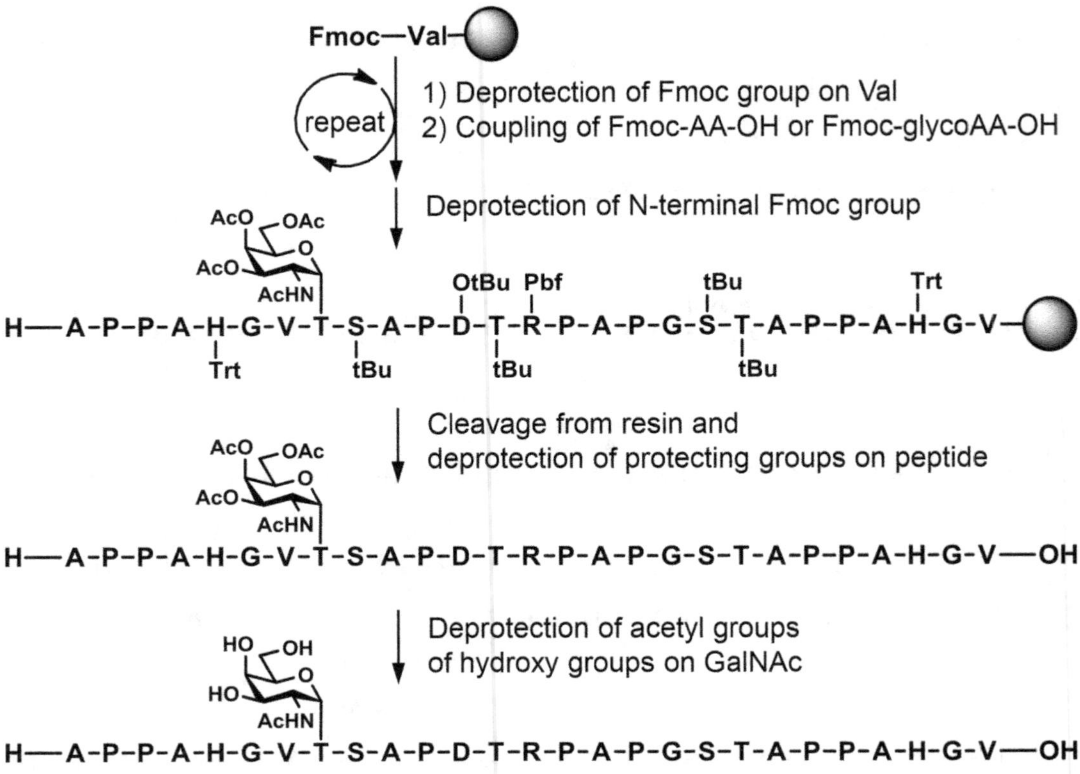

Fig. 3 Concept of synthesis of compound **2**, mono-GalNAc peptide at Thr_8, by SPPS

essential, the use of microwave SPPS machines is recommended because non-microwave methods may require a greater reagent excess or a longer reaction time.

2 Materials

All solutions were prepared using Ultrapure or deionized water with a resistivity of 18 MΩcm at 25 °C and peptide synthesis grade or analytical grade reagents. All reagents were prepared and stored at room temperature unless otherwise specified. Dispose of waste in accordance with the Waste Disposal and Public Cleansing Law.

2.1 Glycopeptide Synthesis by SPPS (Solid-Phase Peptide Synthesis)

1. Preloaded solid-phase carrier. For a MUC1 glycopeptide synthesis, use valine preloaded resin, for example, Fmoc-Val-NovaSyn TGA resin, 0.18 mmol/g.
2. Fmoc-protected amino acid synthons, Fmoc-AA-OH. For a MUC1 glycopeptide synthesis, Fmoc-Gly-OH, Fmoc-His (Trt)-OH, Fmoc-Ala-OH, Fmoc-Pro-OH, Fmoc-Thr(tBu)-OH, Fmoc-Ser(tBu)-OH, Fmoc-Arg(Pbf)-OH, Fmoc-Asp (OtBu)-OH, and Fmoc-Val-OH.

3. Fmoc-protected glycosylated amino acid synthons, Fmoc-glycoAA-OH. For a MUC1 glycopeptide synthesis, Fmoc-Ser (GalNAc(Ac)$_3$-α-D)-OH and Fmoc-Thr(GalNAc (Ac)$_3$-α-D)-OH.
4. *N*,*N*-dimethylformamide (DMF). Peptide synthesis grade is preferred.
5. Dichloromethane for peptide synthesis grade.
6. 20% (v/v) piperidine in dimethylformamide.
7. *N*,*N*-Diisopropylethylamine (DIEA or DIPEA).
8. Coupling reagent: 1-[(1-(cyano-2-ethoxy-2-oxoethylideneaminooxy) dimethylaminomorpholino)] uronium hexafluorophosphate (COMU).
9. Optional; coupling reagent: 1-[bis(dimethylamino)-methylene]-1*H*-benzotriazolium 3-oxide hexafluorophosphate (HBTU).
10. Capping cocktail: a mixture of acetic anhydride/*N*,*N*-diisopropylethylamine (DIEA)/DMF = 5/10/85 (v/v/v).
11. Kaiser test reagent (commercially available).
12. Trifluoroacetic acid (TFA).
13. Cleavage cocktail: a mixture of triisopropylsilane (TIS)/water/TFA = 2.5/2.5/95 (v/v/v).
14. Diethyl ether.
15. Reaction vessel: synthesizer equipped vessel or empty column with filter if syntheses are performed manually. For example, LibraTube®, RT5M100 (Hipep Laboratory, Japan).
16. Shaker: high-speed shaker if syntheses are performed manually. For example, Cute Mixer CM-1000, EYELA.
17. Hot plate, for Kaiser test.
18. Pipettes and micropipettes. Prefer to organic solvent resistance.
19. Stopcock for empty column with filter (Subheading 2.1, **item 15**).
20. Centrifuge.
21. Aspiration bottle.
22. Vortex mixer.
23. Diaphragm pump.
24. Freeze dryer for samples.
25. Pressured nitrogen gas, to blow up solvents.
26. Centrifuge tubes. For example, Corning tube.

2.2 Deprotection of Protecting Groups on Carbohydrate Components

1. Methanol.
2. Sodium methoxide, commercially available as powder or methanol solution.
3. Acetic acid.
4. Nitrogen gas.
5. Ultrapure water or deionized water with a resistivity of 18 MΩcm at 25 °C.
6. 50% acetonitrile solution with 0.1% TFA.
7. pH meter or pH test paper.

2.3 Purification and Analysis Components

1. Eluent A: Ultrapure water or deionized water with 0.1% TFA.
2. Eluent B: acetonitrile with 0.1% TFA.
3. HPLC apparatus.
4. Column: As the first choice, C18 column is recommended. For example, InertSustain C18 (GL Science, Japan).
5. MALDI MS (matrix-assisted laser desorption ionization mass spectrometry) apparatus and plate for measuring molecular weight.
6. Matrix for MALDI-TOF (time-of-flight) MS: 2,5-Dihydroxybenzoic acid (DHB) is prepared as 10 mg/mL in 50% acetonitrile in water with 0.1% TFA.

3 Methods

All procedures were carried out at room temperature unless specified otherwise. The structures of example of synthesis targets, MUC1 glycopeptides, are shown in Fig. 2 and the synthetic scheme of compound **2** is shown in Fig. 3.

3.1 Glycopeptide Synthesis by SPPS (Solid-Phase Peptide Synthesis)

1. Place 150 mg of preloaded resin (*see* **Note 1**) in a reaction vessel (*see* **Note 2**). In this case, the reaction was carried out at the 31.5 μmol scale.
2. Swell resin in DMF (approx. 2 mL) overnight at room temperature, and then filter (*see* **Note 3**).
3. Add 1 mL of 20% piperidine in DMF to the resin (**step 2**) and shake the mixture for 10–30 min at room temperature or for 3 min under microwave heating at 50 °C (*see* **Note 4**), and then filter and wash the resin several times in DMF (*see* **Note 5**).
4. Add 1 mL of the following prepared amino acid coupling cocktail to the resin (**step 3**) (*see* **Note 6**): Fmoc-AA-OH (157 μmol, 5 eq.), COMU (67.5 mg, 157 μmol, 5 eq.), and DIEA (26.8 μL, 157 μmol, 5 eq.), in DMF (1 mL), and shake the mixture for 2 h at room temperature or for 5 min under

microwave heating at 50 °C (*see* **Note 4**). Then, filter and wash the resin several times in DMF. For glycosylated amino acid coupling, the following cocktail was used: Fmoc-glycoAA-OH (47 μmol, 1.5 eq.), COMU (20.2 mg, 47 μmol, 1.5 eq.), and DIEA (8.0 μL, 47 μmol, 1.5 eq.) in DMF (0.7 mL). Then shake the mixture for a longer time, for example, 10–20 min under microwave heating at 50 °C (*see* **Notes 4** and **7**).

5. Add 1 mL of capping cocktail to the resin (**step 4**) to protect the unreacted N-terminal on the resin (*see* **Note 8**) and shake the mixture for 5 min at room temperature and then filter and wash the resin several times in DMF.
6. Repeat **steps 3–5** in Subheading 3.1 until the designed peptides were formed (*see* **Note 9**).
7. Wash the resulting resin in DMF and dichloromethane several times each, and dry under vacuum for at least 1 h (*see* **Note 10**).
8. Add 3 mL of cleavage cocktail to the resin to cleave the peptides from the resin (*see* **Note 11**), shake for 1.5 h at room temperature (*see* **Note 12**), filter, and wash thrice with TFA. Blow nitrogen gas to the combined filtrate to reduce solvent content.
9. Add ice-cold diethyl ether (approx. 10–20 times the volume of the crude mixture) to the crude mixture and centrifuge (5000 × *g* at 4 °C) for 5 min. Repeat this process three times and dry (*see* **Note 13**).
10. Dissolve the resulting residue in water (or in less than 50% acetonitrile) and freeze-dry (*see* **Note 14**).

3.2 Deprotection of Protecting Groups on Carbohydrates

1. Dissolve the resulting crude glycopeptide in methanol (*see* **Note 15**).
2. Add a catalytic amount of NaOMe and maintain pH above 11.5.
3. Occasionally monitor the reaction by HPLC (*see* Subheading 3.3 and **Note 16**).
4. After deprotection was completed, neutralize the reaction mixture with AcOH.
5. Blow nitrogen gas to the resulting crude mixture (**step 4**) to remove the solvent.
6. Dissolve the resulting residue in water (*see* **Note 17**) and freeze-dry.

3.3 Purification and Analysis

1. Analyze and purify the obtained crude glycopeptide (Subheading 3.2, **step 6**) by HPLC with a reverse-phase column (Subheading 2.3, **item 4**), eluents (Subheading 2.3, **items 1** and **2**), and UV detection (Fig. 4 left) (*see* **Note 18**).
2. Collect the fractions and freeze-dry (*see* **Note 19**).

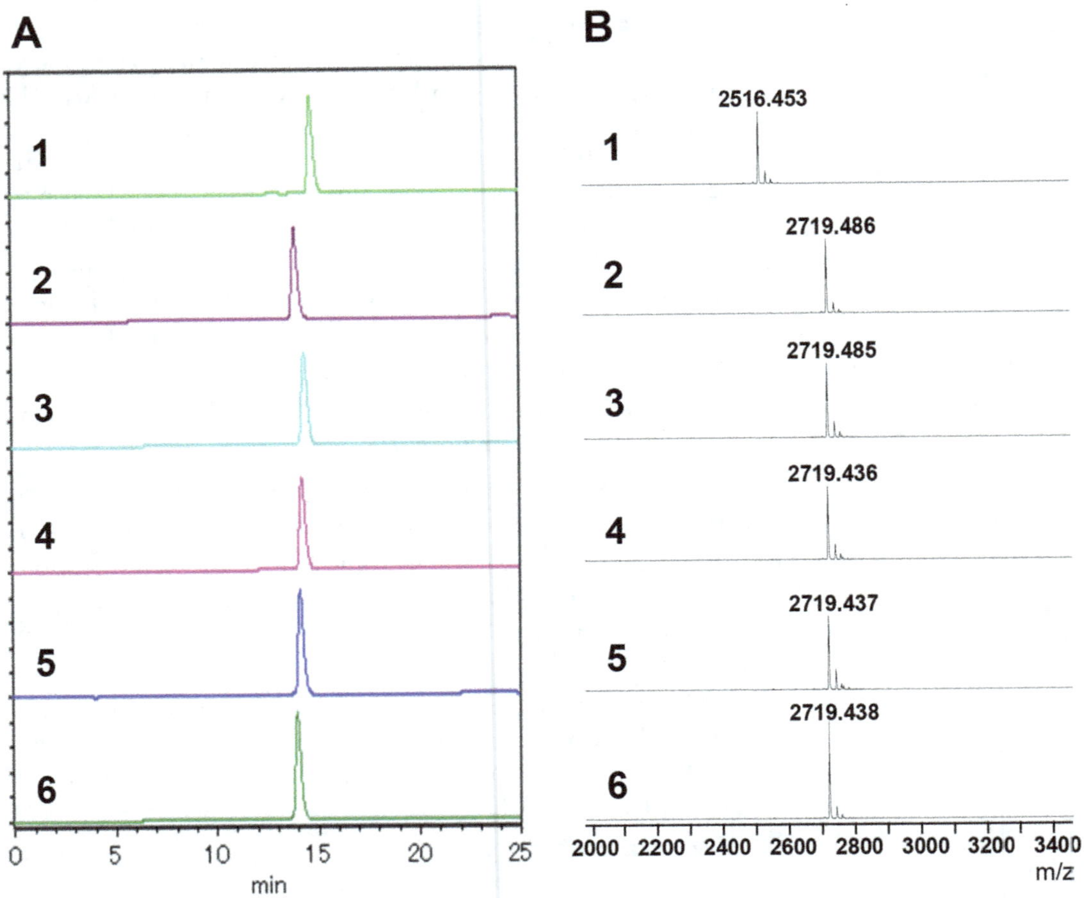

Fig. 4 (**a**) HPLC data of synthesized compounds **1** to **6**. Conditions: column, InertSustain C18 (4.6 mM × 250 mM); elution A/B = 90/10 to 70/30 (v/v) over 20 min where eluent A is H_2O (contains 0.1% TFA) and eluent B is CH_3CN (contains 0.1% TFA), flow rate, 1 mL/min; detect, UV 220 nM; column temperature, 30 °C. (**b**) MALDI-TOF MS data for **1** to **6** by positive mode. Calculated exact mass weights are 2515.2568 ($C_{109}H_{170}N_{34}O_{35}$) for **1** and 2718.3362 ($C_{117}H_{183}N_{35}O_{40}$) for **2** to **6**

3. Perform physical analysis using mass spectrometry (MS) (Fig. 4 right) (*see* **Note 20**), amino acid analysis, etc.
4. Calculate the synthetic yields by the weight balance (*see* **Note 21**).

4 Notes

1. If the resin is not preloaded with amino acids, the loading rate of the first (C-terminal) amino acid can be determined as follows: The first amino acid residue is introduced on the resin (*see* Subheading 3.1, **step 4**) and perform Fmoc group deprotection (*see* Subheading 3.1, **step 3**). Then, measure the absorption of Fmoc piperidine in the filtrate at 301 nm with a

spectrophotometer and calculate the amount of Fmoc from the molar absorption coefficient [$\varepsilon = 7800$ (M^{-1} cm^{-1})], which indicates the amino acid loading rate (mmol/g).

2. Attention should be paid to the resistance temperature of the vessels. Vessel made by polypropylene is recommended.
3. Swelling was achieved by immersion in reaction solvent for at least 1 h at room temperature. Gentle microwave heating results in faster swelling.
4. Microwave-employed SPPS devices are commercially available, e.g., from CEM or Biotage. They may recommend performing more than the 0.1 mmol reaction scale. MWS-1000 (EYELA) allows for easy reaction temperature control. The device protocol was followed as appropriate (coupling reagents, reaction temperature, time, etc.).
5. Thoroughly wash the resin (at least three to five times). Residual piperidine can be induced to cleave the Fmoc group of the synthon in a subsequent coupling reaction.
6. Coupling agents should be prepared immediately before use to minimize the reduction in their activity. COMU solution is not suitable for storage because of the rapid loss in their activity, but HBTU solutions can be stored under refrigeration for 2 weeks. If HBTU is used as a coupling reagent, two equivalents of the base DIEA against HBTU are required, namely, Fmoc-AA-OH (157 μmol, 5 eq.), HBTU (59.7 mg, 157 μmol, 5 eq.), and DIEA (53.6 μL, 315 μmol, 10 eq.) in DMF (1 mL) or Fmoc-glycoAA-OH (47 μmol, 1.5 eq.), HBTU (17.9 mg, 47 μmol, 1.5 eq.), and DIEA (16 μL, 95 μmol, 3 eq.) in DMF (0.7 mL).
7. Fmoc-AA-OH should be used at approximately 5-- 10 equivalents, but Fmoc-glycoAA-OH would be used at only about 1.2–1.5 equivalents because glycosylated amino acids are expensive. In the latter case, the reaction time takes longer. As already mentioned, microwave irradiation is recommended for the coupling of glycosylated amino acids; however, if the couplings are performed at room temperature for 2 h or more, a capping stage (*see* Subheading 3.1, **step 5**) should be performed.
8. This step can be skipped when the coupling proceeded in high yields. If it is desired to follow or there are concerns about coupling reactions (for example, when synthesizing difficult amino acid sequences), a Kaiser test may be performed to check whether any unreacted amino groups remain as follows: A few grains of resin were washed with methanol three times to remove DMF in an Eppendorf tube or a glass test tube and the Kaiser test kit reagent was added. The mixture was then heated at 100 °C for 1 min. If the resin color turns purple, this

indicates that amino groups remain on the resin, and it is therefore recommended to perform the amino acid coupling reaction (Subheading 3.1, **step 4**) again ("double coupling") before acetyl capping.

9. If you finished the peptide synthesis by coupling of final amino acid as in **step 4** in Subheading 3.1 and proceed to cleavage the peptide (Subheading 3.1, **step 8**), you will obtain Fmoc-protected peptides at the N-terminal. If you perform deprotection of the Fmoc group on the final amino acid residue as in **step 3** in Subheading 3.1, you will obtain N-terminal free (amino group) peptides. If further capping is performed as In **step 5** in Subheading 3.1, you will obtain N-terminal acetylated (-NHAc) peptides.
10. After washing the resin, thorough vacuum drying overnight is recommended, as any residual DMF may inhibit the cleavage of the synthesized peptide from the resin.
11. For MUC1 peptide synthesis, 1 to 2 mL of a cleavage cocktail to approximately 10 mg of resin is typically added. When cysteine (Cys) or methionine (Met) is the component of a synthesized peptide, scavengers such as ethanedithiol (EDT), thioanisole, or phenol are required.
12. A longer time is required if less reactive protecting groups, such as the Pbf group on arginine, are introduced.
13. If precipitation did not occur, the amount of ether could be adjusted. If the precipitate still does not appear (e.g., in the case of hydrophobic peptides), it may be omitted.
14. If HPLC analysis of the crude acetyl-protected glycopeptide indicates that the purity is high, it is possible to proceed directly to the stage of deprotection of the protecting groups on carbohydrates (Subheading 3.2). If many impurities are detected or if the HPLC signals overlap with the deacylated product, purification by preparative HPLC is recommended.
15. Adjust to about 1–10 mg/mL. It should be noted that the deprotection materials became more hydrophilic.
16. If the reaction did not proceed well, extra NaOMe is added carefully, as excessive addition may cause ester hydrolysis or peptide epimerization.
17. If the resulting residue did not dissolve in water, add 50% acetonitrile.
18. Thus, 220-nM ultraviolet can detect amide bonds.
19. The pH must be controlled to prevent decomposition, especially in the case of glycopeptides containing sialic acids. Glycosyl bonds may be hydrolyzed under acidic conditions.
20. Ionization of compounds for mass spectrometry (MS), ESI (electrospray ionization), may be available, but MALDI is

recommended for the peptides and the glycopeptides, especially in cases the molecular weight is greater than 500 where strong matrix signals are not disturbed. In MALDI-TOF MS, DHB or CHCA is typically used as matrix. DHB is mainly used for low- to medium-molecular-weight compounds and low-sugar samples and observable *m/z* are typically 207 $[M+H-H_2O]^+$ and 225 $[M+H]^+$ for positive mode and 223 $[M-H]^-$, 447 $[2\ M-H]^-$, and 671 $[3\ M-H]^-$ for negative mode, whereas CHCA is used for macromolecules and sugar-rich samples and observable m/z are typically 172 $[M+H-H_2O]^+$, 190 $[M+H]^+$, 212 $[M+Na]^+$, and 379 $[2\ M+H]^+$ for positive mode and 144 $[M-CO_2H+H-H]^-$ and 188 $[M-H]^-$ for negative mode. The positive mode is often more sensitive; however, the negative mode may be preferred for glycopeptides containing sialic acid.

21. Although the samples were freeze-dried well, some amount of water may remain. If the synthetic glycopeptide is intended for other research uses such as biochemical or medical research, the actual peptide amount should be determined by absorbance.

Acknowledgements

This work was partially supported by AMED under grant numbers JP19ae0101027 and 21mk0101203 and Grants-in-Aid for Scientific Research from the Japan Society for the Promotion of Science (KAKENHI numbers 18580109 and 25293134).

References

1. Matsushita T, Hinou H, Kurogochi M et al (2005) Rapid microwave-assisted solid-phase glycopeptide synthesis. Org Lett 7:877–880
2. Fumoto M, Hinou H, Ohta T et al (2005) Combinatorial synthesis of MUC1 glycopeptides: polymer blotting facilitates chemical and enzymatic synthesis of highly complicated mucin glycopeptides. J Am Chem Soc 127:11804–11818
3. Matsushita T, Hinou H, Fumoto M et al (2006) Construction of highly glycosylated mucin-type glycopeptides based on microwave-assisted solid-phase syntheses and enzymatic modifications. J Org Chem 71:3051–3063
4. Matsushita T, Sadamoto R, Ohyabu N et al (2009) Functional neoglycopeptides: synthesis and characterization of new class MUC1 glycoprotein models having core 2-based *O*-glycan and complex-type *N*-glycan chains. Biochemistry 48:11117–11133
5. Ohyabu N, Hinou H, Matsushita T et al (2009) An essential epitope of anti MUC1 monoclonal antibody KL-6 revealed by focused glycopeptide library. J Am Chem Soc 131:11804–11818
6. Izumi R, Matsushita T, Fujitani N et al (2013) Microwave-assisted solid-phase synthesis of antifreeze glycopeptides. Chem Eur J 19:3913–3920
7. Yoshimura Y, Denda-Nagai K, Takahashi Y et al (2019) Products of chemoenzymatic synthesis representing MUC1 tandem repeat unit with T-, ST- or STn-antigen revealed distinct specificities of anti-MUC1 antibodies. Sci Rep 9: 16641–16652
8. Yokoe T, Ohki Y, Nagashima I et al (2019) Study of glycopeptide synthesis under microwave irradiation at low temperatures towards automated synthesiser. JEMEA J 3:32–39

Chapter 17

MALDI-TOF MS/MS Analysis of Permethylated *O*-Glycan Alditols Released from Mucins

Yukinobu Goso and Makoto Kurihara

Abstract

Mucin glycans are associated with the function of mucin in maintaining mucosal homeostasis. Therefore, the glycomic analysis of mucins is crucial. Matrix-assisted laser desorption/ionization time-of-flight mass spectrometry (MALDI-TOF MS) is one of the most suitable methods for the glycomic analysis of mucin *O*-glycans. In this chapter, we describe methods for analyzing permethylated *O*-glycan alditols released from mucins by MALDI-TOF MS and MALDI-TOF tandem mass spectrometry (MALDI-TOF MS/MS).

Key words Mucins, *O*-Glycan alditols, Permethylation, Mass spectrometry, MALDI-TOF MS, Glycomics

1 Introduction

Mucin is a glycoprotein synthesized by mucosal epithelial cells and highly glycosylated with mucin-type *O*-glycans [1]. Mucin glycans are closely associated with the function of mucin in maintaining mucosal homeostasis [2]. Histochemical studies suggest that the *O*-glycans of mucosal mucins differ among tissues. The distinct roles of mucins in various tissues might be attributed to the differences in mucin *O*-glycans. Therefore, elucidating the glycome of mucin *O*-glycans in each tissue is vital.

To understand the role of tissue-specifically expressed glycans, gene knockout experiments are performed on glycosyltransferases involved in their synthesis. Nakayama and his group reported that deletion of the α-1,4-*N*-acetylglucosaminyltransferase (*A4gnt* knockout), which transfers an α-*N*-acetylglucosamine to the periphery of *O*-glycan of mucins present in gastric gland mucus, resulted in hyperplasia of the pyloric region in the stomach and development of adenocarcinoma [3]. An increase in acidic *O*-gly-

Akihiko Kameyama (ed.), *Mucins: Methods and Protocols*, Methods in Molecular Biology, vol. 2763,
https://doi.org/10.1007/978-1-0716-3670-1_17,

cans was also observed through histochemical examination [4]. In such experiments, understanding the mechanisms by which the *O*-glycans are altered by the deletion of the enzyme is crucial.

Matrix-assisted laser desorption/ionization time-of-flight mass spectrometry (MALDI-TOF MS) is one of the most suitable methods for the glycomic analysis of mucin *O*-glycans. Glycans are excised from mucins with minimal degradation using the alkaline-borohydride treatment and permethylated before mass spectrometry. This process increases the sensitivity of mass spectrometry and facilitates structural analysis of glycans. Furthermore, it contributes to the stability of sialic acid attached to glycans during MS analysis and allows for the analysis of acidic glycans in positive mode [5].

In this chapter, we describe methods to analyze permethylated *O*-glycan alditols released from mucins by MALDI-TOF MS (Fig. 1) and MALDI-TOF MS/MS (Fig. 2) [6]. The results of these knockout experiments are also presented (Fig. 3) [4].

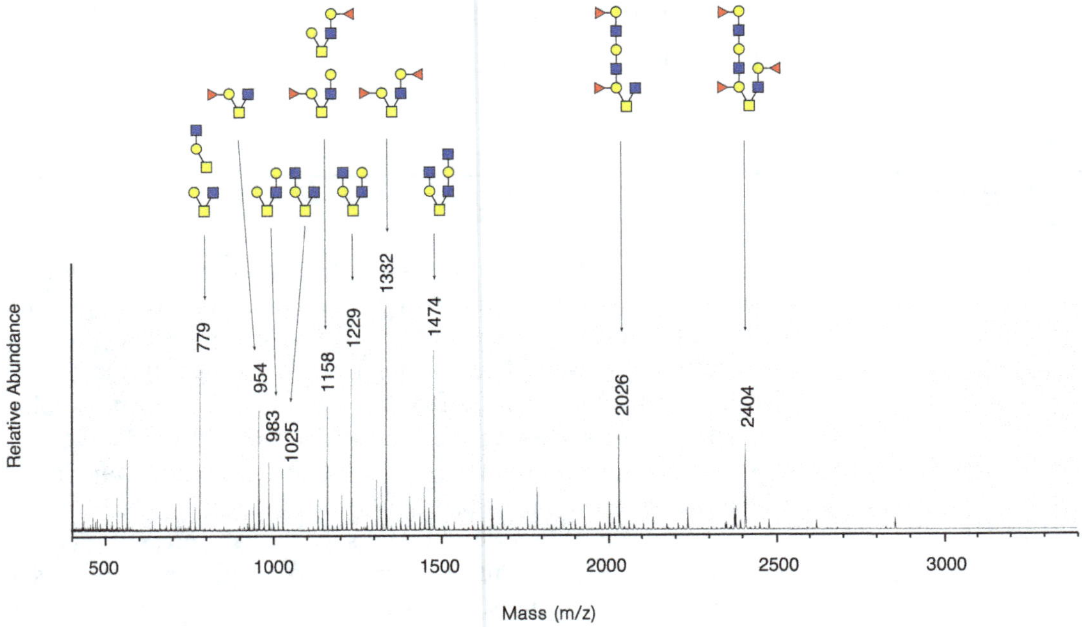

Fig. 1 MALDI-TOF MS of permethylated *O*-glycan alditols obtained from rat gastric mucin. Mucin was treated at 45 °C for 15 h with 50 mM NaOH containing 1 M sodium borohydride. The obtained *O*-glycan alditols were permethylated and analyzed by MALDI-TOF MS. Proposed glycan structures, as determined by MALDI-TOF MS/MS, are illustrated with yellow circles, blue squares, yellow squares, and red triangles denoting galactose, *N*-acetylglucosamine, *N*-acetylgalactosamine, and fucose, respectively. (Reproduced from [6] with permission and some modifications. Copyright 2017 American Chemical Society)

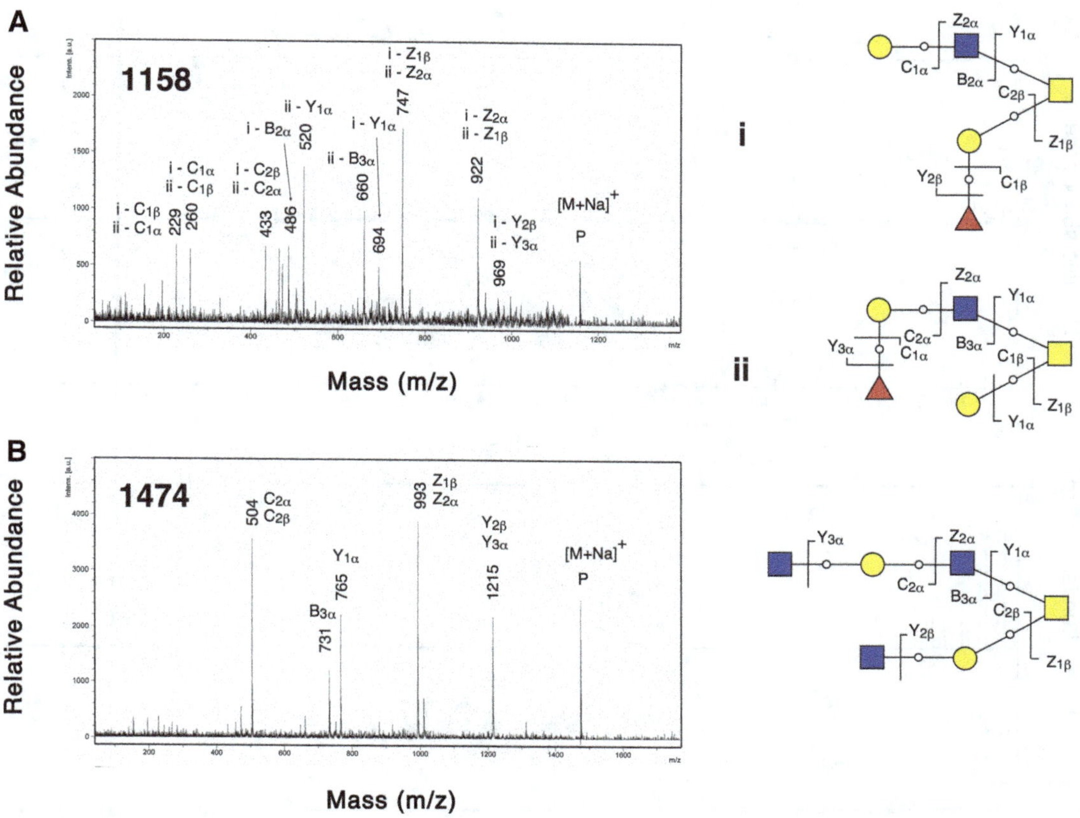

Fig. 2 MALDI-TOF MS/MS of permethylated *O*-glycan alditols indicated in Fig. 1 (mass values: 1158 [A] and 1474 [B]). Glycan structures are illustrated as in Fig. 1, and fragment ions are named according to the proposed nomenclature [9]. Note that two possible structures were identified in the case of the previous *O*-glycan alditols. (Reproduced from [6] with permission and some modifications. Copyright 2017 American Chemical Society)

2 Materials

2.1 Release of O*-Glycans*

1. Purified rat and mouse gastric mucins (*see* **Note 1**).
2. 1.5 mL tube.
3. 50 mM NaOH containing 1 M $NaBH_4$ (prepared at the time of use.).
4. 4 M acetic acid.
5. pH test paper.
6. Dry block heater.
7. Cation exchange resin (H^+-form) (*see* **Note 2**).
8. Empty column cartridges (6 mL).
9. Lyophilizer.
10. Glass test tubes (13 mm × 100 mm).

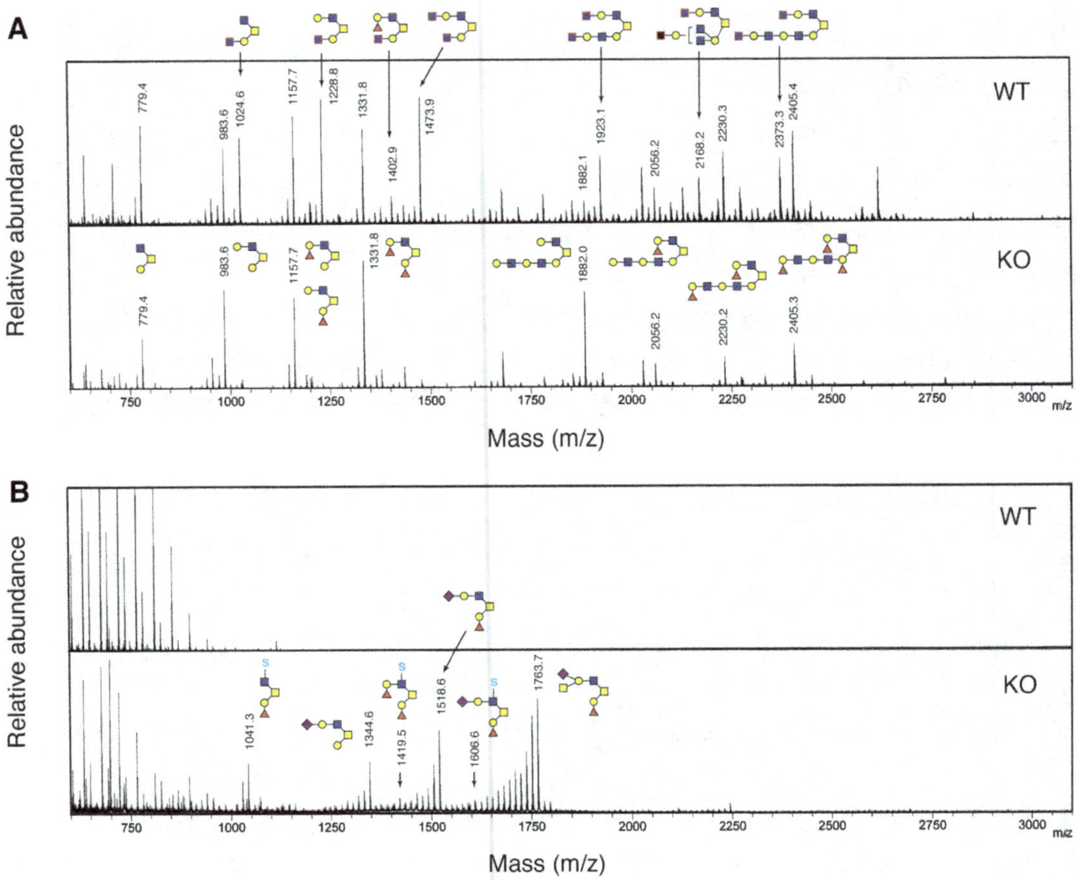

Fig. 3 Analysis of neutral (**a**) and acidic (**b**) *O*-glycan alditols isolated from 10-week-old *A4gnt*+/+ (WT) and *A4gnt*–/– (KO) mice using MALDI-TOF MS and MALDI-TOF MS/MS. Loss of *O*-glycan alditols bearing peripheral α-*N*-acetylglucosamine (blue squares with red border) is observed in *A4gnt*–/– mice. Glycan structures are illustrated as in Fig. 1, except that purple diamond and blue S denote *N*-acetylneuraminic acid and sulfo group, respectively. (Reproduced from [4] with some modifications)

11. Cetrifugal evaporator.
12. Methanol containing 1% acetic acid.
13. Nitrogen gas.
14. Anion exchange resin (acetate-form) (*see* **Note 3**).
15. 0.4 M pyridinium acetate, pH 5.0.

2.2 Permethylation of O-Glycan

1. Screw cap glass tube (13 mm × 100 mm).
2. 50% NaOH aqueous solution.
3. Methanol.
4. Vortex mixer.
5. Dimethyl sulfoxide (DMSO).

6. Ultrasonic bath.
7. Table-top centrifuge.
8. Vacuum desiccator connected to vacuum pump, phosphorus pentoxide (P_2O_5).
9. Methyl iodide.
10. Hood equipped with a ventilation system.
11. Nitrogen gas.
12. Reversed-phase SPE column (1 mL) (*see* **Note 4**).
13. Acetonitrile.
14. 50% acetonitrile.
15. Centrifugal evaporator.

2.3 MALDI-TOF MS and MS/MS Analysis

1. MALDI-TOF MS instrument (*see* **Note 5**).
2. Matrix solution: 50 mg/mL 2,5-dihydrobenzoic acid (DHB) in methanol/water (1:1; v/v).
3. GlycoMod online tool (https://web.expasy.org/glycomod/).
4. GlycoWorkBench, downloadable Java tool (https://glycoworkbench.software.informer.com/).

3 Methods

All solutions are aqueous solutions unless indicated otherwise.

3.1 Release of O-Glycans from Mucin by Alkaline Borohydride Treatment

1. Add the freshly prepared 100 μL of 50 mM NaOH containing 1 M $NaBH_4$ to the mucin sample (150 μg) in the 1.5 mL tube.
2. Incubate the tube at 45 °C for 15 h using a dry block heater.
3. Transfer the sample to a 13 mm × 100 mm tube with a 200 μL of water rinse. After 600 μL of water, add 4 M acetic acid (approximately 60 μL) at 0 °C until pH is lower than 5.0.
4. Pack cation exchange resin columns with 4 mL slurry of Dowex 50Wx4 (H^+-form)/water (1:1 [w/v]) to empty column cartridges (6 mL), and then wash with 20 mL of water.
5. Load the sample to these columns and wash it with 25 mL of water. Collect the eluent from the column.
6. Lyophilize the eluent.
7. Dissolve the sample in 2 mL of water and transfer to a 13 mm × 100 mm tube.
8. Evaporate the solutions at 43 °C using a centrifugal evaporator.
9. Add 1 mL methanol containing 1% (v/v) acetic acid and evaporate at 45 °C with a stream of nitrogen gas using a dry block heater. Repeat this procedure two more times.

10. Pack anion exchange resin columns with 3.3 mL slurry of Toyopearl QAE (acetate-form)/water (1:2.5 [w/v]) to empty column cartridges (6 mL), and wash the columns with 20 mL of water (*see* **Note 6**).
11. Dissolve the glycan samples in 2 mL of water. Pipette the solutions into the columns and wash the column with 20 mL of water. Collect the eluent from the column.
12. Glycan in the eluate is designated as the "neutral glycan."
13. Elute acidic glycans from the columns with 20 mL of 0.4 M pyridinium acetate, pH 5.0.
14. Lyophilize the neutral and acidic glycan solutions.

3.2 Preparation of Permethylation Reagent

Permethylation of *O*-glycans can be performed according to the methods described by Anumula and Taylor [7].

1. Add 50 μL of 50% NaOH in the screw cap tube, followed by 100 μL of methanol, and mix on a vortex mixer (*see* **Note 7**).
2. Add 3 mL of DMSO and mix vigorously on a vortex mixer.
3. Sonicate for 4 min using an ultrasonic bath and then centrifuge at 500 × *g* for 2 min.
4. Remove supernatant by aspiration.
5. Add 3 mL of DMSO and mix vigorously on a vortex mixer.
6. Centrifuge at 500 × *g* for 2 min.
7. Remove supernatant by aspiration.
8. Repeat **steps 5–7** two more times.
9. Add 1 mL of DMSO and mix on a vortex mixer.

3.3 Permethylation of O-Glycans

1. Dry *O*-glycan samples in a screw cap tube in vacuo overnight over P_2O_5.
2. Add 200 μL of DMSO and 4 μL of water.
3. Sonicate for 2 min using an ultrasonic bath.
4. Add the freshly prepared permethylation reagent (200 μL, Subheading 3.2) and methyl iodide (100 μL).
5. Flash the tube with nitrogen gas and cap it.
6. Mix vigorously on a vortex mixer for 5 min.
7. Add 2 mL of water to terminate the reaction and flash with nitrogen gas for 5 min in a hood equipped with a ventilation system to remove residual methyl iodide.
8. Wash the OASIS HLB Cartridge with 2 mL of acetonitrile and then with 4 mL of water.
9. Apply the permethylated glycan solution to the column and wash the column with 3 mL of water.

10. Elute the permethylated glycans with 2.5 mL of 50% (v/v) acetonitrile.
11. Dry the eluate using a centrifugal evaporator.
12. Dissolve the permethylated glycans in a small amount of methanol/water (1:1; v/v).

3.4 MALDI-TOF MS and MALDI-TOF MS/MS Analysis of Permethylated O-Glycans

1. Spot 0.5 μL of the permethylated glycan sample on a stainless steel MALDI plate.
2. Add a 0.5 μL of DHB matrix solution on top of the sample.
3. Dry the spot prior to MS analysis.
4. Acquire mass spectra using the MALDI mass spectrometer operating in positive mode at an accelerating voltage of 20 kV.
5. Perform tandem mass spectrometry in LIFT mode.

3.5 Mass Data Interpretation

1. Connect to the GlycoMod website.
2. Enter *m/z* values of peaks. Fill in the relevant options, such as monoisotopic, positive, Na^+, *O*-linked oligosaccharide/reduced oligosaccharide, and monosaccharide/permethylated.
3. Start GlycoMod. Output is the list of potential components.
4. Perform manual interpretation based on databases of *O*-glycans.
5. Launch GlycoWorkBench.
6. Draw putative structures for the *m/z* value. Set Mass options as [MONO, perMe, Na, 0, redEnd].
7. Input the peak list into the GlycoWorkBench.
8. Using the ANNOTATE tool, match the fragments to the *MS/MS* peak list for all possible structures.
9. Using the FRAGMENT tool, perform a theoretical fragmentation of the putative structures.
10. Identify the most likely structure from **steps 8** and **9**.

4 Notes

1. Rat and mouse gastric mucins were prepared according to the methods of **Ref.** 8. Mucosa was scraped off from the stomach, homogenized in 6 M guanidinium chloride, pH 7.4, containing 2% (v/v) Triton X-100, 50 mM Tris–HCl, 100 mM dithiothreitol, and protease inhibitors, and the mucin was extracted. After alkylating with iodoacetamide, the extract was applied to a Sepharose CL-6B column equilibrated with 4 M guanidinium chloride, pH 7.4, containing 0.5% (v/v) Triton X-100, 50 mM Tris–HCl, and protease inhibitors. The excluded fraction from

the column was further purified by CsCl density gradient centrifugation. The mucin fraction was collected and desalted with ethanol precipitation.

2. Dowex 50Wx4 resin (200–400 mesh, 454 g, Dow Chemical) is prewashed sequentially with 3 L of 2 M NaOH, 4 L of water, 6 L of 3 M HCl, and 2 L of water. Another cation exchange resin may be available.
3. Toyopearl QAE-550C resin (500 mL, TOSOH) is prewashed sequentially with 1 L of 0.5 M HCl, 1.5 L of water, 5 L of 0.5 M NaOH, 2 L of 0.5 M NaOAc, 0.5 L of 0.5 M acetic acid, and 2.5 L of water, and stored at 4 °C. Another anion exchange resin may also be available.
4. OASIS HLB (30 mg/cc) Cartridge (Waters) or equivalent products.
5. A Bruker Autoflex III mass spectrometer or an equivalent equipment.
6. When neutral and acidic glycans are analyzed together, **steps 10–14** can be omitted.
7. Addition of methanol enhanced the dispersion speed of 50% NaOH in DMSO [7].

Acknowledgments

This work was supported by the Japan Society for the Promotion of Science KAKENHI grant 25460934.

References

1. Dekker J, Rossen JWA, Büller HA et al (2002) The MUC family: an obituary. Trends Biochem Sci 27:126–131
2. Corfield AP (2015) Mucins: a biologically relevant glycan barrier in mucosal protection. Biochim Biophys Acta 1850:236–252
3. Karasawa F, Shiota A, Goso Y et al (2012) Essential role of gastric gland mucin in preventing gastric cancer in mice. J Clin Invest 122:923–934
4. Kawakubo M, Komura H, Goso Y et al (2019) Analysis of *A4gnt* knockout mice reveals an essential role for gastric sulfomucins in preventing gastritis cystica profunda. J Histochem Cytochem 67:759–770
5. Morelle W, Michalski J-C (2007) Analysis of protein glycosylation by mass spectrometry. Nat Protoc 2:1585–1602
6. Goso Y, Sugaya T, Ishihara K et al (2017) Comparison of methods to release mucin-type *O*-glycans for glycomic analysis. Anal Chem 89:8870–8876
7. Anumula KR, Taylor PB (1992) A comprehensive procedure for preparation of partially methylated alditol acetates from glycoprotein carbohydrates. Anal Biochem 203:101–108
8. Goso Y, Ishihara K, Kurihara M et al (1999) Rat gastric mucins recognized by monoclonal antibodies RGM21 and HIK1083: isolation of mucin species characteristic of the surface and glandular mucosa. J Biochem 126:375–381
9. Domon B, Costello CE (1988) A systematic nomenclature for carbohydrate fragmentations in FAB-MS/MS spectra of glycoconjugates. Glycoconj J 5:397–409

Chapter 18

Structural Elucidation of Sialylated *O*-Glycan Alditols Obtained from Mucins by Mass Spectrometry

Daigo Tsubokawa, Rei Kawashima, and Takafumi Ichikawa

Abstract

Acidic *O*-glycans having sialic acid and/or sulfate residue are abundantly expressed in intestinal mucins. However, structural elucidation of acidic *O*-glycans is a laborious and time-consuming task due to their large structural variations. Here, we describe a methodology of structural elucidation for sialylated *O*-glycan alditols from intestinal mucins using tandem mass spectroscopy. Methylesterification and mild periodate oxidation of sialylated *O*-glycan alditols assist mass analysis. This description includes the purification process of *O*-glycan alditols for structural analysis.

Key words Mucin, *O*-Glycan alditol, Mass spectrometry, Structural elucidation

1 Introduction

Mucin, a highly *O*-glycosylated glycoprotein with high molecular mass, is a major component of the gastrointestinal mucus and a major factor in the defense mechanism of the gastrointestinal mucosa. Acidic mucins, such as sialomucin and sulfomucin, which link to sialylated and sulfated glycans, respectively, are abundantly produced by intestinal goblet cells [1]. Because sialylated and sulfated glycans are expressed on intestinal mucins during the physiological and pathological alterations in the intestinal mucosa [2–5], structural elucidation of sialylated and sulfated glycans are important for functional analysis of specific acidic mucins [6, 7]. However, structural elucidation of acidic O-glycans is a laborious and time-consuming task due to their large structural variations. Previously, we purified and analyzed specific sialylated glycans associating pathophysiology in the intestinal mucosa [7]. *O*-Glycan alditols were obtained from intestinal mucins and purified by ion-exchange chromatography followed by two-step normal-phase HPLC. The structural elucidation of purified sialylated *O*-glycans was performed by tandem mass spectroscopy. Methylesterification of the sialic acid

Akihiko Kameyama (ed.), *Mucins: Methods and Protocols*, Methods in Molecular Biology, vol. 2763,
https://doi.org/10.1007/978-1-0716-3670-1_18,

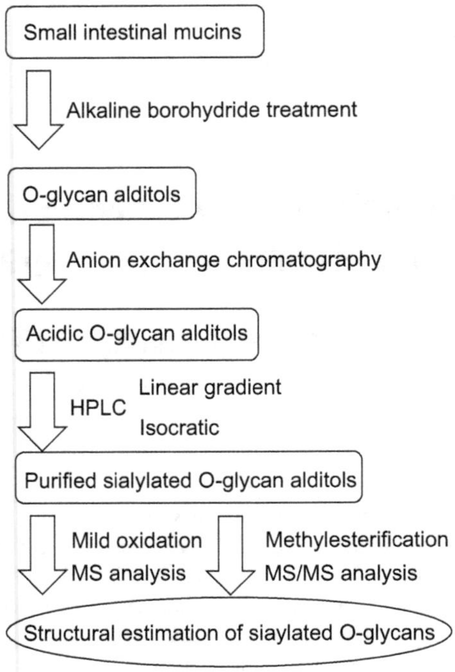

Fig. 1 A strategy of structural estimation of sialylated O-glycans from small intestinal mucins

residue can increase the intensity of its fragment ions [8]. Mild periodate oxidation can cleave the C–C bond of the reducing terminal GalNAc-ol at position C4 and C5, leading to elucidate whether a sialic acid-containing branch is attached at position C3 or C6 of GalNAc-ol [9]. Here, we describe the procedure about purification and structural elucidation of sialylated *O*-glycan alditols from rat small intestinal mucins (Fig. 1).

2 Materials

Prepare all solutions using Ultrapure water and analytical grade reagents. Do not add sodium azide to the reagents. Prepare and store all reagents at room temperature (unless indicated otherwise).

2.1 Mucin Preparation

1. Powdered rat small intestines (*see* **Note 1**).
2. 50 mM Tris–HCl buffer (pH 7.2): Add ~100 mL of water to a 1-L graduated cylinder. Weigh 6.06 g of Tris base and transfer to the cylinder. Mix with water. Add more water until the volume of the mix reaches 800 mL. Mix and adjust the pH to 7.2 with HCl. Add more water until the volume reaches 1 L, and mix the solution.

3. Extraction buffer: 20 mM Na_2 EDTA, 2 mM PMSF, 6 μM pepstatin, 20 mM N-ethylmaleimide in 50 mM Tris–HCl buffer (pH 7.2). Weigh 744 mg of Na_2 EDTA and 250 mg of N-ethylmaleimide, and transfer to a 100 mL graduated cylinder. Add 50 mM Tris–HCl buffer (pH 7.2) until volume reaches 100 mL. Weigh 17.4 mg of PMSF and 0.4 mg of pepstatin, individually transfer to each 1.5 mL microcentrifuge tube, and then add 1 mL of ethanol and methanol, respectively (*see* **Note 2**). Dissolve them using a vortex mixer. Add the ethanol and methanol to the cylinder, and mix the solution. Use immediately after preparation (*see* **Note 3**).
4. 20% (v/v) Triton ×-100: Add 80 mL of water to a 100 mL graduated beaker (*see* **Note 4**). Add 20 mL Triton ×-100 to the beaker, and mix the solution using a stirrer.
5. Polytron homogenizer.
6. Centrifuge (angle rotor 50 mL round bottom).
7. Round-bottom centrifuge tubes (50 mL) (*see* **Note 5**).
8. Shaking incubator.
9. Gel filtration resin (Bio-Gel® A-1.5 m, Bio-Rad Laboratories, Richmond, CA, USA) (*see* **Note 6**).
10. Open glass columns (2.5 × 100 cm).
11. Fraction collector.
12. Elution buffer: 5 mM Na_2 EDTA in 0.5 M NaCl. Add ~2500 mL of water to a 5-L beaker. Weigh 9.30 g of Na_2 EDTA and 147.75 g of NaCl, and transfer to the beaker. Mix with the water. Add more water until the volume reaches 5 L, and mix the solution.
13. Collection tubes.
14. Ethanol.
15. Dialysis tube (flat width 20–30 mm).

2.2 Glycan Preparation

1. 2 M $NaBH_4$ in 100 mM NaOH: Add and mix 90 mL of water and 10 mL 1 M NaOH in a glass beaker. Weigh 7.566 g of $NaBH_4$, transfer to the beaker, and mix with the solution. Use immediately following the preparation.
2. Heat block.
3. Glass tubes with screwcaps (16 × 125 mm).
4. Glacial acetic acid.
5. Open glass columns (1.5 × 10 cm).
6. Strong cation exchange resin (hydrogen form).
7. Strong anion exchange resin (acetate form).
8. Rotary evaporator.

9. Methanol.
10. Gradient former.
11. Peristaltic pump.
12. 0.6 M NaOAc.
13. Fraction collector.
14. Collection tubes.
15. Graphitized carbon columns (bed volume 150 mg and 1 g).
16. 25% (v/v) acetonitrile containing 0.05% (v/v) trifluoroacetic acid.
17. Syringe filters (hydrophilic PTFE, 0.45 μm pore size).
18. Normal-phase HPLC columns (TSKgel® Amide-80, 4.6 × 250 mm, 7.8 × 300 mm, Tosoh Bioscience).
19. 50% (v/v) acetonitrile.
20. Microsyringe for HPLC.
21. HPLC eluate A: 80% (v/v) acetonitrile in 2.5 mM NaH_2PO_4.
22. HPLC eluate B: 30% (v/v) acetonitrile in 2.5 mM NaH_2PO_4.
23. HPLC system with high-pressure gradient pumps.

2.3 Glycan Analysis

1. Dimethyl sulfoxide (DMSO).
2. Methyl iodide.
3. Glass vials with caps (1 mL).
4. 50% (v/v) acetonitrile.
5. 2,5-Dihydroxybenzoic acid (2,5-DHB) solution: Weigh 10 mg of 2,5-DHB and transfer to a 1.5 mL microcentrifuge tube. Add 500 μL of acetonitrile and dissolve the 2,5-DHB, and then add 500 μL of water into the tube. Store at −20 °C.
6. Mild periodate oxidation solution: 0.06 mM $NaIO_4$ in 40 mM imidazole buffer (pH 6.5). Add ~10 mL of water to a 100 mL graduated cylinder. Weigh 272 mg of imidazole and transfer to the cylinder. Mix with the water. Add more water until the volume of the mix reaches 80 mL. Mix and adjust the pH to 6.5 with HCl. Add more water until the volume reaches 100 mL, and mix the solution. Weigh 1.25 mg of $NaIO_4$ and transfer to a 1.5 mL microcentrifuge tube and add 1 mL of 40 mM imidazole buffer to the microcentrifuge tube (pH 6.5). Dissolve the reagent using a vortex mixer. Put immediately the tube on ice. Use immediately after preparation (*see* **Note 3**).
7. Butane-diol reagent: Add 98.54 μL of water and 1.46 μL butane-2,3-diol to a 1.5 mL microcentrifuge tube. Mix the solution using a vortex mixer. Put immediately the tube on ice. Use immediately after preparation (*see* **Note 3**).
8. Graphitized carbon columns (150 mg).

9. 25% (v/v) acetonitrile.
10. 25% (v/v) acetonitrile containing 0.05% (v/v) trifluoroacetic acid.
11. Mass spectrometer: MALDI-TOF mass spectrometer (Autoflex III TOF/TOF instrument, Bruker Daltonik GmbH, Bremen, Germany).

3 Methods

Carry out all procedures at room temperature unless specified otherwise.

3.1 Preparation of Small Intestinal Mucins

1. Weight 1.5 g powdered small intestines, and transfer to a centrifuge tube (*see* **Note 7**).
2. Add 10 mL of 50 mM Tris–HCl buffer (pH 7.2) into the centrifuge tube, and mix small intestines with the solution. (*see* **Note 8**).
3. Stand the centrifuge tube on boiled water for 5 min, and then cool on ice.
4. Add 10 mL of extraction buffer into the centrifuge tube.
5. Homogenize the small intestines in the centrifuge tube on ice for 8 min (shafts revolutions, 13,000 rpm) (*see* **Note 9**).
6. Add 2.2 mL of 20% Triton ×-100 to the centrifuge tube, and mix the solution using a vortex mixer.
7. Incubate the centrifuge tube by a shaking incubator (200–300 rpm) at 37 °C for 1 h.
8. Centrifuge at 10,000 × *g* for 30 min at 4 °C.
9. Collect supernatant from the centrifuge tube, and store at 4 °C (*see* **Note 10**).
10. Apply 25 mL of supernatant to a gel filtration column (bed volume 300 mL) (*see* **Note 11**), elute with 300 mL of elution buffer, and collect the effluent in collection tubes using a fraction collector (7 mL each).
11. Pool the void fractions (Fr. 8–12) as mucin to a glass beaker (*see* **Note 12**).
12. Add three volumes of ethanol to the glass beaker, gently mix the solution using a stirring rod, and keep at 4 °C overnight.
13. Transfer the mixture to centrifuge tubes, and centrifuge at 10,000 × *g* for 30 min at 4 °C.
14. Discard the supernatant, and then dissolve the precipitate with water.

15. Dialyze the solution against over 50 volumes of water overnight (*see* **Note 13**).
16. Lyophilize the dialyzed solution, and store at 4 °C (*see* **Note 14**).

3.2 Purification of Glycans

3.2.1 Preparation of O-Glycan Alditols

1. Dissolved lyophilized mucins that has been prepared in Subheading 3.1 in water (20 mg/mL), and transfer to glass tubes (1 mL each).
2. Add 1 mL of 2 M $NaBH_4$ in 100 mM NaOH to each tube, cap the tube securely, and mix the solution using a vortex mixer.
3. Incubate the glass tubes using a heat block at 50 °C for 24 h (*see* **Note 15**).
4. Pool and transfer the reaction mixture to a glass beaker on ice. Add glacial acetic acid to neutralize the reaction mixture (*see* **Note 16**), and then add an equal volume of water to the glass beaker.
5. Apply 40 mL of the reaction mixture to a cation exchange column (bed volume 25 mL) (*see* **Note 17**), wash with 40 mL of water, and collect the flow-through fraction. Then, dry up the solution using a rotary evaporator.
6. Add methanol, and dissolve the residues. Evaporate the solution. Repeat this addition and evaporation procedure three times (*see* **Note 18**).
7. Add 20 mL of water, and dissolved the residues. Store at 4 °C before use (*see* **Note 10**).

3.2.2 Preparation of Acidic O-Glycan Alditols

1. Apply 10 mL of the stored sample that has been prepared in Subheading 3.2.1 to an anion exchange column (bed volume 15 mL) (*see* **Note 19**).
2. Wash with 30 mL of water, elute with a linear gradient of 80 mL of 0–0.6 M NaOAc (*see* **Note 20**), and collect the effluent in collection tubes using a fraction collector (2 mL each) (*see* **Note 21**) .
3. Pool the eluted fractions (Fr. 25–45) as acidic glycans (*see* **Note 12**).
4. Apply ~20 mL of the eluted fractions to a graphitized carbon column (1 g) for desalting (*see* **Note 22**), wash with 10 mL of water, and elute with 10 mL of 25% (v/v) acetonitrile containing 0.05% (v/v) trifluoroacetic acid.
5. Lyophilize the eluted solution, and store at 4 °C (*see* **Note 23**).

3.2.3 First-Step HPLC

1. Dissolve the lyophilized sample that has been prepared in Subheading 3.2.2 in 50% (v/v) acetonitrile (10 mg/mL), and then filter the sample by a syringe filter.

2. Inject ~100 μL of the filtered sample to a HPLC system equipped with normal-phase HPLC columns (7.8 × 300 mm × 2 columns) (*see* **Note 24**).
3. Elute with a 30-min linear gradient of HPLC eluate A/B (75/25 to 50/50) at a flow rate of 2.0 mL/min (*see* **Note 25**).
4. Monitor the effluent solution at 210 nm. Collect each glycan fraction in collection tubes (*see* **Note 26**).
5. Lyophilize each glycan fraction, and then dissolved in water (~2 mg/mL).
6. Apply ~1 mL of the fraction to a graphitized carbon column (150 mg) (*see* **Note 27**), wash with 2 mL of water, and elute with 2 mL of 25% (v/v) acetonitrile containing 0.05% (v/v) trifluoroacetic acid.
7. Lyophilize the eluted solution, and store at 4 °C (*see* **Note 28**).

3.2.4 Second-Step HPLC

1. Dissolve the lyophilized sample that has been prepared in Subheading 3.2.3 in 50% (v/v) acetonitrile (~2 mg/mL), and then filter the sample using a syringe filter.
2. Inject ~75 μL of the filtered sample to a HPLC system equipped with normal-phase HPLC columns (4.6 × 250 mm × 2 columns) (*see* **Note 29**).
3. Elute with a 200 min isocratic condition of HPLC eluate A/B (88/12) at a flow rate of 1.0 mL/min (*see* **Note 25**).
4. Monitor the effluent solution at 210 nm. Collect each glycan fraction in collection tubes (*see* **Note 30**).
5. Lyophilize each glycan fraction, and then dissolve in water (~2 mg/mL).
6. Apply ~1 mL of the fraction to a graphitized carbon column (150 mg) (*see* **Note 27**), wash with 2 mL of water, and elute with 2 mL of 25% (v/v) acetonitrile containing 0.05% (v/v) trifluoroacetic acid.
7. Lyophilize the eluted solution, and store at 4 °C (*see* **Notes 28** and **38**) (Fig. 2).

3.3 Glycan Analysis

3.3.1 MS/MS Analysis of Sialylated O-Glycan Alditols

1. Dissolve the lyophilized sample that has been prepared in Subheading 3.2.4 in water (~10 mg/mL).
2. Add 5 μL of the sample to a glass vial, and dry (*see* **Note 31**).
3. Add 1 μL of DMSO to the glass vial, and dissolve the residues by pipetting.
4. Add 1 μL of methyl iodide to the glass vial, and mix the solution by pipetting. Cap the glass vial, and incubate for 2 h at room temperature (*see* **Note 32**).

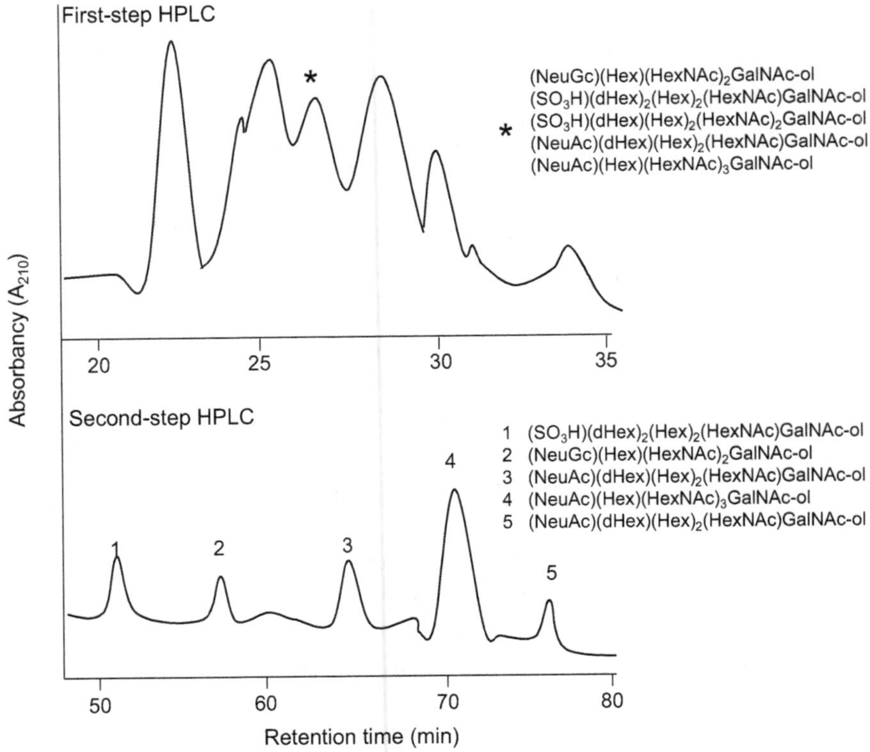

Fig. 2 Purification of acidic O-glycans by two-step normal-phase HPLC. Fast separation can be performed by the first-step linear gradient condition. Purified acidic O-glycans can be obtained by the second-step isocratic condition

5. Dry the solution under a gentle stream of nitrogen, and then dissolve the residues by 1 μL of 50% acetonitrile by pipetting.
6. Deposit 1 μL of 2,5-DHB solution onto a MALDI target plate. Deposit 1 μL of the sample solution over the spot of 2,5-DHB, and mix by pipetting. Allow the spot to dry.
7. Acquire a reading of the MS/MS spectrum (*see* **Note 33**) (Fig. 3).

3.3.2 MS Analysis of Sialic Acid-Containing Branch

1. Transfer the lyophilized sample (approx. 10 mg/mL) that has been prepared in Subheading 3.2.4 to a 0.5 mL microcentrifuge tube, and stand on ice.
2. Add 7 μL of ice-cold mild periodate oxidation solution to the microtube, and dissolve the lyophilized sample by pipetting.
3. Stand the microtube on ice for 5 min in dark (*see* **Note 34**).
4. Add 0.5 μL of butane-diol regent to the microtube, and mix the solution by pipetting.
5. Stand the microtube on ice for 40 min (*see* **Note 35**).

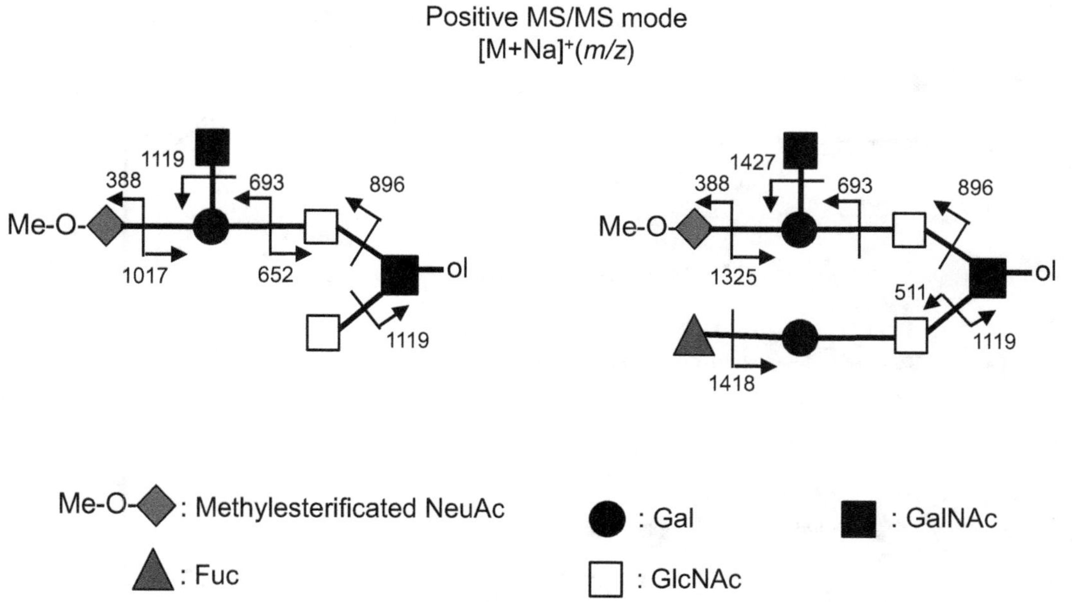

Fig. 3 The fragment ions of sialylated O-glycans following methylesterification can be detected by MS/MS analysis in positive ion mode

6. Add 0.2 mL of water to the microtube, and mix the solution by pipetting.
7. Apply the sample to a graphitized carbon column (150 mg) (*see* **Note 27**), wash with 2 mL of water, and elute with 2 mL of 25% (v/v) acetonitrile followed by 2 mL of 25% (v/v) acetonitrile containing 0.05% (v/v) trifluoroacetic acid (*see* **Note 36**).
8. Lyophilize each eluted solution, and then dissolve each residue by 5 μL of water.
9. Deposit 1 μL of 2,5-DHB solution onto a MALDI target plate. Deposit 2 μL of the sample solution over the spot of 2,5-DHB, and mix by pipetting. Allow the spot to dry.
10. Acquire a reading of the MS spectrum (*see* **Note 37**) (Fig. 4).

4 Notes

1. Rat small intestines are successively cut open, washed with chilled saline, and frozen in deep freezer, and then lyophilized followed by splitting longitudinally with sharp scissors until powderization.
2. This process should be carried out for dissolutions of PMSF and pepstatin.
3. The solution is for a single use only.

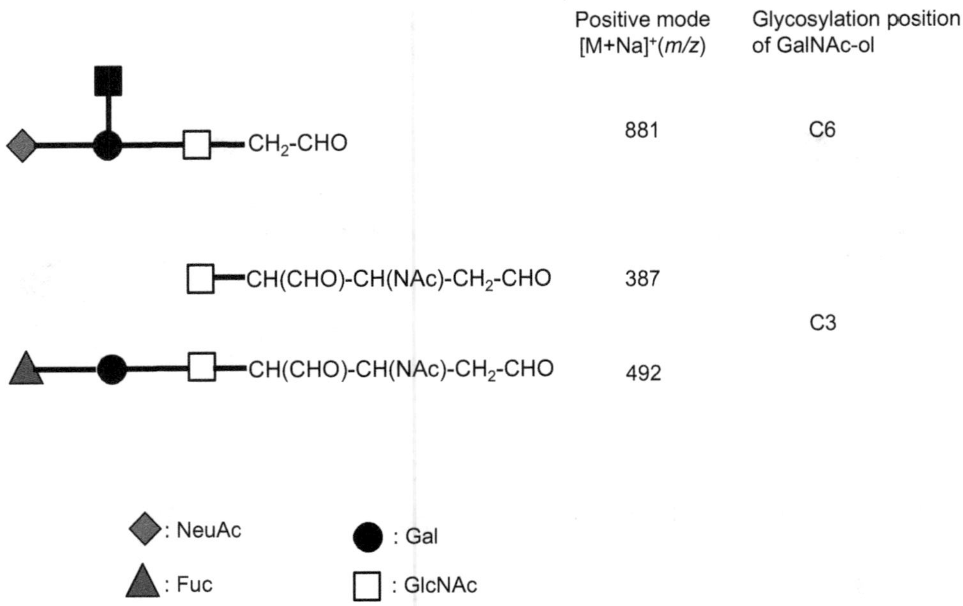

Fig. 4 The glycosylation branches with or without sialic acid following mild periodate oxidation can be detected by MS analysis in positive ion mode

4. Triton ×-100 should be added to the beaker following water to prevent bubbling. Preheating of water can faster dissolve Triton ×-100.
5. Round bottom is better than conical bottom for homogenization of tissues.
6. A commercially available gel filtration agarose resin (fractionation range: 10–1500 kDa) can be used.
7. One to one and a half gram dry weight of powdered small intestines can be obtained from a male 8-week-old Wistar rat. We usually extract from 1.5 g × 6 tubes of powdered small intestines at one experiment.
8. A micro-spatula is useful to mix in a centrifuge tube.
9. Firstly, set shaft revolutions to 3000 ~ 5000 rpm in a few seconds for gentle mixing, and then set to 13,000 rpm.
10. Long-term storage over 1 week should be performed less than −20 °C.
11. A gel filtration column should be equilibrated with three bed volumes of elution buffer before applying a sample.
12. The fractions can be monitored by hexose measurement [10].
13. Unnecessary small molecules can be removed.
14. Approximately 100 mg dry weight of small intestinal mucin may be obtained from a male 8-week-old Wistar rat.

15. Mix the solutions by a vortex and loose the caps to escape hydrogen gas per hour as possible.
16. Add acetic acid dropwise. The addition of acid may produce the drastic release of hydrogen gas.
17. A cation exchange column should be washed with three bed volumes of water before a sample applying. Sodium and peptides can be removed by a cation exchange chromatography.
18. Borate should be evaporated through this procedure.
19. An anion exchange column should be equilibrated with three bed volumes of 1 M NaOAc and then washed with three bed volumes of water before applying a sample.
20. The linear gradient eluent can be made by a gradient former with 40 mL water and 40 mL 0.6 M NaOAc. The eluent should be mixed by a magnetic stirrer.
21. The flow rate should be adjusted at 1.0 mL/min using a peristaltic pump for reproducibility.
22. A graphitized carbon column (1 g) should be washed with 5 mL of 25% (v/v) acetonitrile containing 0.05% (v/v) trifluoroacetic acid and 10 mL of water before applying a sample.
23. Approximately 2 mg dry weight of acidic *O*-glycan alditols may be obtained from 100 mg dry weight small intestinal mucin.
24. The HPLC columns should be washed with HPLC eluate B for 10 min and then equilibrated with HPLC eluate A/B (75/25) for 10 min before sample injection.
25. HPLC eluate should be degassed for 5 min in an ultrasonic bath.
26. Glycan fractions may be obtained in retention times of 15–35 min.
27. A graphitized carbon column (150 mg) should be washed with 1 mL of 25% (v/v) acetonitrile containing 0.05% (v/v) trifluoroacetic acid and 2 mL of water before applying a sample.
28. Acidic glycan alditols contained in the fractions can be detected by MALDI-TOF/MS in negative mode.
29. The HPLC columns should be washed with HPLC eluate B for 10 min and then equilibrated with HPLC eluate A/B (88/12) for 10 min before sample injection.
30. Glycan fractions may be obtained in retention times for 30–200 min. Purified alditols can be obtained as a single component by second-step HPLC. Since glycan isomers can be separated by this process, MS/MS analysis would be performed accurately.
31. Spray-drying under a gentle stream of nitrogen is better than air-drying.

32. The COOH-terminal residue of sialic acid can be methylesterificated by this process.
33. Sialylated alditol and its fragment ions can be detected as the corresponding methylesterificated alditols by MALDI-TOF/MS/MS in positive mode.
34. An aluminum foil is useful to cover samples.
35. Excess periodate can be destroyed by this process.
36. Sialic acid-containing branches can be eluted with 25% acetonitrile containing 0.05% trifluoroacetic acid. Neutral branches, which do not have acidic residues, can be eluted with 25% acetonitrile.
37. The branchial ions can be detected by MALDI-TOF/MS in positive mode.
38. Amino sugar analysis [11] and NMR spectroscopy [12] of purified alditols would help in the determination of glycan composition and glycosylation position, respectively [6, 7, 13].

Acknowledgements

The authors express our sincere appreciation to S. Sugawara for her technical assistance.

References

1. Holmén Larsson JM, Thomsson KA, Rodríguez-Piñeiro AM et al (2013) Studies of mucus in mouse stomach, small intestine, and colon. III. Gastrointestinal Muc5ac and Muc2 mucin O-glycan patterns reveal a regiospecific distribution. Am J Physiol Gastrointest Liver Physiol 305:G357–G363
2. Tsubokawa D, Ishiwata K, Goso Y et al (2015) Induction of Sd(a)-sialomucin and sulfated H-sulfomucin in mouse small intestinal mucosa by infection with parasitic helminth. Exp Parasitol 153:165–173
3. Iwai T, Ichikawa T, Goso Y et al (2009) Effects of indomethacin on the rat small intestinal mucosa: immunohistochemical and biochemical studies using anti-mucin monoclonal antibodies. J Gastroenterol 44:277–284
4. Saegusa Y, Ichikawa T, Iwai T et al (2008) Changes in the mucus barrier of the rat during 5-fluorouracil-induced gastrointestinal mucositis. Scand J Gastroenterol 43:59–65
5. Kawashima R, Fujimaki M, Ikenoue Y et al (2016) Influence of an elemental diet on 5-fluorouracil-induced morphological changes in the mouse salivary gland and colon. Support Care Cancer 24:1609–1616
6. Tsubokawa D, Goso Y, Sawaguchi A et al (2007) A monoclonal antibody, PGM34, against 6-sulfated blood-group H type 2 antigen, on the carbohydrate moiety of mucin. FEBS J 274:1833–1848
7. Tsubokawa D, Goso Y, Kawashima R et al (2012) The monoclonal antibody HCM31 specifically recognises the Sd (a) tetrasaccharide in goblet cell mucin. FEBS Open Bio 2:223–233
8. Powell AK, Harvey DJ (1996) Stabilization of sialic acids in N-linked oligosaccharides and gangliosides for analysis by positive ion matrix-assisted laser desorption/ionization mass spectrometry. Rapid Commun Mass Spectrom 10:1027–1032
9. Chai W, Stoll MS, Galustian C et al (2003) Neoglycolipid technology: deciphering information content of glycome. Methods Enzymol 362:160–195
10. Dubois M, Gilles KA, Hamilton JK et al (1956) Colorimetric method for determination of

sugars and related substances. Anal Chem 28: 350–356

11. Ishihara K, Kameyama J, Hotta K (1993) Development of an HPLC method to estimate hexosamines and its application to determine mucin content in rat and human gastric mucosa. Comp Biochem Physiol B 104:781–786
12. Kamerling JP, Vliegenthart JFG (1992) High-resolution ^{1}H-nuclear magnetic resonance spectroscopy of oligosaccharides released from mucin-type O-glycoproteins. Biol Magn Reson 10:1–287
13. Ishihara K, Kurihara M, Goso Y et al (1996) Peripheral alpha-linked N-acetylglucosamine on the carbohydrate moiety of mucin derived from mammalian gastric gland mucous cells: epitope recognized by a newly characterized monoclonal antibody. Biochem J 318:409–416

Chapter 19

Differential Glycoform Analysis of MUC1 Derived from Biological Specimens Using an Antibody-Overlay Lectin Microarray

Atsushi Matsuda, Patcharaporn Boottanun, Sachiko Koizumi, Misugi Nagai, and Atsushi Kuno

Abstract

The association between altered glycosylation of MUC1 and various disease events has sparked significant interest. However, analytical technologies to investigate the disease-related glycoforms of endogenous MUC1 in blood and tissue specimens are limited. Therefore, we devised a reliable technique for differential analysis of endogenous MUC1 glycoforms based on an antibody-assisted lectin microarray. Its highly sensitive detection aids in analyzing soluble MUC1 from relatively small amounts of serum via a simple enrichment process. Micro−/macro-dissection of the MUC1-positive region is combined with glycoform analysis of the membrane-tethered MUC1. Thus, we have optimized the protocol for sample qualification using immunohistochemistry, sample pretreatment for tissue sections, protein extraction, purification via immunoprecipitation, and the antibody-overlay lectin microarray, which are sequentially essential for differential glycoform analysis of endogenous MUC1.

Key words Lectin microarray, MUC1, Glycan profiling, Tissue section, Serum

1 Introduction

Glycosylation is a posttranslational modification that plays a vital role in various protein biological functions of proteins. Protein glycosylation, including *N*-linked and *O*-linked glycans, changes specifically according to the progression of certain diseases. Recent technological innovations, such as mass spectrometry and microarrays, have facilitated the analysis of glycosylation from small amounts of a specific protein bearing *N*- and *O*-linked glycosylation in clinical specimens, improving our understanding of the relationship between glyco-alterations and their biological significance. Notably, the lectin microarray (LMA) has contributed to the differential glycoform analysis of target protein biomarkers using a relatively large number of clinical specimens [1]. In the case of

Akihiko Kameyama (ed.), *Mucins: Methods and Protocols*, Methods in Molecular Biology, vol. 2763,
https://doi.org/10.1007/978-1-0716-3670-1_19,

membrane proteins as candidate molecular targets, in-depth mass spectrometry-based glycopeptide analysis serves as an immunogen map to create a unique antibody recognizing both the glycan and peptide components of a "discovered" disease-specific glycopeptide sequence.

Table 1[a] presents tissue analyses of rare technical difficult cases, including mucin 1 (MUC1).

MUC1 is a typical mucin that carries heavily *O*-linked glycosylated tandem repeats in the N unit with *N*-linked glycan at its C terminus. The aberrant expression and glycosylation of MUC1 are implicated in various aspects of cancer, including proliferation, invasion, and metastasis [33, 34]. They have gained attention as targets for the development of biomarkers and molecular-targeted agents. The aberrant MUC1 with altered glycosylation, such as CA15–3 and KL-6, is clinically used as a diagnostic marker for some cancers and diseases. The cancer-associated Tn-glycoform of the membrane-tethered MUC1 could serve as a potential neoantigen for engineered CAR T cell targeting [35]. However, obtaining evidence to rigorously define MUC1 glycoform diversity in clinical specimens is time-consuming owing to the aforementioned structural characteristics and the difficulty of purifying endogenous MUC1 [36, 37]. Accordingly, we designed and validated a novel technique for MUC1 glycan profiling from clinical specimens based on antibody-overlay LMA [4, 14]. Combined with laser capture microdissection (LCM), our novel MUC1 glycoform analysis technique (Fig. 1) aided in efficiently comparing MUC1 glycoforms in cancerous and surrounding noncancerous areas of the same tissue section. The protocol comprised five steps: sample qualification using immunohistochemistry (IHC), fragment collection for one of the serial tissue sections, protein extraction, purification using immunoprecipitation (IP), and antibody-overlay LMA. Every step should be handled through rational manipulations for reliable data; therefore, we have provided a step-by-step description of the protocol with abundant technical notes. Alternatively, the analytical strategy could be applied to serological MUC1 to detect disease-specific glycoforms as serological biomarkers, with some modification in the pretreatment step.

2 Materials

Prepare all solutions using Ultrapure water and analytical-grade reagents. Do not add sodium azide to the reagents.

2.1 IHC of MUC1

1. Anti-sialylated MUC1 monoclonal antibody (MY.1E12) (*see* **Note 1**).
2. Antigen retrieval buffer: 10 mM sodium citrate buffer (pH 6.0).

Table 1
Endogenous glycoprotein glycan profiling using antibody-overlay LMA analysis

Target glycoprotein	Sample type		Detection method	References
	Tissue			
Podoplanin	FFPE testis tissue from a patient with seminoma	3.6 mm^3	ALP	Kuno et al. [2]
PSA, MME	Frozen OCT-embedded prostate tissue from prostate cancer	200 μg total protein	ALP	Li et al. [3]
MUC1	FFPE tissue section from CCA and PDAC patients	0.0125 mm^3	ALP	Matsuda et al. [4]
Basigin	Frozen tissue from PDAC patients	0.1 mm^3	ALP	Wagatsuma et al. [5]
ERC/MSLN	FFPE tissue from patients with epithelioid mesothelioma	0.04 mm^3	ALP	Fujihira et al. [6]
	Serum and plasma			
Fetuin-A	Sera of patients with pancreatic cancer	100 μL	DL	Kuwamoto et al. [7]
AGP	Sera of patients with chronic hepatitis C	0.5 μL	ALP	Kuno et al. [8]
Mac-2 binding protein	Sera of patients with chronic hepatitis C and healthy volunteers	2.0 μL	ALP	Kuno et al. [9]
CPN2, CSF1R, SPARCL1, ICOSLG, PIGR	Sera of HCC patients and healthy volunteers	10 μL	ALP	Kaji et al. [10]
IgA1	Sera of patients with monoclonal immunoglobulin deposition disease associated with monoclonal IgA	[b]	ALP	Narimatsu et al. [11]
Anti-thyroglobulin IgG (TgAb)	Sera from patients with Hashimoto's thyroiditis	10 μg of purified TgAb	DL	Yuan et al. [12]
α-1-Antitrypsin (A1AT)	Sera from patients with lung cancer and benign pulmonary disease	[b]	DL	Liang et al. [13]
MUC1	Sera of CCA patients	5 μL	ALP	Matsuda et al. [14]
MMP-3	Sera of patients with rheumatoid arthritis (RA) and synovial fluid of patients with RA and osteoarthritis	40 μL	ALP	Takeshita et al. [15]
HBV particle	Sera of a patient with HBV infection	250 ng HBV particles	ALP	Wagatsuma et al. [16]
IgG	Plasma of small lung nodules from patients	5 μL	DL	Zou et al. [17]

(continued)

Table 1
(continued)

Target glycoprotein	Sample type		Detection method	References
IgG	Plasma of patients with HIV infection	[b]	DL	Giron et al. [18]
IgG4	Sera of patients with IgG4-related disease	[b]		Hu et al. [19]
IgG	Sera of patients with primary biliary cholangitis	200 μL	ALP	Zheng et al. [20]
NRG1	Sera of patients with papillary thyroid cancer	[b]	ALP	Liu et al. [21]
M2BP	Sera of patients with liver cirrhosis	1.0 μL	ALP	Noro et al. [22]
Clusterin	Sera from patients with prostate cancer and benign prostate disease	50 μg/mL total proteins	ALP	Kazuno et al. [23]
IgG, IgM	Sera from patients with HCC, hepatitis B cirrhosis, and chronic hepatitis B, and healthy volunteers	200 μL	ALP	Zhang et al. [24]
IgG	Sera from patients with peripheral artery disease	200 μL	ALP	Li et al. [25]
IgG	Sera from patients with abdominal aortic aneurysm	200 μL	ALP	Li et al. [26]
IgA1	Sera from patients with IgA nephropathy	5 μL	ALP	Uenoyama et al. [27]
IgG	Sera from patients with dermatomyositis	[b]	ALP	Li et al. [28]
IgG	Sera from patients with Takayasu arteritis	200 μL	ALP	Liu et al. [29]
	Other			
Transferrin (Tf)	Cerebrospinal fluid	10 ng of purified Tf	ALP	Futakawa et al. [30]
L1CAM	Bile of patients with CC and hepatolithiasis	100 μL	ALP	Matsuda et al. [31]
PSA, PSMA	Seminal PSA, recombinant human PSMA	2–200 ng/100 μL of enriched PSA	Tyramide-assisted ALP	Meany et al. [32]

[a] Reproduced from [1] with some modifications
[b] Optimized sample dilution or immunoprecipitated/purified samples

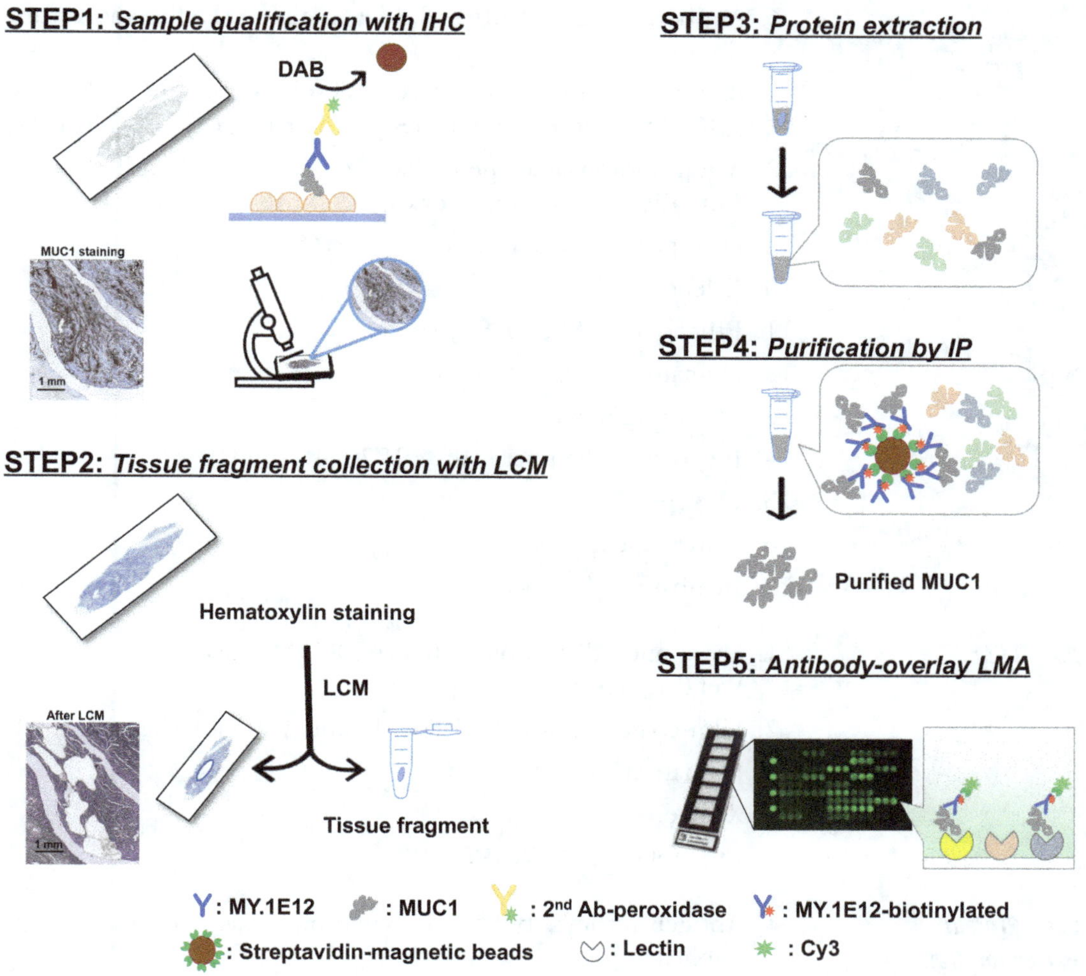

Fig. 1 Overview of LCM targeting of MUC1-positive cells in tissue sections and glycan profiling with MY.1E12-overlay LMA from serum and dissected tissue fragments. Step 1: Tissue section qualification using IHC; the tissue sections are stained using MY.1E12 antibody to observe MUC1-positive areas under a microscope. Step 2: Tissue fragment collection using LCM; the tissue sections were stained with hematoxylin, and MUC1-positive areas similar to those observed using IHC were collected via LCM. Step 3: The obtained tissue fragments were processed following the protein extraction step. Step 4: Crude proteins were immunoprecipitated using MY.1E12 antibody; subsequently, the purified MUC1 was subjected to Step 5 – antibody-overlay LMA. (Reproduced with some modifications from [4, 14] with permission. Copyright 2017 United States & Canadian Academy of Pathology and copyright 2015 American Chemical Society). Disclaimer: free PNG images of lab objects were obtained from www.labicons.com

3. Autoclave.
4. Endogenous peroxidase activity quenching buffer: methanol containing 0.3% hydrogen peroxide.
5. Blocking buffer: 10 mM sodium phosphate, 150 mM NaCl (PBS) containing 2% horse serum (pH 7.2).

6. Antibody diluent buffer: PBS containing 1% bovine serum albumin (pH 7.2).
7. Biotinylated horse anti-mouse IgG solution in VECTASTAIN ABC Kit (Vector Laboratories, Burlingame, CA) (*see* **Note 2**).
8. Avidin-biotinylated peroxidase complex in VECTASTAIN ABC Kit (Vector Laboratories) (*see* **Note 2**).
9. DAB peroxidase substrate.
10. Xylene.
11. Ethanol (70, 80, 90, 95, 100%).
12. Hematoxylin.
13. Mounting medium.
14. Incubation chamber (195 × 172 × 48 [mm]).
15. PBS (pH 7.2).
16. Water repellent pen.
17. Incubation chamber.

2.2 LCM

1. Laser microdissection system 6000DM (Leica Microsystems, Wetzlar, Germany).
2. Film-coated glass slide (PEN Membrane, Leica Microsystems).
3. Hematoxylin.
4. Xylene.
5. Ethanol (70, 80, 90, 95, 100%).

2.3 Protein Extraction from Dissected Tissue Fragments

1. Antigen retrieval buffer: the same buffer as that described in Subheading 2.1.
2. Extraction buffer: PBS containing 0.5% NP-40 (pH 7.4).
3. Sonicator.
4. Microtube (0.5 mL).
5. Refrigerated microcentrifuge.

2.4 IP of MUC1

1. Biotinylated MY.1E12 (250 μg/mL) (*see* **Note 3**).
2. Streptavidin-conjugated magnetic beads (Dynabeads™ MyOne™ Streptavidin C1, Thermo Fisher Scientific, Waltham, MA).
3. PBS containing 1% Triton-X100 in (PBSTx).
4. PBS containing 0.2% SDS.
5. Tube mixer with heating and cooling block.
6. Magnet stand.
7. Microtube (1.5 mL).

2.5 Antibody-Overlay LMA

1. Biotinylated MY.1E12 (25 μg/mL): the same antibody as that described in Subheading 2.4.
2. Lectin-immobilized microarray slide (LecChip®; Precision System Science Co., Matsudo, Japan).
3. LMA scanner (GlycoStation; Mx, Yokohama, Japan).
4. Masking solution: human serum polyclonal IgG (10 mg/mL).
5. Cy3-labeled streptavidin (GE Healthcare Amersham, Little Chalfont, UK) 50 μg/mL solution.
6. PBSTx.
7. Incubation chamber.

3 Methods

3.1 IHC Staining with Anti-MUC1 Antibody

1. Deparaffinization of formalin-fixed paraffin-embedded (FFPE) tissue sections with xylene and rehydration using an alcohol gradient (70–100%), followed by submersion in tap water.
2. For antigen retrieval, soak the tissue sections in the antigen retrieval buffer and autoclave for 10 min at 115 °C.
3. Wash tissue sections with PBS for 5 min at room temperature.
4. Incubate tissue sections with the endogenous peroxidase activity quenching buffer for 15 min.
5. Block nonspecific reactions by soaking in the blocking buffer for 30 min at room temperature with gentle shaking in an incubation chamber (*see* **Note 4**).
6. Remove the blocking reagent from tissue sections and apply the MY.1E12 solution at a final concentration of 1 μg/mL in the antibody diluent buffer onto the tissue section.
7. Incubate for 1 h at room temperature by shaking gently in an incubation chamber.
8. Remove the antibody solution, and wash the tissue section thrice with PBS for 5 min each.
9. Drop the biotinylated horse anti-mouse IgG solution onto the tissue sections.
10. Incubate for 30 min at room temperature by shaking gently in an incubation chamber.
11. Wash for 5 min in PBS.
12. Incubate for 30 min with the prepared avidin-biotinylated peroxidase complex in an incubation chamber.
13. Wash for 5 min in PBS.

14. Add the appropriate volume of DAB peroxidase substrate onto the inside area marked with a water-repellent pen and incubate until the desired stain intensity develops.
15. Rinse in tap water.
16. Counterstaining with hematoxylin.
17. Rinse with running water.
18. Fix the stained tissue section with the mounting medium.

3.2 Tissue LCM

1. A portion of the FFPE serial tissue sections (5 μm thickness) used for staining, as mentioned above (Subheading 3.1), are mounted on a film-coated glass slide.
2. Deparaffinization with xylene and rehydration using an alcohol gradient (70–100%), followed by submersion in tap water.
3. Staining with hematoxylin.
4. Wash the tissue sections with running water, air-dry (*see* **Note 5**), and proceed to the LCM step.
5. Set the tissue section and 0.5 mL microtube in the LCM (*see* **Note 6**).
6. Microdissect MY.1E12-positive cells via microscopic observation from an area of at least 2.5 mm^2.
7. Collect dissected tissue fragments into microtubes.
8. After confirming that tissue fragments have been collected (Fig. 2a), cover, spin down, and store them at room temperature until the protein extraction process.

3.3 Protein Extraction

1. Add 0.2 mL of the antigen retrieval buffer into the microtubes where tissue fragments were collected (Fig. 2b).
2. Heat denature at 95 °C for 60 min for antigen retrieval.
3. Centrifugation at 20,000 × *g* for 5 min at 4 °C.

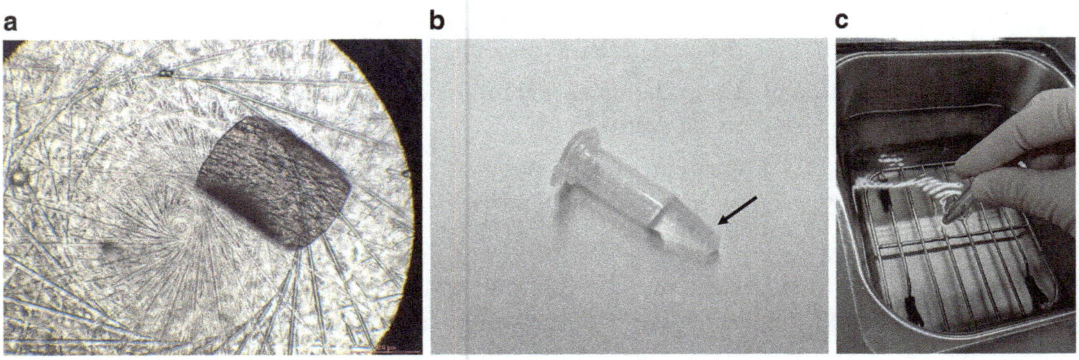

Fig. 2 Tissue fragment collection using LCM and protein extraction. (**a**) A tissue fragment on the lid of a 0.5 mL protein LoBind tube after cutting using LCM. (**b**) Tissue fragment in 0.2 mL of the antigen retrieval buffer. (**c**) Protein extraction is performed by sonicating the tissue fragments thrice for 3 s each

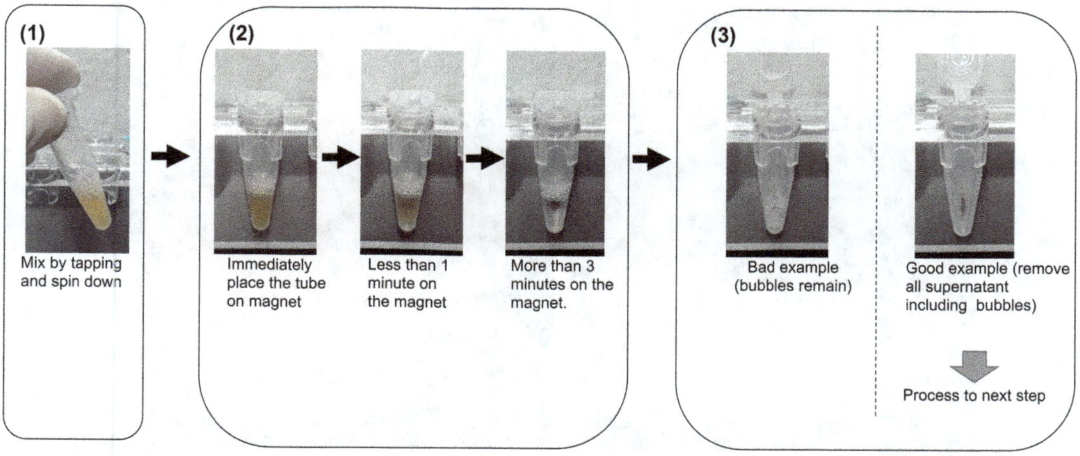

Fig. 3 Schematic overview of the IP protocol with crucial points that should be carefully performed. (1) After adding PBSTx, gently mix by tapping and briefly spinning down the sample during washing. (2) Immediately place the tube on a magnet stand for at least 3 min or until the solution is clear. (3) Remove all supernatant including the bubbles

4. Wash the tissue fragments with 0.2 mL of PBS after discarding the supernatant.
5. Centrifugation at 20,000 × *g* for 5 min at 4 °C.
6. Carefully discard the supernatants.
7. Add 20 μL of extraction buffer to the tissue fragments.
8. Gently sonicate the tissue fragments thrice for 3 s each (Fig. 2c).
9. Incubate the microtubes for 1 h on ice (*see* **Note 7**).
10. Centrifugation at 20,000 × *g* for 5 min at 4 °C.
11. Collect the supernatants as protein extraction samples.
12. Store the prepared samples below −20 °C until further use.

3.4 MUC1 IP

A schematic overview of the IP protocol is shown in Fig. 3.

1. To prepare the reaction beads, add biotinylated MY.1E12 (500 ng) and streptavidin magnetic bead solution (10 μL) to a 1.5 mL microtube (*see* **Note 8**).
2. Incubate for 1 h at 4 °C (*see* **Note 9**).
3. Collect reaction beads using a magnetic stand and discard the supernatants.
4. Wash the beads thrice with 200 μL of PBSTx, discard all supernatants, and process reaction beads for the next step.
5. Add 40 μL of sample solution to the beads (*see* **Note 10**).
6. Incubate at 4 °C for 1 h (*see* **Note 11**).

Fig. 4 After applying the sample, the LecChip was placed into the humid incubation chamber and incubated at 20 °C overnight with gentle shaking

7. Discard the supernatants and wash the beads thrice with 200 μL of PBSTx.
8. Add 10 μL of PBS containing 0.2% SDS to the magnetic beads and suspend.
9. Heat denature at 95 °C for 10 min.
10. Bring the tubes to room temperature and add 10 μL of streptavidin magnetic beads to the tubes (*see* **Note 12**).
11. Incubate for 30 min at 4 °C and collect the supernatants (*see* **Note 13**).

3.5 Antibody-Overlay LMA

All reaction steps in this section are performed in an incubation chamber, as shown in Fig. 4.

1. Bring the LecChip to room temperature and open the LecChip packaging.
2. Wash each well thrice with 60 μL of PBSTx.
3. Dilute each sample with PBSTx to the desired concentration and apply it to the LecChip (*see* **Note 14**).
4. Gently place the LecChip into the incubation chamber and incubate overnight at 20 °C with gentle shaking.
5. After incubation, add 2 μL of human serum IgG (10 mg/mL) as a masking solution for free lectins to each well.
6. Incubate at 20 °C for 30 min with gentle shaking.
7. Wash each well thrice with 60 μL of PBSTx.
8. Apply 60 μL of biotinylated MY.1E12 solution (100 ng) in PBSTx to each well and incubate at 20 °C for 1 h with gentle shaking.

9. Wash each well thrice with 60 μL of PBSTx.
10. Apply 60 μL of Cy3-labeled streptavidin solution in PBSTx to each well and incubate at 20 °C for 30 min with gentle shaking.
11. Wash each well thrice with 60 μL of PBSTx.
12. After the final wash, fill the wells with 60 μL of PBSTx.
13. Scan the LecChip using a microarray scanner.

4 Notes

1. Among the many anti-MUC1 antibodies available, this chapter describes protocols using the most proven widely used antibody, anti-sialylated MUC1 (MY.1E12) [38, 39], for MUC1 glycan profiling.
2. To enhance weak signals in IHC, we recommend a kit containing an avidin-biotin-peroxidase complex; subsequently, prepare the solution according to the manufacturer's protocol before use.
3. This biotinylated antibody is prepared using a Biotin-Labeling Kit-NH_2 (Dojindo Laboratories, Kumamoto, Japan).
4. In this step, surround the tissue with a water-repellent pen beforehand to minimize the amount of solution used.
5. After staining, air-dry the slide overnight or warm it at 37 °C to save time. The staining slide should be dried thoroughly; if the tissue is still moist, it cannot be cut off via LCM.
6. For small tissue fragments, 10–60 μL of MilliQ may be added to the tube prior to collection to prevent the tissues from dispersing due to static electricity. In this case, protein extraction should be performed on the same day.
7. Crude protein extracts can be stored at −30 °C until further use. For large specimens, simultaneous Cy3-labeling of multiple stored batches helps reduce inter-experimental variation and eases the experimental process.
8. The magnetic beads should be washed twice with 5 volumes of PBSTx in advance (10 μL magnetic beads with 50 μL of PBSTx).
9. Incubate at 1400 rpm in a tube mixer with heating and cooling block functions.
10. The sample solution for this reaction should be diluted twofold for the tissue fragment protein extract and fourfold for the serum (10 μL of serum). If necessary, the sample can be pre-treated (precleared) to absorb nonspecifically binding proteins, which can potentially react with streptavidin-magnetic beads; removing nonspecific components is recommended prior to the actual IP procedure.

11. Incubate at 1400 rpm in a tube mixer with heating and cooling block functions. To obtain a high sample yield, this step can be performed overnight.
12. Add the unconjugated streptavidin magnetic beads to the extracted solution to deplete the biotinylated antibody co-precipitated via heat denaturing.
13. Incubate at 1400 rpm in a tube mixer with heating and cooling block functions.
14. Following the recommended loading volume for LecChip, 5–20 μL of the immunoprecipitated sample (equivalent to 2.5–10 μL as serum volume) was filled up to 60 μL with PBSTx and applied to the chip well, respectively (Fig. 5). The significant lectin signals are almost saturated at 5 μL of normal

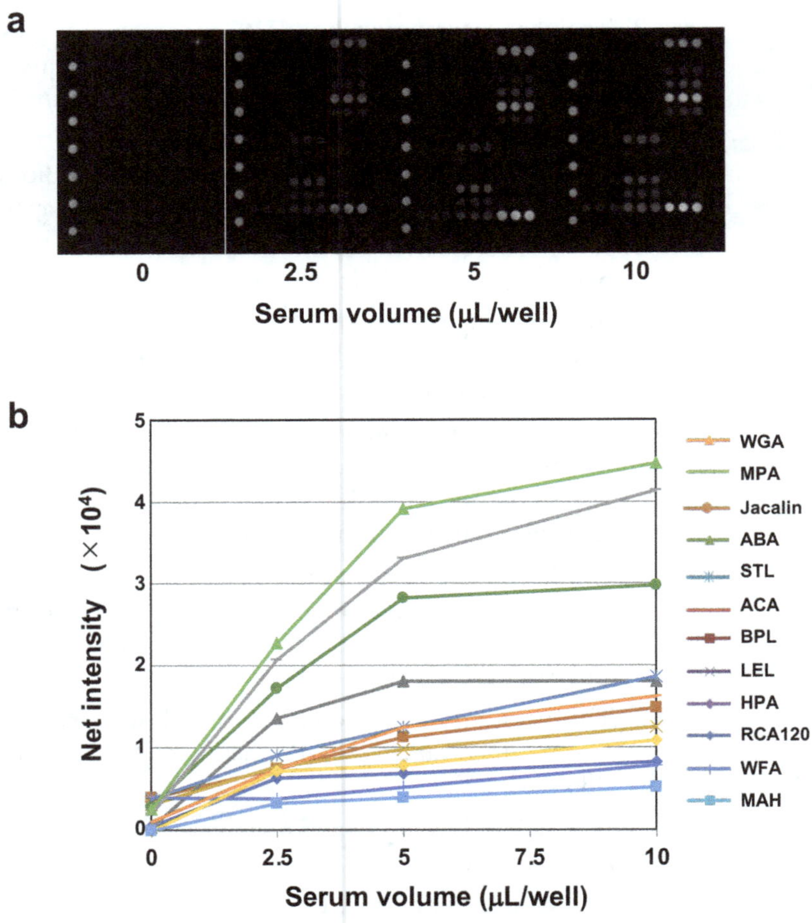

Fig. 5 Glycan profiling of diluted serum MUC1 with MY.1E12-overlay LMA. (**a**) Scanning images of glycan profiling with MY.1E12-overlay LMA of MUC1; 0 (negative control), 5, 10, and 20 μL of the elution solution of MUC1 purified from normal control serum (equivalent to 2.5, 5, 10 μL of serum). (**b**) Significant signal intensities of 13 lectins binding to serum MUC1. (Reproduced from [14] with permission. Copyright 2015 American Chemical Society)

control serum. Therefore, for reliable MUC1 detection in serum specimens, the recommended amount for loading onto the LMA slide is ≤5 μL in this section.

15. The protein extraction protocol from tissue fragments (Subheading 3.3) follows that of a previous report [40].
16. The antibody-overlay LMA protocol (Subheading 3.5) follows that of a previous report [2].

Acknowledgments

The authors would like to acknowledge Prof. J. Shoda (formally the University of Tsukuba) and Prof. J. Hirabayashi (formally AIST) for providing us with the opportunity for MUC1 glycan analysis. We would like to thank Editage (www.editage.com) for English language editing.

References

1. Kuno A, Matsuda A, Unno S et al (2014) Differential glycan analysis of an endogenous glycoprotein: toward clinical implementation – from sample pretreatment to data standardization. Methods Mol Biol 1200:265–285
2. Kuno A, Kato Y, Matsuda A et al (2009) Focused differential glycan analysis with the platform antibody-assisted lectin profiling for glycan-related biomarker verification. Mol Cell Proteomics 8:99–108
3. Li Y, Tao SC, Bova GS et al (2011) Detection and verification of glycosylation patterns of glycoproteins from clinical specimens using lectin microarrays and lectin-based immunosorbent assays. Anal Chem 83:8509–8516
4. Matsuda A, Higashi M, Nakagawa T et al (2017) Assessment of tumor characteristics based on glycoform analysis of membrane-tethered MUC1. Lab Invest 97:1103–1113
5. Wagatsuma T, Nagai-Okatani C, Matsuda A et al (2020) Discovery of pancreatic ductal adenocarcinoma-related aberrant glycosylations: a multilateral approach of lectin microarray-based tissue glycomic profiling with public transcriptomic datasets. Front Oncol 10:338
6. Fujihira H, Takakura D, Matsuda A et al (2021) Bisecting-GlcNAc on Asn388 is characteristic to ERC/mesothelin expressed on epithelioid mesothelioma cells. J Biochem 170: 317–326
7. Kuwamoto K, Takeda Y, Shirai A et al (2010) Identification of various types of α2-HS glycoprotein in sera of patients with pancreatic cancer: possible implication in resistance to protease treatment. Mol Med Rep 3:651–656
8. Kuno A, Ikehara Y, Tanaka Y et al (2011) Multilectin assay for detecting fibrosis-specific glyco-alteration by means of lectin microarray. Clin Chem 57:48–56
9. Kuno A, Ikehara Y, Tanaka Y et al (2013) A serum "sweet-doughnut" protein facilitates fibrosis evaluation and therapy assessment in patients with viral hepatitis. Sci Rep 3:1065
10. Kaji H, Ocho M, Togayachi A et al (2013) Glycoproteomic discovery of serological biomarker candidates for HCV/HBV infection-associated liver fibrosis and hepatocellular carcinoma. J Proteome Res 12:2630–2640
11. Narimatsu Y, Kuno A, Ito H et al (2014) IgA nephropathy caused by unusual polymerization of IgA1 with aberrant N-glycosylation in a patient with monoclonal immunoglobulin deposition disease. PLoS One 9:e91079
12. Yuan S, Li Q, Zhang Y et al (2015) Changes in anti-thyroglobulin IgG glycosylation patterns in Hashimoto's thyroiditis patients. J Clin Endocrinol Metabol 100:717–724
13. Liang Y, Ma T, Thakur A et al (2015) Differentially expressed glycosylated patterns of α-1-antitrypsin as serum biomarkers for the diagnosis of lung cancer. Glycobiology 25:331–340
14. Matsuda A, Kuno A, Nakagawa T et al (2015) Lectin microarray-based sero-biomarker verification targeting aberrant O-linked glycosylation on mucin 1. Anal Chem 87:7274–7281

15. Takeshita M, Kuno A, Suzuki K et al (2016) Alteration of matrix metalloproteinase-3 O-glycan structure as a biomarker for disease activity of rheumatoid arthritis. Arthritis Res Ther 18:112
16. Wagatsuma T, Kuno A, Angata K et al (2018) Highly sensitive glycan profiling of hepatitis B viral particles and a simple method for Dane particle enrichment. Anal Chem 90:10196–10203
17. Zou X, Yao F, Yang F et al (2019) Glycomic signatures of plasma IgG improve preoperative prediction of the invasiveness of small lung nodules. Molecules 25:28
18. Giron LB, Papasavvas E, Azzoni L et al (2020) Plasma and antibody glycomic biomarkers of time to HIV rebound and viral setpoint. AIDS 34:681–686
19. Hu C, Zhang P, Li L et al (2021) Assessing serum IgG4 glycosylation profiles of IgG4-related disease using lectin microarray. Clin Exp Rheumatol 39:393–402
20. Zeng X, Li S, Tang S et al (2021) Changes of serum IgG glycosylation patterns in primary biliary cholangitis patients. Front Immunol 12:669137
21. Liu WL, Cao YM, Liao T et al (2021) Multiple lectin assays in detecting glycol-alteration status of serum NRG1 in papillary thyroid cancer. Transl Cancer Res 10:3218–3224
22. Noro E, Matsuda A, Kyoutou T et al (2021) N-glycan structures of Wisteria floribunda agglutinin-positive Mac2 binding protein in the serum of patients with liver fibrosis†. Glycobiology 31:1268–1278
23. Kazuno S, Fujimura T, Fujime M, Miura Y, Ueno T (2021) O-glycosylated clusterin as a sensitive marker for diagnosing early stages of prostate cancer. Prostate 81:170–181
24. Zhang Y, Zhang S, Liu J et al (2022) Identification of serum glycobiomarkers for Hepatocellular Carcinoma using lectin microarrays. Front Immunol 13:973993
25. Li S, Meng J, Xu F et al (2022) IgG glycosylation profiling of peripheral artery diseases with lectin microarray. J Clin Med 11:5727
26. Li S, Meng J, Lv Y et al (2022) Changes in serum IgG glycosylation patterns for abdominal aortic aneurysm patients. J Cardiovasc Dev Dis 9:291
27. Uenoyama Y, Matsuda A, Ohashi K et al (2022) Development and evaluation of a robust sandwich immunoassay system detecting serum WFA-reactive IgA1 for diagnosis of IgA nephropathy. Int J Mol Sci 23:5165
28. Li X, Bai J, Li S et al (2022) Association of the serum IgG glycosylation with disease activity of anti-transcription intermediary factor 1 gamma positive dermatomyositis. Clin Exp Rheumato 41:230
29. Liu L, Li J, Yang Y, Hu C, Tian X (2023) Altered glycosylation profiles of serum IgG in Takayasu arteritis. Eur J Med Res 28:69
30. Futakawa S, Nara K, Miyajima M et al (2012) A unique N-glycan on human transferrin in CSF: a possible biomarker for iNPH. Neurobiol Aging 33:1807–1815
31. Matsuda A, Kuno A, Matsuzaki H et al (2013) Glycoproteomics-based cancer marker discovery adopting dual enrichment with Wisteria floribunda agglutinin for high specific glycodiagnosis of cholangiocarcinoma. J Proteome 85:1–11
32. Meany DL, Hackler L Jr, Zhang H, Chan DW (2011) Tyramide signal amplification for antibody-overlay lectin microarray: a strategy to improve the sensitivity of targeted glycan profiling. J Proteome Res 10:1425–1431
33. Hollingsworth MA, Swanson BJ (2004) Mucins in cancer: protection and control of the cell surface. Nat Rev Cancer 4:45–60
34. Kufe DW (2009) Mucins in cancer: function, prognosis and therapy. Nat Rev Cancer 9:874–885
35. Posey AD Jr, Schwab RD, Boesteanu AC et al (2016) Engineered CAR T cells targeting the cancer-associated TN-glycoform of the membrane mucin MUC1 control adenocarcinoma. Immunity 44:1444–1454
36. Storr SJ, Royle L, Chapman CJ et al (2008) The O-linked glycosylation of secretory/shed MUC1 from an advanced breast cancer patient's serum. Glycobiology 18:456–462
37. Andersch-Björkman Y, Thomsson KA, Holmén Larsson JM, Ekerhovd E, Hansson GC (2007) Large scale identification of proteins, mucins, and their O-glycosylation in the endocervical mucus during the menstrual cycle. Mol Cell Proteomics 6:708–716
38. Takeuchi H, Kato K, Denda-Nagai K et al (2002) The epitope recognized by the unique anti-MUC1 monoclonal antibody MY.1E12 involves sialyl alpha 2-3galactosyl beta 1-3N-acetylgalactosaminide linked to a distinct threonine residue in the MUC1 tandem repeat. J Immunol Methods 270:199–209
39. Kokubu R, Ohno S, Kuratani H et al (2022) O-Glycan-dependent interaction between MUC1 glycopeptide and MY.1E12 antibody by NMR, molecular dynamics and docking simulations. Int J Mol Sci 23:143
40. Matsuda A, Kuno A, Ishida H et al (2008) Development of an all-in-one technology for glycan profiling targeting formalin-embedded tissue sections. Biochem Biophys Re Commun 370:259–263

Chapter 20

ISOGlyP: *O*-Glycosylation Site Prediction Using Peptide Sequences and GALNTs

Luisa Gracia Mazuca and Jonathon E. Mohl

Abstract

Mucin-type *O*-glycosylation is one of the most common posttranslational modifications of proteins. The abnormal expression of various polypeptide GalNAc-transferases (GALNTs) which initiate and define sites of *O*-glycosylation is linked to many cancers and other diseases. Many current *O*-glycosylation prediction programs utilize *O*-glycoproteomics data obtained without using the transferase isoform(s) responsible for the glycosylation. With 20 different GALNTs in humans, having the ability to predict and interpret *O*-glycosylation sites in terms of specific GALNT isoforms is invaluable.

To fill this gap, ISOGlyP (isoform-specific *O*-glycosylation prediction) has been developed. Using position-specific enhancement values generated based on GalNAc-T isoform-specific amino acid preferences, ISOGlyP predicts the propensity that a site would be glycosylated by a specific transferase. ISOGlyP gave an overall prediction accuracy of 70% against in vivo data, which is comparable to that of the NetOGlyc4.0 predictor. Additionally, ISOGlyP can identify the known effects of long- and short-range prior glycosylation and can generate potential peptide sequences selectively glycosylated by specific isoforms. ISOGlyP is freely available for use at https://ISOGlyP.utep.edu. The code is also available on GitHub (https://github.com/jonmohl/ISOGlyP).

Key words GalNAc-T, GALNT, Mucin-type *O*-glycosylation, *O*-Glycosylation prediction, PTM prediction

1 Introduction

Mucin-type *O*-glycan biosynthesis is initiated by the transfer of *N*-acetylgalactosamine (GalNAc) from UDP-GalNAc to the hydroxyl groups of Ser or Thr residues in a polypeptide, catalyzed by a large family of polypeptide N-acetylgalactosaminyltransferases (GALNTs). In humans, 20 isoforms have been identified. Multiple members of the GALNT family have also been identified in *Drosophila*, *Caenorhabditis elegans*, and other single and multicellular organisms [1]. Several GALNTs show close sequence orthologues

Akihiko Kameyama (ed.), *Mucins: Methods and Protocols*, Methods in Molecular Biology, vol. 2763,
https://doi.org/10.1007/978-1-0716-3670-1_20,

across species suggesting that a number of them may be responsible for biologically significant functions, which have been conserved during evolution. Structurally, the GALNTs consist of an N-terminal catalytic domain tethered by a short linker to a C-terminal ricin-like lectin domain [2].

Initial studies on the GALNTs have revealed a wide range of peptide and glycopeptide substrate properties. For example, GALNT7 and T10 prefer substrates previously modified with *O*-linked GalNAc on nearby Ser/Thr residues, hence having so-called glycopeptide or filling-in activities [3–5]. The catalytic domain of GALNT10, interestingly, has been shown to be solely responsible for its near absolute glycopeptide specificity [6]. In contrast, isoforms such as GALNT2 and T4 possess altered preferences against glycopeptide substrates [7–10], while others, GALNT1 and T2, can be partially inhibited by neighboring glycosylation [5, 11, 12]. These latter transferases, preferring non-glycosylated over glycosylated substrates, have been called early or initiating transferases.

ISOGlyP [13] utilizes transferase-specific random peptide-derived data [14–20] allowing for the prediction of sites modified by a specific GALNT. This ability compliments over more general prediction software that were trained without regard to the GALNTs, such as NetOGlyc 4.0 [21]. Here we describe how to utilize the ISOGlyP site to identify both GALNT-specific sites and how to design peptides that are specifically modified by one or more GALNTs.

2 Methods

1. Go to the ISOGlyP (Fig. 1) site at https://isoglyp.utep.edu/.
2. Have the protein/peptide sequence of interest ready in FASTA format. For an example, download this sequence in FASTA format: https://rest.uniprot.org/uniprotkb/Q9GZV9.fasta. The sequence corresponds to the fibroblast growth factor 23 (FGF23, Uniprot accession: Q9GZV9).

2.1 Mucin-Type O-Glycosylation Prediction

The FASTA sequence(s) can be inputted by pasting the sequence into the text box area, uploading a FASTA file, or by providing the Uniprot reference numbers. If prior *O*-GalNAc glycosylation is known on either an S or T, replace the S and T by $ and +, respectively. After the input sequence(s) have been entered, the next steps are to select the desired prediction parameters, such as which GALNTs. Finally, after input sequences are submitted for processing, a status page may appear with the ISOGlyP displaying the results on a new webpage once the predictions are done.

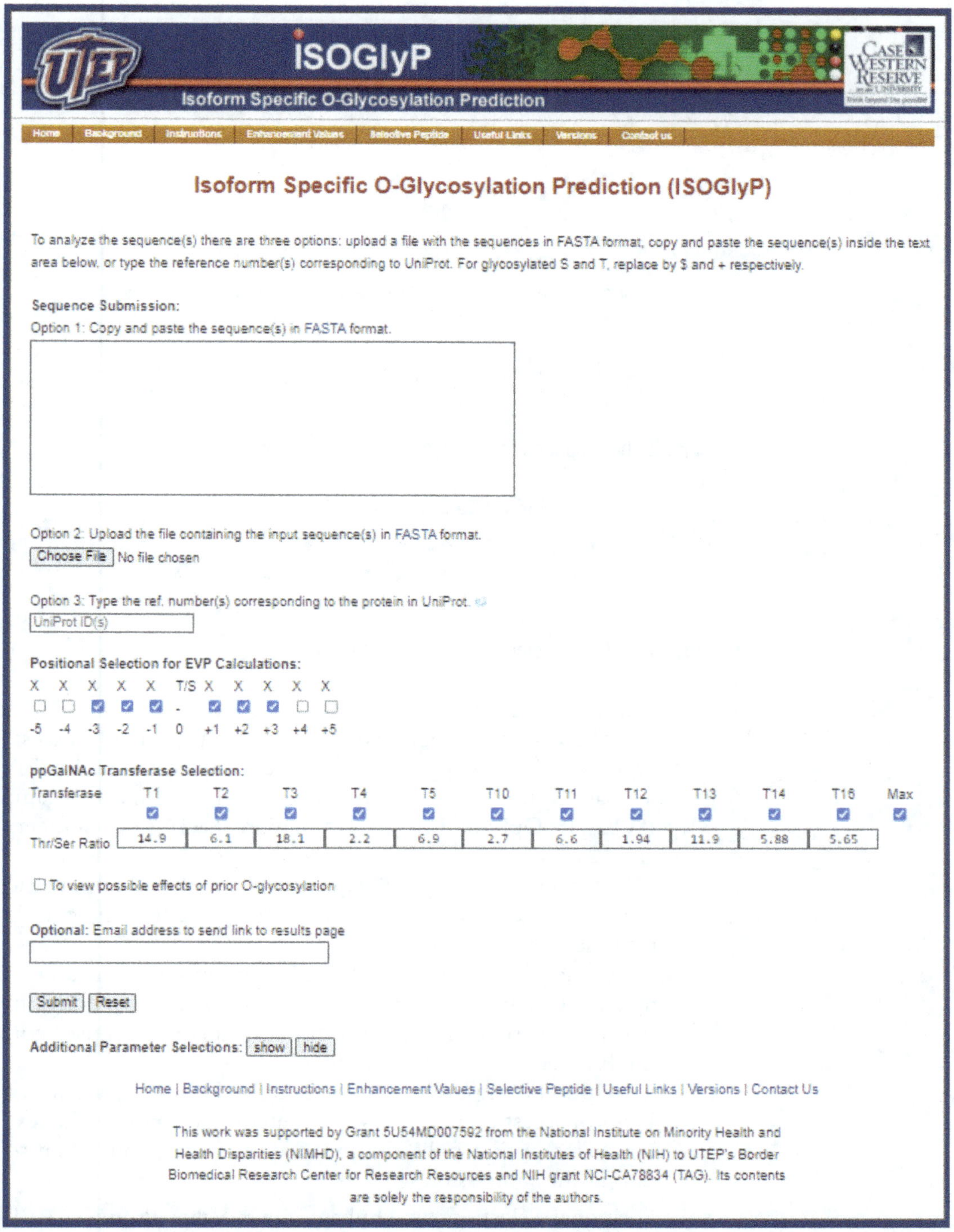

Fig. 1 ISOGlyP's homepage

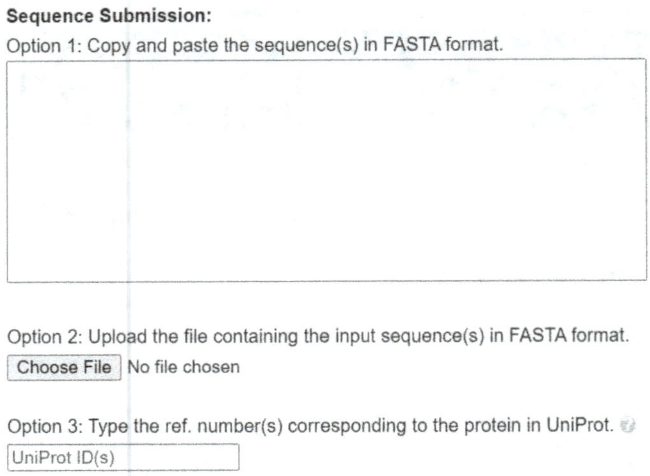

Fig. 2 Three options for imputing a sequence/sequences

Fig. 3 Positional selection for EVP calculations

1. Sequence Upload

   There are three options to submit the peptide of interest. Option 1 is to copy and paste the sequence(s) in FASTA format. Option 2 is to upload from your local machine into ISOGlyP. Option 3 is to list Uniprot accession numbers (Fig. 2).

   Option 1: Copy/paste sequence(s) – With this option, the user will either directly type or copy and paste the amino acids into the box for processing.

   Option 2: Uploading a file – This option allows the user to upload sequence(s) saved in a text document for prediction. Every single sequence saved in the text document must be in the FASTA format for processing.

   Option 3: Input Uniprot accession – Copy or type a list of Uniprot accession number(s), separated by spaces into the text box. This option pulls the protein sequences from Uniprot during the submission.

2. Positional Selection for (Enhancement Value Product) EVP Calculations

   Prediction constraints: The user can choose prediction constraints by checking the boxes below the XXXXX-T/S-XXXXX sequence. Although enhancement values are available for +/− 5 residues, we believe that utilizing a +/−3 X residue window will reduce "noise" in the calculation (Fig. 3).

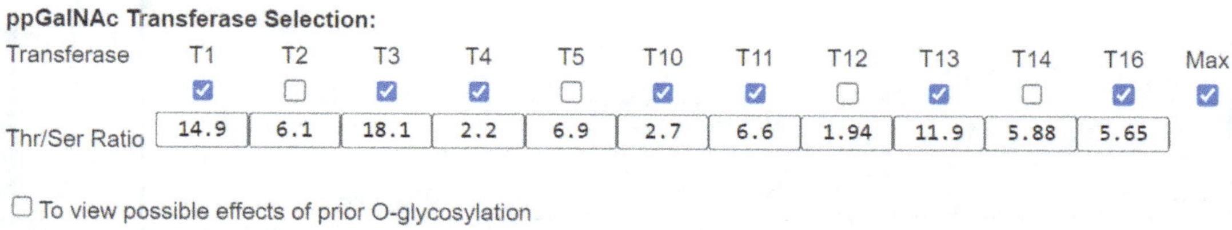

Fig. 4 GALNT selection

3. GALNT Selection

   Select isoform: The GALNT isoform enhancement values to use are chosen by checking the boxes below the GALN-Transferases heading. To date we have data for GALNT1, T2, T3, T5, T10, T11, T12, T13, T14, and T16. The MAX option returns the max EVP from the selected GALNT isoforms. As data for additional GALNT isoforms are obtained, their values will be added to the site (Fig. 4) (*see* **Note 1**).

   Thr/Ser ratio: By changing this weighting parameter, it allows for global changes of the EVP for each serine site. Values calculated from the enhancement value tables will be multiplied by the inserted weighting factor (see **Note 2**).

   Prior *O*-glycosylation: Check the box to view any possible effects of prior *O*-glycosylation in the results.

4. Additional Parameters

   Enter email address: This step is optional. The user may enter an email address to receive a link to the results page.

   Cys/Thr values: Cys and Thr value enhancement settings are made below the C/T values' pull downs. As there are no determined enhancement values for Cys or Thr, their default values are chosen to be 1.0. However, since these residues are homologous to Ser structurally, there is an option to set their values to those of Ser.

5. Press Submit

   The uploaded sequences will be submitted for processing by clicking on the submit button, which will transfer the user to a new window displaying either the prediction status page if the job is still being processed or the results page.

6. View Results

   The results page gives the residue number, flanking sequence, and the EVP values of user-selected GALNTs (Fig. 5). The higher the EVP value, the greater likelihood of that position being glycosylated. Note that the results can be exported in CSV file (*see* **Note 3**).

2.2 Selective Peptide Identification

1. To determine peptides that are predicted to be glycosylated by a specific GALNT(s) while excluding peptides that are also positive for unwanted GALNT(s), choose the "Selective Peptide" tab from the main navigation bar.

ISOGlyP Results

To view with possible prior O-glycosylation effects, click here.

Enhancement Value Version: 20200516

Amino Acid Positions Selected: - - X X X S/T X X X - - -

Thr/Ser Ratio: [14.9, 6.1, 18.1, 2.2, 6.9, 2.7, 6.6, 1.94, 11.9, 5.88, 5.65]

Download: Results or Fasta_Submission

Sequence	S/T	Position	Pattern	T1	T2	T3	T4	T5	T10	T11	T12	T13	T14	T16	Max
>sp|Q9GZV9|FGF23_HUMAN Fibroblast growth factor 23 OS=Homo sapiens OX=9606 GN=FGF23 PE=1 SV=1															
	S	15	VCALCSVCSMS	0.09	0.21	0.05	0.08	0.34	0.09	0.66	0.09	0.76	0.04	0.19	
	S	18	LCSVCSMSVLR	0.08	0.15	0.03	0.07	0.58	0.04	0.40	0.10	0.70	0.04	0.13	
	S	20	SVCSMSVLRAY	0.02	0.03	0.01	0.04	0.14	0.05	0.05	0.11	0.15	0.00	0.03	
	S	29	AYPNASPLLGS	0.04	0.24	0.02	0.03	0.10	0.06	0.69	0.04	0.34	0.02	0.11	
	S	34	SPLLGSSWGGL	0.04	0.14	0.01	0.03	0.40	0.12	0.60	0.05	0.77	0.04	0.19	
	S	35	PLLGSSWGGLI	0.18	0.77	0.06	0.03	1.14	0.15	1.12	0.05	1.17	0.12	0.44	1.17
	T	44	LIHLYTATARN	0.80	0.24	0.83	0.79	0.85	0.90	0.75	0.98	1.14	0.21	0.29	1.14
	T	46	HLYTATARNSY	1.20	0.42	1.65	0.33	0.57	1.42	0.71	0.77	1.01	0.43	0.39	1.65
	S	50	ATARNSYHLQI	0.01	0.01	0.00	0.01	0.01	0.03	0.02	0.01	0.06	0.00	0.01	
	T	68	GAPHQTIYSAL	0.46	0.26	0.30	0.52	0.20	0.22	0.29	0.27	0.49	0.08	0.14	
	S	71	HQTIYSALMIR	0.02	0.02	0.01	0.02	0.06	0.07	0.07	0.05	0.28	0.00	0.02	
	S	77	ALMIRSEDAGF	0.03	0.05	0.01	0.02	0.08	0.10	0.15	0.02	0.13	0.02	0.05	
	T	86	GFVVITGVMSR	0.68	0.25	0.66	0.91	0.23	0.35	0.14	0.31	1.27	0.23	0.32	1.27
	S	90	ITGVMSRRYLC	0.04	0.02	0.02	0.02	0.10	0.41	0.03	0.19	0.20	0.01	0.03	
	S	105	GNIFGSHYFDP	0.01	0.05	0.00	0.01	0.04	0.07	0.03	0.01	0.10	0.01	0.02	
	T	119	RFQHQTLENGY	0.25	0.05	0.12	0.26	0.04	0.20	0.06	0.15	0.27	0.02	0.05	
	S	129	YDVYHSPQYHF	0.21	0.28	0.12	0.04	0.43	0.16	0.30	0.21	1.00	0.05	0.19	1.00
	S	137	YHFLVSLGRAK	0.05	0.05	0.11	0.04	0.28	0.06	0.20	0.08	0.27	0.00	0.03	
	S	155	NPPPYSQFLSR	0.03	0.06	0.00	0.02	0.04	0.08	0.10	0.03	0.17	0.00	0.02	
	S	159	YSQFLSRRNEI	0.01	0.01	0.02	0.02	0.07	0.20	0.02	0.11	0.06	0.00	0.01	
	T	171	LIHFNTPIPRR	2.41	1.10	2.69	3.07	4.93	0.33	2.96	3.68	1.40	0.29	0.65	4.93
	T	178	IPRRHTRSAED	0.29	0.16	1.77	0.22	0.38	0.74	0.32	1.22	0.41	0.52	0.31	1.77
	S	180	RRHTRSAEDDS	0.03	0.06	0.01	0.01	0.07	0.11	0.22	0.07	0.16	0.06	0.09	
	S	185	SAEDDSERDPL	0.05	0.05	0.00	0.01	0.02	0.18	0.20	0.09	0.15	0.15	0.07	
	T	200	PRARMTPAPAS	5.04	2.79	6.90	12.07	11.69	3.65	10.64	19.87	2.30	1.40	3.67	19.87
	S	205	TPAPASCSQEL	0.08	0.18	0.12	0.04	0.34	0.09	0.71	0.08	0.86	0.07	0.12	
	S	207	APASCSQELPS	0.05	0.13	0.01	0.01	0.07	0.07	0.20	0.04	0.29	0.03	0.10	
	S	212	SQELPSAEDNS	0.10	0.38	0.01	0.02	0.10	0.13	0.97	0.05	0.38	0.29	0.78	
	S	217	SAEDNSPMASD	0.06	0.07	0.01	0.10	0.20	0.08	0.54	0.22	0.51	0.11	0.17	
	S	221	NSPMASDPLGV	0.07	0.29	0.01	0.02	0.13	0.15	0.61	0.02	0.27	0.03	0.09	
	T	234	GGRVNTHAGGT	0.39	0.12	0.36	0.24	0.32	0.32	0.26	0.39	0.78	0.47	0.22	
	T	239	THAGGTGPEGC	0.41	0.93	0.14	0.26	0.36	3.88	1.13	0.15	0.93	5.16	3.56	5.16

Fig. 5 ISOGlyP results

2. General Parameters

   These parameters are useful to customize the search (Fig. 6).

   Glycosylation site amino acid: Select between threonine, serine, or both Ser/Thr.

   Transferases: Select the transferase isoform(s) desired to glycosylate the sequence(s) (i.e., Selected-For) and the transferase isoforms desired to not glycosylate the sequence (i.e., Selected-Against). Also select the EVP cutoff values for the Selected-For (default value of >1.2 EVP) and Selected-Against (default value <0.8 EVP) transferases.

   Positional selection for EVP calculations: Select the range of EV positions (up to +/− 5 residues of the acceptor) utilized in the analysis.

   Email: This step is optional. The user may enter an email address to receive a link to the results page.

3. Sequence Submission/Random Peptide Creation

   Sequence submission: This option serves to search a given peptide/protein sequence for isoform-specific sites of likely glycosylation. Here the user inputs (or uploads) the protein/peptide sequence of interest in FASTA format (into the left box) and submits by pressing the submit button at the bottom (Fig. 7).

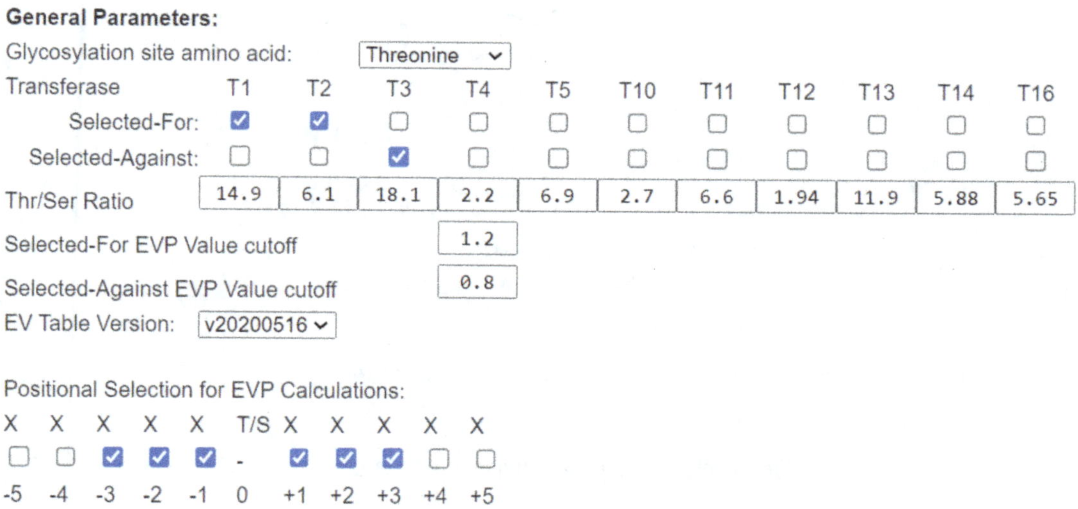

Fig. 6 Selective peptide general parameters

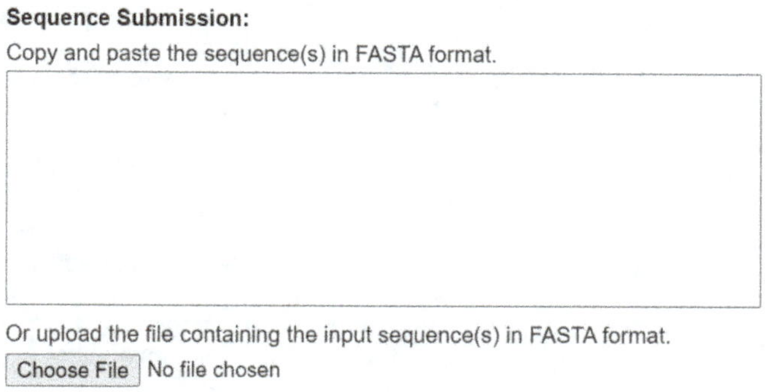

Fig. 7 Sequence submission for the selective peptide tool

Random peptide creation: This option produces a series of random peptide sequences that will likely be selectively glycosylated by the given transferase. Options can be filled in to guide the creation of peptides (Fig. 8); *see* Fig. 9 as an example.

4. Results page: The results are returned as a list of Thr and Ser residues which meet the preset criteria of the user (Fig. 10).

3 Notes

1. Enhancement value tables: ISOGlyP uses experimentally determined transferase-specific enhancement values (EVs) for each GalNAc-T isoform to determine the likelihood of glycosylation of a Ser or Thr residue in a given protein or peptide sequence. Currently, EVs, obtained from random peptide-derived amino acid preferences, for 11 of the 20 GalNAc-T isoforms are available [14–19]. The EVs represent the propensity for a

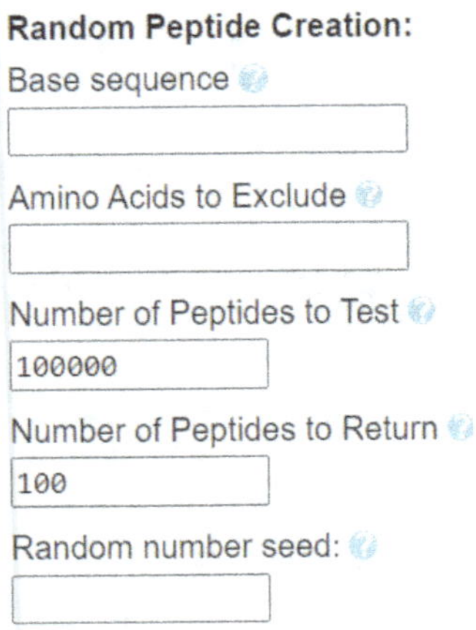

Fig. 8 Selective peptide – random peptide creation

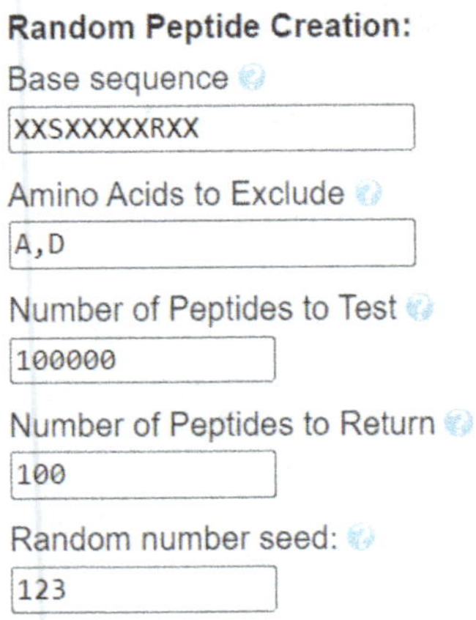

Fig. 9 Selective peptide – random peptide creation with sample input

given amino acid residue flanking a given glycosite, to be found after glycosylation by a given GalNAc-T isoform.

2. Thr/Ser ratio: Ser acceptors are glycosylated at lower rates than Thr acceptors by the GALNTs [20]; therefore, to account for the lower glycosylation rates of Ser acceptors, their EVP values are further divided by the experimentally determined Thr/Ser rate ratios obtained for each transferase isoform [20]. The Thr/Ser rate ratios for the GALNT isoforms (T1, T2, T3, T4, T5, T10, T11, T12, T13, and T16) were obtained by comparing the rates of glycosylation of Ser and Thr sites with random and fixed sequence peptide substrates [20]. These values appear on the ISOGlyP home page. These Thr/Ser rate values range from ~2 to ~18 depending on the GALNTGalNAc-transferase (GALNT) subfamily [20]. Note that for GALNT14 its Thr/Ser ratio was taken as the average

ISOGlyP Selective Peptide Results

Transferase	T1	T2	T3	T4	T5	T10	T11	T12	T13	T14	T16
Selected-For	1	1	0	0	0	0	0	0	0	0	0
Selected-Against	0	0	1	0	0	0	0	0	0	0	0
Thr/Ser Ratio	14.9	6.1	18.1	2.2	6.9	2.7	6.6	1.94	11.9	5.88	5.65

Positions:	-5	-4	-3	-2	-1	X	+1	+2	+3	+4	+5
	-	-	+	+	+	T/S	+	+	+	-	-

100000 randomly generated peptides using the sequence ---XXSXXXXX as a framwork.

Seed used in random number generator: 123

Enhancement Value Version: 20200516

Download: Results or Fasta_Of_Results

Peptide Number	Sequence	Positive EVP	Negative EVP
1	-EPSPGP	5.86	0.41
2	-VPSPGP	5.71	0.76
3	-GPSPLP	5.17	0.19
4	-RPSPPP	4.51	0.73
5	-GPSVVP	4.40	0.28
6	-FPSPMP	4.31	0.47
7	-QPSVGP	4.16	0.55
8	-LPSPPP	4.07	0.42
9	-GPSPPY	3.89	0.13
10	-GPSEQP	3.59	0.10
11	-GPSMEP	3.56	0.11
12	-HPSPMP	3.54	0.47
13	-YPSEGP	3.49	0.34
14	-YPSMGP	3.49	0.56
15	-LPSPEP	3.43	0.24
16	-EPSVGP	3.34	0.26
17	-PPSPFP	3.32	0.21
18	-NPSPQP	3.28	0.54
19	-YPSPGY	3.22	0.38
20	-KPSPMP	3.22	0.28

Fig. 10 Selective peptide – random peptide creation results

of that of GALNT2 and T16 as these transferases are in the same GALNT subfamily [20].

3. Interpreting EVP values: EVP values reflect relative rates of glycosylation. The higher the value, the faster and more likely a site would be glycosylated by a particular isoform. An EVP value of 1 would indicate the transferase perceives the sequence as relatively neutral, i.e., not inhibited nor enhanced, but nevertheless likely to be glycosylated. An EVP value greater than 1 would suggest a higher rate or likelihood of glycosylation; therefore, a value of 2 would suggest a twofold rate of glycosylation. Very large EVPs would suggest exceptional sites. EVP values less than 1 would suggest the transferase does not prefer that site – but still could conceivably glycosylate the site if given enough time or transferase. Nevertheless, an EVP value of 0.2 would suggest a very poor site not likely to be glycosylated. Keep in mind that we are simply multiplying the EV values to obtain the product – at the present time we do not know if some positions might be more important (weighted) than others. Further studies will be required to address this issue. Also note that the EVP values do not take into account end

effects; therefore, based on our experience, predictions within 3–5 residues of the N- or C- terminal of a peptide may be too high. Finally, at the present time, the EVP values calculated by ISOGlyP for a Ser or Thr residue in the same flanking peptide sequence are identical and do not reflect the intrinsic lower rates of Ser glycosylation compared to Thr glycosylation. At the present time few systematic studies have been performed on T1 and T2 and none with the other isoforms, quantifying this difference. Therefore, we recommend that the EVP values for Ser residues be roughly decreased by a factor of approximately 10 when comparing to the EVP values for Thr residues.

Acknowledgement

This work was supported by grant 5U54MD007592 from the National Institute on Minority Health and Health Disparities (NIMHD), a component of the National Institutes of Health (NIH) to UTEP's Border Biomedical Research Center for Research Resources and NIH grant NCI-CA78834 (TAG) for the data used for the enhancement tables and initial site. The contents are solely the responsibility of the authors.

References

1. Bennett EP, Mandel U, Clausen H et al (2012) Control of mucin-type O-glycosylation: a classification of the polypeptide GalNAc-transferase gene family. Glycobiology 22:736–756
2. Fritz TA, Raman J, Tabak LA (2006) Dynamic association between the catalytic and lectin domains of human UDP-GalNAc: polypeptide alpha -N-acetylgalactosaminyltransferase-2. J Biol Chem 281:8613–8619
3. Bennett EP, Hassan H, Hollingsworth MA et al (1999) A novel human UDP-N-acetyl-D-galactosamine:polypeptide N- acetylgalactosaminyltransferase, GalNAc-T7, with specifici*ty for partial* G*al*NAc-glycosylated acceptor substrates. FEBS Lett 460:226–230
4. Cheng L, Tachibana K, Zhang Y et al (2002) Characterization of a novel human U*DP-GalNAc tra*nsferase, pp-GalNAc-T10. FEBS Lett 531:115–121
5. Pratt MR, Hang HC, Ten Hagen KG et al (2004) Deconvoluting the functions of polypeptide *N*-a-Acetylgalactosaminyltransferase family members by Glycopeptide substrate profiling. Chem Biol 11:1009–1016
6. Perrine CL, Ganguli A, Wu P et al (2009) The glycopeptide preferring polypeptide- GalNAc transferase-10 (GALNT10), involved in mucin-type O-glycosylation, has a unique GalNAc-O-Ser/Thr binding site in its catalytic domain not found in GALNT1 or T2. J Biol Chem 284:20387–20397
7. Hassan H, Reis CA, Bennett EP et al (2000) The lectin domain of UDP-N-acetyl-D-galactosamine: polypeptide N-acetylgalactosaminyltransferase-T4 directs its glycopeptide specificities. J Biol Chem 275:38197–38205
8. Hanisch FG, Reis CA, Clausen H et al (2001) Evidence for glycosylation-dependent activities of polypeptide N-acetylgalactosaminyltransferases rGalNAc-T2 and -T4 on mucin glycopeptides. Glycobiology 11:731–740
9. Wandall HH, Irazoqui F, Tarp MA et al (2007) The lectin domains of polypeptide GalNAc-transferases exhibit carbohydrate-binding specificity for GalNAc: lectin binding to GalNAc-glycopeptide substrates is required for high density GalNAc-O-glycosylation. Glycobiology 17:374–387
10. Raman J, Fritz TA, Gerken TA et al (2008) The catalytic and lectin domains of UDP- GalNAc :

Polypeptide alpha-N-Acetylgalactosaminyltransferase function in concert to direct glycosylation site selection. J Biol Chem 283: 22942–22951

11. Gerken TA, Gilmore M, Zhang J (2002) Determination of the site-specific oligosaccharide distribution of the O-glycans attached to the porcine submaxillary mucin tandem repeat: further evidence for the modulation of O-glycan side chain structures by peptide sequence. J Biol Chem 277:7736–7751
12. Gerken TA, Tep C, Rarick J (2004) Role of peptide sequence and neighboring residue glycosylation on the substrate specificity of the uridine 5′-diphosphate-a-*N*-acetylgalactosamine: polypeptide *N*-acetylgalactosaminyl transferases T1 and T2: kinetic modeling of the porcine and canine submaxillary gland mucin tandem repeats. Biochemistry 43: 9888–9900
13. Mohl JE, Gerken TA, Leung MY (2021) ISOGlyP: de novo prediction of isoform-specific mucin-type O-glycosylation. Glycobiology 31: 168–172
14. Gerken TA, Raman J, Fritz TA et al (2006) Identification of common and unique peptide substrate preferences for the UDP-GalNAc: polypeptide alpha -N-acetylgalactosaminyltransferases T1 & T2 (GALNT1 & T2) derived from oriented random peptide substrates. J Biol Chem 281:32403–32416
15. Gerken TA, Ten Hagen KG, Jamison O (2008) Conservation of peptide acceptor preferences between drosophila and mammalian polypeptide-GalNAc transferase orthologue pairs. Glycobiology 18:861–870
16. Gerken TA, Jamison O, Perrine CL et al (2011) Emerging paradigms for the initiation of mucin-type protein O-glycosylation by the polypeptide GalNAc transferase (GALNT) family of glycosyltransferases. J Biol Chem 286:14493–14507
17. Perrine C, Ju T, Cummings RD et al (2009) Systematic determination of the peptide acceptor preferences for the human udp-gal:glycoprotein-alpha-galnac beta 3 galactosyltransferase (t-synthase). Glycobiology 19:321–328
18. de las Rivas M, Paul Daniel EJ, Coelho H et al (2018) Structural and mechanistic insights into the catalytic-domain-mediated short-range glycosylation preferences of galnac-t4. ACS Central Sci 4:1274–1290
19. Festari MF, Trajtenberg F, Berois N et al (2017) Revisiting the human polypeptide galnac-t1 and t13 paralogs. Glycobiology 27: 140–153
20. Paul Daniel EJ, de las Rivas M, Lira-Navarrete E et al (2020) Ser and Thr acceptor preferences of the GalNAc-Ts vary among isoenzymes to modulate mucin-type O-glycosylation. Glycobiology 30:910–922
21. Steentoft C, Vakhrushev SY, Joshi HJ et al (2013) Precision mapping of the human o-galnac glycoproteome through simplecell technology. EMBO J 32:1478–1488

Part IV

Molecular Biology

Chapter 21

Assessment of Mucin-Associated Gene Expression Levels on the Ocular Surface

Jun Shoji and Satoru Yamagami

Abstract

The ocular surface is covered with a mucus layer. The mucin-associated genes expressed in the ocular surface cells include *MUC1*, *MUC4*, *MUC5AC*, and *MUC16*. Impression cytology is useful for collecting specimens from the ocular surface, their histological examination, and measuring mucin-associated gene expression levels. The expression of mucin-associated gene levels was assessed by quantitative polymerase chain reaction. The expression levels of these mucin-associated genes are potential biomarkers for ocular surface diseases, including dry eye disease.

Key words Impression cytology, MUC gene, Ocular surface, Real-time polymerase chain reaction

1 Introduction

The ocular surface, formed by the conjunctiva, cornea, and tear film, is a tissue in direct contact with the external environment and therefore has biological defense functions [1]. The tear film comprises two layers: a lipid layer secreted by the meibomian gland and a mucoaqueous layer secreted by the lacrimal gland (Fig. 1). Secretory mucins, including MUC5AC, are secreted by goblet cells in the conjunctival epithelium and the lowest layer of the mucoaqueous layer, providing lubrication between the tear fluid and the corneal and conjunctival epithelium [2, 3]. In addition, mucous trapping is a function of secretory mucin, by which foreign substances and microorganisms entering from the external environment are entrapped and eliminated as ocular discharge; this is one of the defense mechanisms of the ocular surface [4]. The epithelial layer consists of the corneal and conjunctival epithelia and membrane-associated mucins. Membrane-associated mucins, including MUC1, MUC4, and MUC16, present in contact with the microvilli of corneal and conjunctival epithelium, are the distinguishing components of the mucosal glycocalyx.

Akihiko Kameyama (ed.), *Mucins: Methods and Protocols*, Methods in Molecular Biology, vol. 2763,
https://doi.org/10.1007/978-1-0716-3670-1_21,

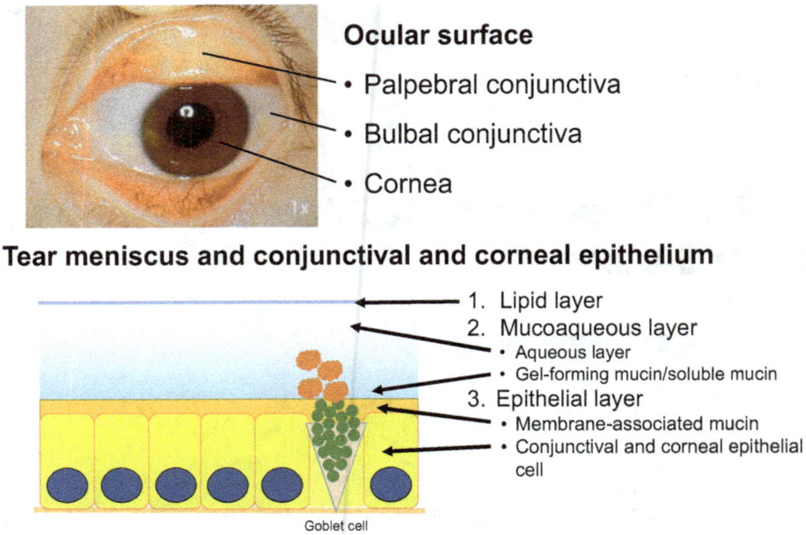

Fig. 1 Ocular surface structure Conjunctival and corneal surfaces are covered with a tear film consisting of lipid and mucoaqueous layers. The lipid layer is the superficial layer of the tear film on the ocular surface. The mucin layer, including gel-forming and soluble mucins, has a decreasing concentration gradient from the epithelium toward the aqueous layer

Membrane-associated mucins maintain wetness, stabilize the tear film, and create a physical barrier on the ocular surface [5].

Ocular surface diseases, including Sjögren syndrome, Stevens–Johnson syndrome, and ocular cicatricial pemphigoid, are characterized by abnormalities such as a decrease in goblet cells and squamous metaplasia in the conjunctival epithelium [6]. Impression cytology is a valuable tool for diagnosing ocular surface disease. It is a cellular biopsy method that allows observation of pathological changes in the conjunctival and corneal epithelium. Furthermore, quantitative reverse-transcription polymerase chain reaction (qRT-PCR) can measure mRNA expression in conjunctival and corneal epithelium and inflammatory cells, including impression cytology specimens [7].

Abnormalities in mucin-related gene expression in the conjunctival and corneal epithelium have been reported in ocular surface diseases. In Sjögren's syndrome, a dry eye disease, MUC5AC expression is reduced due to histopathological abnormalities on the ocular surface with decreased goblet cells in the conjunctival epithelium [8, 9]. In adult patients with chronic atopic keratoconjunctivitis, another dry eye disease, expression levels of MUC5AC and MUC16 are decreased on the ocular surface [10, 11]. Therefore, measurement of ocular surface mucin-related gene expression is a potential diagnostic test for dry eye-related diseases. Here, we present a method for measuring mucin-related gene expression in specimens collected by the impression cytology method.

2 Materials

All instruments used in the experiments should be sterile and RNase-free.

2.1 Impression Cytology

1. Schirmer's test paper or nitrocellulose membrane (*see* **Notes 1** and **2**).
2. RNA stabilization reagent (*see* **Note 3**).
3. Ophthalmic anesthetics (*see* **Note 4**).
4. Forceps (*see* **Note 5**).
5. Glass rod.
6. Microtube (1.5 mL) for preserving Schirmer's test papers (Fig. 2).

2.2 Messenger RNA Extraction

1. Micropipettes and micropipette tips (*see* **Note 6**).
2. Microtubes (1.5 mL).
3. Lysis buffer: Buffer RLT (Qiagen; *see* **Note 7**).
4. β-Mercaptoethanol.
5. Vortex mixer.

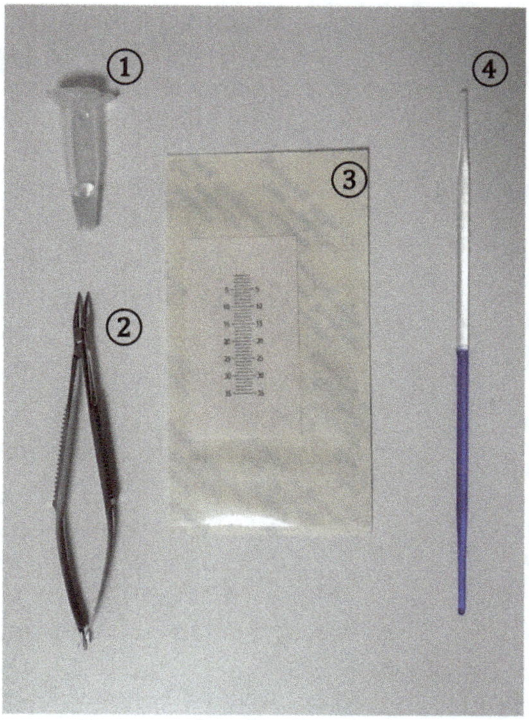

Fig. 2 Instruments used for impression cytology ① Microtube (1.5 mL) containing RNA stabilization reagent for preserving Schirmer's test papers, ② Beaupre forceps, ③ Schirmer's test paper, ④ glass rod

6. Microcentrifuge (with rotor for 2 mL tubes).
7. Messenger RNA extraction kit: RNeasy® Mini Kit (Qiagen).
8. cDNA reverse transcription kit: High-Capacity cDNA Reverse Transcription Kit with RNase inhibitor (Applied Biosystems).
9. Thermal cycler (*see* **Note 8**).

2.3 Real-Time RT-PCR

1. Micropipettes and micropipette tips (*see* **Note 9**).
2. Microtubes (1.5 mL) for PCR reagent preparation.
3. TaqMan® Gene Expression Assay (Applied Biosystems): MUCl, Hs00159367_ml; MUC4, Hs00366414_ml; MUC5AC, Hs00873651_ml; MUC16, Hs01065189_ml; GAPDH, Hs99999905_ml (*see* **Note 10**).
4. TaqMan® Gene Expression Master Mix (Applied Biosystems).
5. qPCR plate and real-time PCR plate seal (*see* **Note 11**).
6. Real-time PCR system.

3 Methods

3.1 Impression Cytology

The 5 mm tip of Schirmer's test paper should be used for impression cytology. After local anesthesia with ophthalmic eye drops, apply Schirmer's test paper to the upper tarsal or bulbar conjunctiva using forceps (Fig. 3a, c, d), and press it gently using a glass rod or forceps; then, remove and preserve it in RNA stabilization reagent at 4 °C until mRNA extraction (*see* **Note 12**).

3.2 Messenger RNA Extraction

1. To isolate cellular components from Schirmer's test paper, prepare a microtube containing 350 μL of a 200:2 mixture of RLT buffer and β-mercaptoethanol.
2. Place the Schirmer's test paper in the microtube and vortex for 15 s.
3. After spinning down using a centrifuge, remove the Schirmer's test paper.
4. Extract total RNA from each specimen using the RNeasy® Mini Kit following the manufacturer's instructions (*see* **Note 13**).
5. Next, reverse transcribe the RNA to complementary DNA (cDNA) using a High-Capacity cDNA Reverse Transcription Kit according to the manufacturer's protocol. Reverse transcription can be carried out using a thermal cycler.

3.3 Real-Time RT-PCR

1. As in Table 1, dispense 2× TaqMan Gene Expression Master Mix, 20× TaqMan Gene Expression Assay probe (predesigned primer/probe), sample cDNA, and RNase-free H_2O into microtubes according to the manufacturer's protocol.

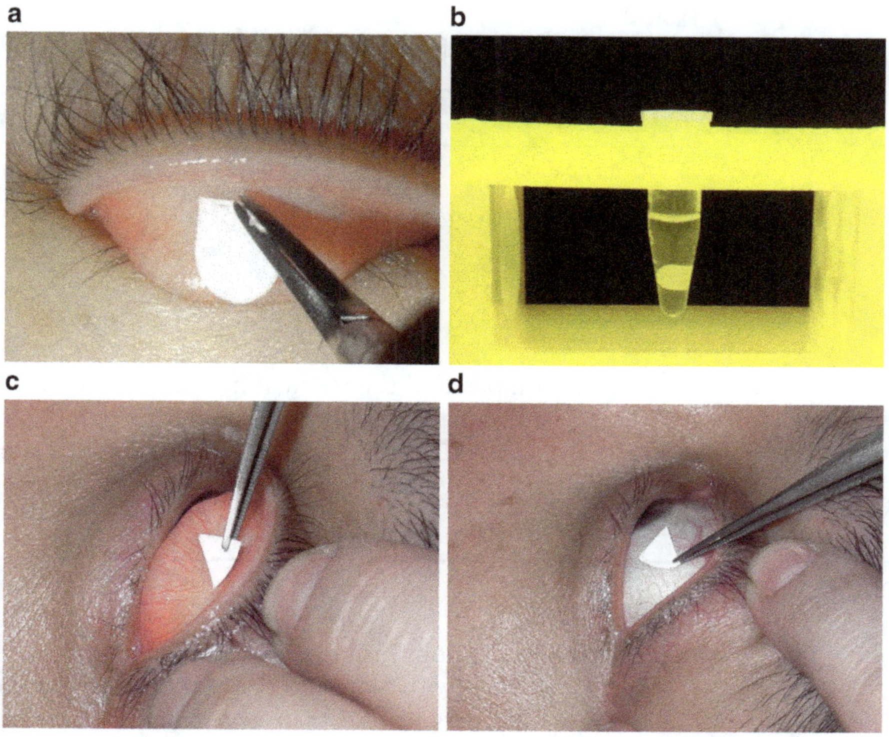

Fig. 3 Techniques for performing impression cytology Using forceps, the Schirmer's test paper is pressed against the conjunctiva and peeled off. The collection is performed in the sitting position with a slit-lamp microscope or a supine position with an operating microscope. (**a**) In the sitting position, the specimen is collected from the palpebral conjunctiva. (**b**) Microtubes containing RNA stabilization reagent to preserve Schirmer's test paper. (**c**) In the supine position, the specimen is collected from the upper palpebral conjunctiva. (**d**) In the supine position, the specimen is collected from the superior bulbar conjunctiva

Table 1
Composition of real-time PCR solution

PCR mixture components	Amount (μL)
20× TaqMan® gene expression probe	1.0
cDNA template + RNase-free water	9.0
2× TaqMan® gene expression master mix	10.0
Total volume	20.0

2. Mix the contents by inverting the tube several times.
3. Dispense 20 μL of the PCR solution into each well of the reaction plate. Seal the plate with a real-time PCR film.
4. Analyze the samples using the real-time PCR system. Example of PCR conditions is shown in Table 2.

Table 2
Examples of PCR conditions in the Step One Plus™ real-time PCR system

	Temperature (°C)	Time	Cycle
UDG incubation	50	2 min	Hold
AmpliTaq gold®, up enzyme activation	95	10 min	Hold
Denature	95	15 s	40
Anneal/extend	60	1 min	

5. Obtain comparative threshold (Ct) values. Target Ct values should be normalized to those of glyceraldehyde 3-phosphate dehydrogenase (*GAPDH*) from the same sample. Data should be analyzed using the ΔΔCt method.

4 Notes

1. Impression cytology using nitrocellulose membrane collects more cells than with Schirmer test strips. However, Schirmer test strips are less painful than nitrocellulose membranes for patients and are therefore recommended for collecting specimens for repeated clinical testing.
2. The Schirmer test strips used for specimen collection are separated at the 5 mm tip. We used the tear production measuring strips (AYUMI Pharmaceutical Corporation, Tokyo, Japan) as the Schirmer test paper.
3. Our recommended RNA stabilization reagent is RNAlater® (Qiagen, Hilden, Germany).
4. Ophthalmic anesthetics used as eye drops include lidocaine hydrochloride, oxybuprocaine hydrochloride, proxymetacaine hydrochloride, and tetracaine hydrochloride.
5. We recommend Beaupre forceps (Inami & Co., Ltd., Tokyo, Japan) to grasp the Schirmer test paper (*see* Fig. 2).
6. We recommend 0.1–1000 μL micropipettes and micropipette tips.
7. RLT buffer is recommended for collecting and lysing cells from Schirmer's test paper.
8. We used OpenArray™ Plate Compression Pad for Dual Flat Block GeneAmp™ PCR System 9700 (Applied Biosystems) as a thermal cycler.
9. We recommend 1–1000 μL micropipettes and micropipette tips.

10. Primers/probes can be searched under "Search Applied Biosystems TaqMan Assays" at https://www.thermofisher.com/jp/ja/home/life-science/pcr/real-time-pcr/real-time-pcr-assays/taqman-gene-expression.html. Expression levels of different cytokines and chemokines on the ocular surface can be measured using different primers.
11. We used 96-Well Fast Plate (Catalog #38800, Sorenson™ BioScientific) for our qRT-PCR.
12. Specimens can be collected by a technique similar to Schirmer's test instead of impression cytology. Briefly, a Schirmer's test strip is placed in the lower conjunctival sac, removed after 3 min, and the 5 mm tip is cut off.
13. Messenger RNA extraction can also be performed with an automated nucleic acid extractor. We have obtained the same results with automated extraction using magREAD® (Precision System Science).

Acknowledgements

We sincerely appreciate Mrs. Akiko Tomioka for her technical assistance. We are also grateful to our collaborators, Dr. Noriko Inada, Dr. Yukiko Tonozuka, Dr. Akira Hirota, Dr. Mariko Horinaka, and Dr. Rumi Adachi, for their hard work in researching the clinical application of this technique.

References

1. Lemp MA, Blackman HJ (1981) Ocular surface defense mechanisms. Ann Ophthalmol 13: 61–63
2. Pflugfelder SC, Stern ME (2020) Biological functions of tear film. Exp Eye Res 197: 108115
3. Swamynathan SK, Wells A (2020) Conjunctival goblet cells: ocular surface function, disorders that affect them, and the potential for their regeneration. Ocul Surf 18:19–26
4. Shoji J, Sakimoto T, Kitano S et al (1988) Trapping role of mucin. Nippon Ganka Gakkai Zasshi 92:2038–2047. (in Japanese)
5. Ablamowicz AF, Nichols JJ (2016) Ocular surface membrane-associated mucin. Ocul Surf 14:331–341
6. Pflugfelder SC, de Paiva CS (2017) The pathophysiology of dry eye disease: what we know and future directions for research. Ophthalmology 124:S4–S13
7. Shoji J (2020) Ocular allergy test and biomarkers on the ocular surface: clinical test for evaluating the ocular surface condition in allergic conjunctival diseases. Allergol Int 69:496–504
8. Argüeso P, Balaram M, Spurr-Michaud S et al (2002) Decreased levels of the goblet cell mucin MUC5AC in tears of patients with Sjögren syndrome. Invest Ophthalmol Vis Sci 43: 1004–1011
9. Shoji J, Inada N, Tomioka A et al (2020) Assessment of mucin-related gene alterations following treatment with rebamipide ophthalmic suspension in Sjögren's syndrome-associated dry eye. PLoS One 15:e0242617
10. Dogru M, Matsumoto Y, Okada N et al (2008) Alterations of the ocular surface epithelial MUC16 and goblet cell MUC5AC in patients with atopic keratoconjunctivitis. Allergy 63: 1324–1334
11. Horinaka M, Shoji J, Tomioka A et al (2021) Alterations in mucin-associated gene expression on the ocular surface in active and stable stages of atopic and vernal keratoconjunctivitis. J Ophthalmol 31:9914786

Chapter 22

Methylation-Specific Electrophoresis

Seiya Yokoyama, Kei Matsuo, and Akihide Tanimoto

Abstract

Methylation of CpG sites in the promoter region of genomic DNA is an important epigenetic modification that plays a critical role in gene regulation, particularly in gene silencing. Epigenetic abnormalities, along with genetic alterations, are implicated in carcinogenesis and cancer progression. Numerous studies have investigated the role of epigenetics in cancer using various tools to assess DNA methylation. However, conventional analysis methods for DNA methylation require a large amount of DNA but lack higher sensitivity, making them unsuitable for analysis of samples with high heterogeneity, such as tumor tissues. In this study, we introduce a novel electrophoresis method named "methylation-specific electrophoresis (MSE)," which utilizes a denaturing gradient acrylamide gel. We demonstrate the applicability of the MSE method for DNA methylation analysis of the mucin gene as an example.

Key words DNA methylation pattern, Tumor heterogeneity, Clinical sample

1 Introduction

The investigation of gene expression in clinical specimens is essential for clinical cancer research [1]. However, obtaining a sufficient quantity and quality of RNA or protein for analysis using clinical specimens can be challenging. On the other hand, epigenetics, including DNA methylation, plays an important role in gene regulation, particularly in gene silencing, and hypomethylation enhances promoter transcriptional activity [2, 3]. Recently, we have demonstrated that mucin gene expression is regulated by DNA methylation as an epigenetic control in cancer cell lines and clinical specimens [4]. Furthermore, abnormal epigenetic alterations occur during the initial carcinogenesis and progression of malignancy [5, 6]. Therefore, a method for detecting alterations in methylation status is a valuable tool for early cancer diagnosis, assessing tumor response to targeted therapy, and predicting prognosis.

Akihiko Kameyama (ed.), *Mucins: Methods and Protocols*, Methods in Molecular Biology, vol. 2763,
https://doi.org/10.1007/978-1-0716-3670-1_22,

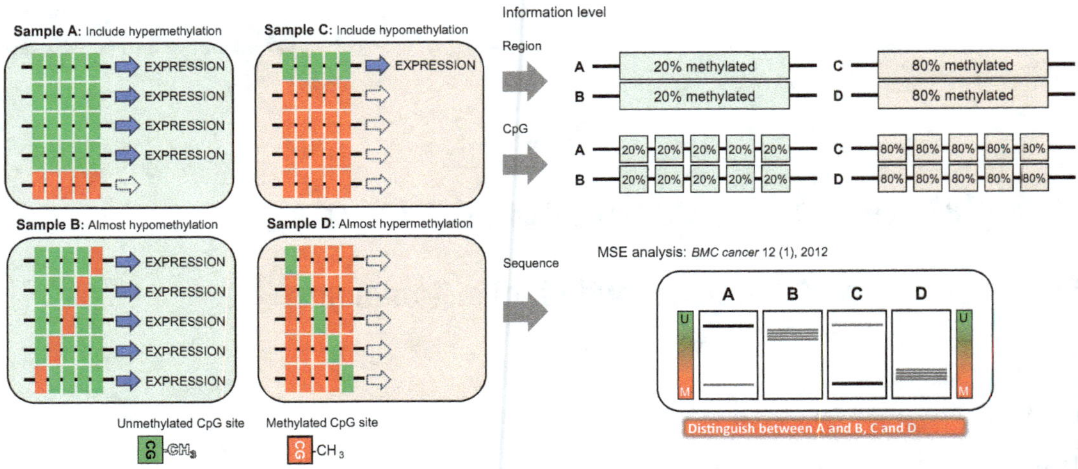

Fig. 1 The importance of methylated CpG site sequential pattern analysis in high-diversity samples such as clinical specimens. Left side: The scheme depicts crude samples with various methylation statuses. Right side: The analysis results of various methods, which provide different information depths, using these crude samples

Several methods for analyzing DNA methylation have been developed, which typically require a large amount of high-quality and low-heterogeneous DNA, such as that obtained from cultured cells and subcutaneous tumor-bearing animal models. However, DNA extracted from clinical samples, including resected tissue, biopsy specimens, and cytological specimens, often exhibits higher heterogeneity. Clinical samples obtained from urine sediment, saliva, sputum, bronchial washing fluid, bile, and pancreatic juice also contain DNA, but the quantity and quality are often insufficient for methylation analysis [7–11].

In DNA methylation analysis using heterogeneous clinical samples, it is necessary to detect differences among heterogeneous diversity of methylated CpG site sequential patterns (referred to as samples A to D) as shown in Fig. 1. Sample A exhibits predominantly complete unmethylation with the minor presence of a complete methylation. Sample B shows mostly hypomethylation patterns. Sample C displays predominantly complete methylation with the minor presence of a complete unmethylation. Sample D demonstrates mostly hypermethylation patterns. These samples may exhibit varying expression levels of the target gene. However, a conventional methylation-specific PCR method would show 20% methylation in samples A and B and 80% methylation in samples C and D, despite the presence of different methylation CpG site sequential patterns in these samples. Similarly, the method of CpG site individual analysis like mass-array and pyrosequencing would show that each CpG sites are 20% methylated in samples A and B, respectively, and 80% methylated in samples C and D, respectively. Consequently, these conventional methods fail to distinguish the actual methylation status and the potential variations in target gene expression among these samples. This disadvantage poses challenges to the development of DNA methylation markers. Therefore, it is crucial to consider information regarding methylation CpG site sequential patterns to evaluate the DNA methylation status accurately.

To distinguish the methylation CpG site sequential patterns, we developed a method of methylation-specific electrophoresis (MSE) [12]. DNA consists of two strands, known as the antisense and sense strands, composed of polynucleotides. The double-stranded structure is maintained through hydrogen bonding between adenine and thymine, as well as between guanine and cytosine. The stability of this double-stranded structure is reinforced by a higher GC content, because G-C base pairs have three hydrogen bonds, while A-T base pairs have two. To analyze DNA methylation, DNA is treated with sodium bisulfite. Sodium bisulfite selectively converts unmethylated cytosine residues into uracil through deamination, while methylated cytosines (including 5-methylcytosine and 5-hydroxymethylcytosine) remain unchanged. The sodium bisulfite-treated DNA is then subjected to PCR amplification, during which the uracils are converted into thymines. As a result, all originally unmethylated cytosines are converted to thymines, while methylated cytosines remain as cytosines. Those steps to a distinct nucleotide difference between methylated and unmethylated cytosines, although the original DNA sequence is preserved. In the MSE method, amplicons obtained from bisulfite-treated DNA, which have the same length and nearly identical sequences except for the different methylated CpG sites, can be distinguished based on their altered electrophoretic mobility. The separation is achieved by subjecting the DNA samples to polyacrylamide gels containing a linear gradient of DNA denaturants, a mixture of formamide and urea. The partially melted double-stranded DNA molecules with different CpG methylation patterns exhibit varying degrees of stability, resulting in differences in their migration during electrophoresis. By exploiting the relationship between sequence and the robustness of DNA under denaturing conditions, the MSE method enables the differentiation of amplicons based on their methylated CpG sequences. Thus, these samples are distinguished as different methylation statuses (Fig. 1). In this section, we provide an example of the MSE analysis for MUC2 gene methylation.

2 Materials

Prepare all solutions using Ultrapure water and analytical grade, nuclease- and protease-free reagents. Prepare and store all reagents at room temperature (unless indicated otherwise).

2.1 DNA Sample Preparation

1. DNA extraction from human tissue samples can be performed using commercially available kits or established protocols specifically designed for a broad range of solid tissues. The choice of extraction method may depend on factors such as the tissue type and sample size.
2. DNA conversion of unmethylated cytosines to uracil can be performed using bisulfite or enzymatic treatment commercially available kits.
3. DNA purification on each step can be performed using commercially available kits or established protocols such as gel extraction.
4. High-fidelity PCR polymerase.
5. Primer design: All the PCR primers are designed to refer to bisulfite-treated sequence, i.e., all cytosines without a methylated CpG site convert thymine. Second PCR primers are designed to be optimal sample fragment sizes (100 bp to 400 bp) to include a CpG site as little as possible. Add clamp sequence (5′-CGCCCGCCGCGCGCGGCGGGCGGGGCG GGGGCACGGGGGG-3′) on the 5′ end of the second PCR forward primer.

   For example, primer sequences to the MUC2 promoter amplify the following:

   Nested forward primer, 5′-TTTGGGGTTAGGTTTGGAAG-3′; nested reverse primer, 5′-ACCTTCTTCAAAATAAAA CAACC-3′;

   target forward primer, 5′-CGCCCGCCGCGCGCGGCGGG CGGGGCGGGGGCACGGGGGG TTTTAGAGTTTGG GTTTTAG-3′; target reverse primer, 5′- -TAACCTAAATACC AACACAC-3′
6. 2× gel loading dye: Mix 0.25 mL of 2% bromophenol blue, 0.25 mL of 2% xylene cyanol, 7 mL of glycerol, and 2.5 mL of distilled H_2O.
7. PCR system.

2.2 Electrophoresis

1. Acrylamide gel: 40 (w/v)% acrylamide/bis mixed solution (37.5: 1). Weigh 38.93 g of acrylamide and 1.07 g of bis–acrylamide and mix them and add distilled H_2O to 100 mL. Filter the solution through a Whatman No. 1 and store it at 4 °C.
2. 50× TAE buffer: Mix 242.0 g of Tris base, 57.1 mL of acetic acid (glacial), and 100 mL of 0.5 M EDTA. Adjust pH 8.0 and dilute to 600 mL with distilled H_2O, and then autoclave for 20–30 min. Store at room temperature.
3. Glycerol.
4. 10% ammonium peroxodisulfate (APS) solution.

5. Tetramethylethylenediamine (TEMED).
6. Denaturant: formamide (deionized) and urea.
7. Gradient gel former.
8. Large-gel (20 cm × 20 cm) electrophoresis chamber with temperature control.
9. 20 cm × 20 cm glass plate for slab gel preparation.
10. Power supply.
11. Running buffer (TAE buffer): 40 mM Tris–acetic acid, 1 mM EDTA, pH 7.5.
12. Higher-sensitivity DNA gel stains such as an intercalating agent that binds DNA.
13. Reference marker: 100-bp ladder marker (*see* **Note 1**).

3 Methods

Carry out all procedures at room temperature unless specified otherwise.

3.1 Sample Preparation

1. DNA extraction was performed from fresh tissue blocks measuring approximately 2 × 2 × 2 mm, as well as from deparaffinized 10 μm FFPE tissue sections, residual material from 1 mL liquid-based cytology (LBC) specimens, 0.5 mL pancreatic juice, or 0.5 mL bile juice. The extracted DNA underwent quality checks to assess fragmentation, purity, and concentration. The recovered DNA was quality checked in fragmentation performed by electrophoresis and purity performed OD spectrum measurement and concentration. The sample has no over 400-bp fragments or less than 1 μg/L fails for further analysis.
2. The recovered DNA converts unmethylated cytosines to uracil by bisulfite or enzymatic treatment.
3. Target regions were amplified by the nested PCR approach using a bisulfite DNA sample. PCR amplicons were purified after all PCR steps.

   For **example, MUC2 PCR step**

Nested PCR	Target PCR
Initial denaturation step, 95 °C 2 min	Initial denaturation step, 95 °C 2 min
Start cycle step (30 cycle)	Start cycle step (30 cycle)
Denaturing, 98 °C 20 s	Denaturing, 98 °C 20 s
Annealing, 59 °C 15 s	Annealing, 51 °C 15 s
Extending, 72 °C 15 s	Extending, 72 °C 15 s
End of cycle step	End of cycle step

(continued)

Table 1
Composition of the gel stock solution

Denaturing (%)	0%	10%	20%	30%	40%	50%
40% acrylamide/bis soln	12.5 mL	12.5 mL	12.5 mL	12.5 mL	12.5 mL	12.5 mL
50× TAE buffer	1 mL	1 mL	1 mL	1 mL	1 mL	1 mL
Formamide	0 mL	2 mL	4 mL	6 mL	8 mL	10 mL
Urea	0.0 g	2.1 g	4.2 g	6.3 g	8.4 g	10.6 g
Glycerol[a]	0 mL	(5 mL)	(5 mL)	(5 mL)	(5 mL)	(5 mL)
dH_2O	q.l.	q.l.	q.l.	q.l.	q.l.	q.l.
Final	50 mL	50 mL	50 mL	50 mL	50 mL	50 mL

[a]If using a high concentration of denaturant gel, add glycerol in the solution

Final extending, 72 °C 1 min	Final extending, 72 °C 1 min

To verify the presence of PCR products of the expected size, perform agarose gel electrophoresis using approximately 2–3 μL of the PCR product. If MSE is not immediately performed, the purified PCR product can be stored at −20 °C or lower for stability.

4. Add an equal volume of 2× gel loading dye to the sample.

3.2 Gel Preparation

1. Unpolymerized acrylamide is toxic. Wear rubber gloves when handling it.
2. Fifty milliliters of acrylamide gel solution containing two types (low and high; conditions in Table 1) of denaturant concentrate preparation. A high concentration of denaturant gel solution contains 1/10 vol glycerol (*see* **Note 2**).

   For example, **gradient gel suitable for MUC2 analysis.**
   Acrylamide/bis concentration: 8%.
   Denaturing concentration gradient: 25–35%.

	Low denaturing gel soln	High denaturing gel soln
40% acrylamide/bis soln	12.5 mL	12.5 mL
50× TAE buffer	1 mL	1 mL
Formamide	5 mL	7 mL
Urea	5.25 g	7.35 g
Glycerol	0 mL	5 mL
Final	50 mL	50 mL

3. Add 100 μL of 10%(w/v) APS solution and 10 μL of TEMED solution to both the denaturant concentration gel solutions just before setting in a gradient gel former (*see* **Note 3**).
4. Make the denaturing gradient gel using a gel former. Put the gel solution into the glass plate gently and carefully (*see* **Notes 4**, **5**, and **6**). Inject the gel solution into the gel cassette until it reaches the upper edge of the plate. Carefully insert the comb into the gel solution, making sure to prevent the entry of air bubbles.
5. Let the gel polymerize for approximately 120 min (*see* **Note 7**).
6. After polymerization, remove the comb, then wash them well with TAE buffer at least three times using an injector, etc.

3.3 Electrophoresis

1. Fill the electrophoresis tank with 7 L of running buffer.
2. Warm up the running buffer until the tank reaches 62 °C.
3. Set the constructed gel in the electrophoresis tank and then wait 10 to 30 min until it settles down at 62 °C (*see* **Note 8**).
4. Apply prepared samples by mixing an equivalent of 200 ng of the PCR product with the same quantity of 2× loading dye (see **Note 9**). When injecting the sample into the well, take maximum care not to let the sample diffuse.
5. Once the temperature has stabilized, electrophorese at 60 V for approximately 17 h (*see* **Note 10**).

3.4 Staining

1. Remove the smaller glass plate from the gel by gently lifting the spacer along the edge. Take caution and proceed slowly to avoid damaging the gel or causing any distortion. Then rinse for 5 min with the gel shaken gently with the TAE buffer and discard the buffer.
2. Soak the gel in a stain solution (400 mL TAE buffer containing DNA gel stains such as an intercalating agent) for 30–60 min with gentle shaking.
3. To generate an image of the electrophoretic profile, please follow the instructions provided in the operation manual for the specific imaging equipment being used. Ideally, a large-size gel image scanner is recommended for obtaining high-resolution images. However, if a scanner is not available, you can use a transilluminator in combination with a digital camera that has a manual exposure function. Adjust the camera settings appropriately to capture a clear and well-exposed image of the gel (*see* **Note 11**).

   For example**, MSE result of the MUC2 promoter region**

   The MUC2 analysis results are presented in Fig. 2. The LS174T cell line was utilized as a low methylation control, while HPAF2 served as a high methylation control. The

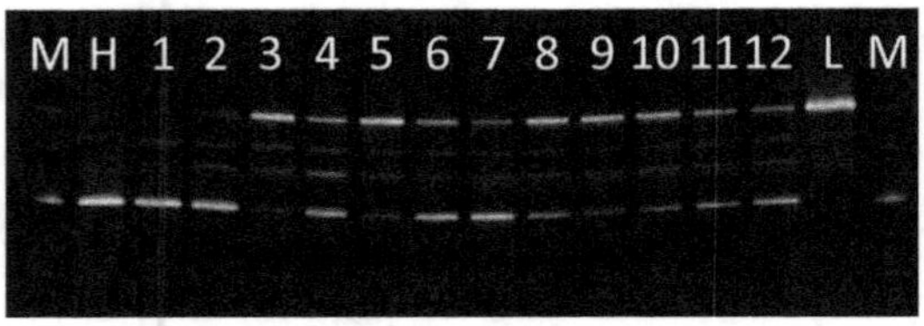

Fig. 2 Result of the MUC2 promoter region. Separation on denaturant gradient (25–35%) acrylamide gel run at constant 60 V for 17 h in 1× TAE buffer. Lane M, 100-bp ladder marker; lane H, high methylation control; lane 1–12, human pancreatic tissues; lane L, low methylation control

methylation statuses of the LS174T and HPAF2 cell lines have been previously measured in other studies using methods such as massARRAY® and other techniques. In the case of a sample with a completely unmethylated pattern, the electropherogram displays a band resembling the low methylation control, positioned at the upper position. Conversely, if the sample contains a completely methylated pattern, the electropherogram will show a band similar to the high methylation control, positioned at the lower position. Therefore, in Fig. 2, samples 3 and 8 exhibit a high intensity of the upper band, suggesting a hypomethylation status. Conversely, samples 1 and 2 show either no or low intensity of the upper band and show high intensity of the lower band, indicating a hypermethylation status. When calculating the hypomethylation index to compare the methylation status between the different gels, the ratio of the intensity of the completely unmethylated band to the intensity of the entire band of sample is calculated. Then, this ratio is normalized using a control sample.

4 Notes

1. Prepare marker lanes to compare the movement patterns between the different gels.
2. Denaturing 10% acrylamide gel solution (target size is 100–300 bp). If the target region size is 200–400 bp, choose 8% acrylamide gel.
3. Once TEMED is added, the polymerization of acrylamide begins immediately. Therefore, it is important to work efficiently.
4. Chill the gel stock solution: The polymerization of acrylamide progresses rapidly at higher temperatures. To control the polymerization process, it is advisable to thoroughly cool the gel stock solution in advance. This helps to maintain a consistent and uniform polymerization throughout the gel.

5. Avoid temperature differences: Large temperature differences between the glass plate and the gel stock solution can lead to uneven polymerization and distortion of the denaturing gradient. It is recommended to minimize temperature variations and ensure that the glass plate and stock solution are at similar temperatures before gel casting.
6. Perform gel casting in a low-temperature room: If possible, perform the gel casting process in a low-temperature environment to keep the glass plate at a lower temperature. This helps maintain the stability of the gel and reduces the risk of distortion. However, keep in mind that gel polymerization may take longer in a low-temperature room. After the gel has solidified to a certain degree, it can be returned to room temperature.
7. Prepare gel at the time of use, and it cannot be stored.
8. Set the temperature of the gel apparatus to approximately 2 °C higher than the intended electrophoresis temperature. This is done to compensate for the drop in buffer temperature that can occur during the sample application process. Wait until the temperature stabilizes at the desired level before proceeding with the electrophoresis, ensuring that the buffer and gel are at the appropriate temperature for optimal separation of the samples.
9. It is recommended to avoid using the wells at the ends of the gel since they can potentially cause distortion during electrophoresis.
10. If you are experiencing low resolution in the gel image, increasing the concentration of acrylamide in the gel to 10% or more, adjusting the denaturing gradient, and/or extending the migration time during electrophoresis can help optimize the separation of DNA fragments based on their sequence differences.
11. The minigel imaging equipment and Polaroid cameras may have limitations in terms of resolution and may not provide optimal image quality.

Acknowledgements

This study was supported, in part, by a grant from Grants-in-Aid for Scientific Research on Scientific Research (C; 15 K21247, 18 K07326, and 21 K07222, to S. Yokoyama) from the Ministry of Education, Science, Sports, Culture and Technology, Japan, by the Kodama Memorial Foundation, Japan (to S. Yokoyama), by the grant from the Shinnihon Foundation of Advanced Medical Treatment Research (to S. Yokoyama), and by the Pancreas Research Foundation of Japan (to S. Yokoyama).

References

1. Kononen J, Bubendorf L, Kallioniemi A et al (1998) Tissue microarrays for high-throughput molecular profiling of tumor specimens. Nat Med 4(7):844–847. https://doi.org/10.1038/nm0798-844
2. Ushijima T, Okochi-Takada E (2005) Aberrant methylations in cancer cells: where do they come from? Cancer Sci 96(4):206–211. https://doi.org/10.1111/j.1349-7006.2005.00035.x
3. Ushijima T (2005) Detection and interpretation of altered methylation patterns in cancer cells. Nat Rev Cancer 5(3):223–231. https://doi.org/10.1038/nrc1571
4. Yamada N, Kitamoto S, Yokoyama S et al (2011) Epigenetic regulation of mucin genes in human cancers. Clin Epigenetics 2:85–96
5. Yokoyama S, Higashi M, Kitamoto S et al (2016) Aberrant methylation of MUC1 and MUC4 promoters are potential prognostic biomarkers for pancreatic ductal adenocarcinomas. Oncotarget 7:42553–42565
6. Yokoyama S, Hamada T, Higashi M et al (2020) Predicted prognosis of patients with pancreatic cancer by machine learning. Clin Cancer Res 26:2411–2421
7. Bastian PJ, Palapattu GS, Lin X et al (2005) Preoperative serum DNA GSTP1 CpG Island hypermethylation and the risk of early prostate-specific antigen recurrence following radical prostatectomy. Clin Cancer Res 11(11):4037–4043. https://doi.org/10.1158/1078-0432.CCR-04-2446
8. Fujiwara K, Fujimoto N, Tabata M et al (2005) Identification of epigenetic aberrant promoter methylation in serum DNA is useful for early detection of lung cancer. Clin Cancer Res 11(3):1219–1225
9. Klump B, Hsieh CJ, Dette S et al (2003) Promoter methylation of INK4a/ARF as detected in bile-significance for the differential diagnosis in biliary disease. Clin Cancer Res 9(5):1773–1778
10. Matsubayashi H, Canto M, Sato N et al (2006) DNA methylation alterations in the pancreatic juice of patients with suspected pancreatic disease. Cancer Res 66(2):1208–1217. https://doi.org/10.1158/0008-5472.CAN-05-2664
11. Yokoyama S, Iwaya H, Akahane T et al (2022) Sequential evaluation of MUC promoter methylation using next-generation sequencing-based custom-made panels in liquid-based cytology specimens of pancreatic cancer. Diag Cytopath 50(11):499–507
12. Yokoyama S, Kitamoto S, Yamada N et al (2012) The application of methylation specific electrophoresis (MSE) to DNA methylation analysis of the 5′ CpG Island of mucin in cancer cells. BMC Cancer 12:67

Chapter 23

Expression Analysis of Genes Corresponding to Mucins and Their Glycans from Cervical Tissue Using RNA Sequencing

Sean Fair and Laura Abril-Parreño

Abstract

In the female reproductive tract, mucin proteins are the main component of mucus secreted by cervical goblet cells and play an essential role in many biological functions. They act as a medium for lubrication and mainting a cervical mucosal barrier against ascending pathogens from the vagina while also allowing sperm migration. The expression of mucin genes as well as the levels of *O*-glycosylation changes across the oestrous cycle. Detection and characterization of mucins and their glycans is important to understand the interface between the external and the internal environment, as the cervical epithelium represents the first line of defense against infections of the upper reproductive tract. Advances in the field of molecular biology have made possible to study differences in mucin and glycan gene expression which can help to understand impeded sperm transport as well as variation in the susceptibility to infection. This chapter discusses procedures relevant for both animals and humans on how to recover cervical tissue and perform a gene expression analysis of genes corresponding to mucins and their glycans using RNA sequencing.

Key words MUC genes, Glycogenes, RNA-sequencing, Glyco-biomarkers, Transcriptome

1 Introduction

Cervical goblet epithelial cells secrete large amounts of cervical mucus, which is a viscoelastic fluid that allows the entry of sperm to the upper female reproductive tract while also inhibiting ascending pathogens from the vagina. In the lead up to ovulation when levels of estrogen are high, mucus production and hydration increase leading to a corresponding decrease in mucus viscosity [1]. This is driven by mucin gene expression giving rise to a peak in mucin production around the time of ovulation, which also promotes mucus hydration [1] and allows sperm migration [2]. Based on transcriptomic data, two categories of mucins have been identified – gel-forming mucins and membrane-bound mucins, including the following 21 mucin genes that have been

Akihiko Kameyama (ed.), *Mucins: Methods and Protocols*, Methods in Molecular Biology, vol. 2763,
https://doi.org/10.1007/978-1-0716-3670-1_23,

identified in the human body: *MUC1, MUC2, MUC3A, MUC3B, MUC4, MUC5AC, MUC5B, MUC6, MUC7, MUC8, MUC9, MUC12, MUC13, MUC14, MUC15, MUC16, MUC17, MUC19, MUC20, MUC21,* and *MUC22* [3]. The gel-forming mucins are on the epithelial cell layer and the individual mucin subunits polymerize via end-to-end disulfide bonds to form larger macromonomers resulting in linear structures [4]. However, other studies have proposed the possibility of branched structures for specific mucins [5]. In the cervix, the secreted gel-forming mucins are *MUC2, MUC5AC, MUC5B*, and *MUC6* [6, 7].

The second category of mucins, the membrane-bound mucins, are extended molecules that traverse the epithelium to the surface. These are composed of an N-terminal extracellular domain, a single transmembrane domain, and a C-terminal cytoplasmic domain. In most transmembrane mucins (except MUC4), the glycosylated extracellular domain is non-covalently associated with the transmembrane domain of the sea urchin sperm protein, enterokinase and agrin (SEA) domain [8]. Several membrane-bound mucins are expressed in the reproductive tract epithelium (MUCs 1, 3A, 3B, 4, 12, 13, and 16) [7]. Some of these mucins have been found in the human (*MUC1* and *MUC6*) [9] and in the bovine cervix (*MUC1, MUC4, MUC16*, and *MUC20*) [10]. *MUC1* and *MUC4* play an important biological role in cell-cell and cell-extracellular matrix interactions [11].

Mucin core proteins are produced in the rough endoplasmic reticulum of the cell and are later transported to the Golgi apparatus where glycosyltransferases add sugar residues to the hydroxyl groups of threonine and serine, in a process known as *O*-glycosylation [12]. During mucin biosynthesis, *O*-glycosylation is initiated with the addition of GalNAc to a serine or threonine residue on the mucin protein backbone, resulting in GalNAc-O-Ser/Thr or Tn antigen structures. This first step can be catalyzed by up to 20 GalNAc transferases (ppGalNAcTs). The next step is the attachment of core 1, 2, and 3 structures. Core 1 (also called T antigen) is synthesized by C1GalT1 and consists of Galβ1–3GalNAc structures. By the action of C2GnT1-3 enzymes, core 1 can be converted to core 2, resulting in Galβ1–3(GlcNAcβ1-6)GalNAc structures. Core 3 is synthetized by the β3GnT6 enzyme, which results in a GlcNAcβ1-3GalNAc structure. Core 3 can be converted to core 4 by the action of GCNT3, resulting in a GlcNAcβ1-3 [GlcNAcβ1-6]GalNAc structure. Once the core structures are attached, 30 or more distinct glycosyltransferases sequentially add other sugar moieties, resulting in the completed *O*-glycan structure. Most of the *O*-glycans have peripheral additions that carry sialic acid and/or fucose by the action of sialyltransferases and fucosyltransferases, respectively [13].

The aforementioned process reflects that glycans are produced through complex circuits that involve coordinated expression of

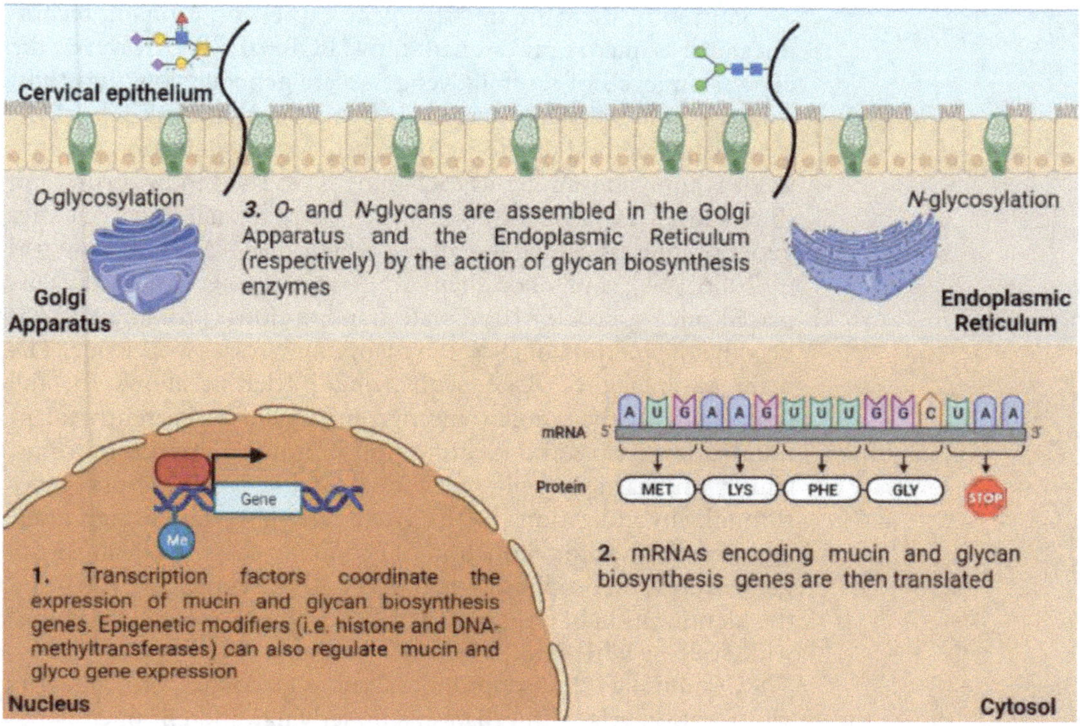

Fig. 1 Diagram of the glyco-genomic regulation. Transcription factors and epigenetic modifiers can change the polygenetic circuits at the transcriptional level, while miRNAs do so at the translational level. *O*- and *N*-glycosylation patterns emerge from the expression and/or repression of different enzymes (i.e., glycosyltransferases)
This picture has been created with BioRender.com

many different genes (Fig. 1). To date, more than 200 genes have been found; these are available and listed as glycogenes in the GlycoGene Database (GGDB, https://acgg.asia/db/ggdb/). Evaluating the expression levels of glycogenes is also a useful step in studying the biological phenomena involving biomarkers. Using RNA sequencing, Abril-Parreño et al. [6] showed differences in cervical gene expression between ewe breeds with differences in the ability of thawed sperm to traverse the cervix. For example, ewe breeds with the lowest cervical sperm transport (Suffolk) had lower expression of ST3 beta-galactoside alpha-2,3-sialyltransferase 3 (*ST3GAL3*) and ST3 beta-galactoside alpha-2,3-sialyltransferase 4 (*ST3GAL4*) compared to ewe breeds with good cervical sperm transport (Norwegian White Sheep and Fur). The proteins encoded by these genes catalyze the transfer of sialic acid from CMP-sialic acid to galactose-containing substrates. However, in the same study they observed higher expression of ST6 beta-galactoside alpha-2,6-sialyltransferase 2 (*ST6GAL2*) and higher expression of ST6 beta-galactoside alpha-2,6-sialyltransferase 1 (*ST6GAL1*) in Suffolk compared to Fur ewes.

Currently, there are multiple gene expression profiling techniques such as microarray or quantitative PCR (qPCR). However, the expression levels of several glycogenes are generally low, and thus, some of them are barely detectable by the microarray. While qPCR is a reliable method due to its high accuracy and detection sensitivity, it is time consuming as each gene is assessed individually, and thus, studies are normally limited to a relatively small panel of target genes of interest. The initial development of RNA sequencing was in 2006 [14], and since then, RNA sequencing has rapidly displaced microarrays for large-scale transcriptional profiling and has become the method of choice in many laboratories worldwide. The main advantage of RNA sequencing is that it allows for full sequencing of the whole transcriptome including currently unannotated transcripts and identification of novel transcript isoforms, while microarrays only profile predefined transcripts/genes through hybridization [10]. There are also other significant advances such as the reduction of economic costs [15], single-cell gene expression profiling by single-cell RNA sequencing [16], and the identification of transcription factor binding sites on a genome-wide scale by ChIP-seq [17].

Combining transcriptomics and glycomics to identify glycosylation-related mechanisms underlying different phenotypes and pathogenesis is a relatively new area of glycobiology. Adding RNA-sequencing data to glycomics provide a more in-depth understanding of the regulation of glycosylation mediated by low-abundance genes. Going forward the combination of systems biology and artificial intelligence will provide fresh insights into the complex, interleaved biosynthesis and functional role of glycans in various biological contexts. Here we describe a method to extract RNA from animal cervical tissue and to assess gene expression using RNA sequencing and associated bioinformatic pipelines. This is relevant for researchers studying the glycobiology of the human or animal cervix and its secretions as it relates to reproductive health.

2 Materials

2.1 Cervical Tissue Collection

1. Dulbecco's PBS.
2. RNaseZap.
3. Petri dishes.
4. Cryogenic vials.
5. Liquid nitrogen.
6. Refrigerator (−80 °C).

2.2 RNA Extraction from Cervical Tissue

1. RNA-free water (Thermo Fisher Scientific, Waltham, Massachusetts, USA).
2. RNaseZap.
3. 70% ethanol.
4. Ethanol.
5. Gloves.
6. 50 mL conical tube.
7. Small polystyrene box (approx. 30 cm length × 20 cm width).
8. Liquid nitrogen.
9. Fume hood.
10. 10.Chloroform.
11. Trizol.
12. Homogenizer.
13. Refrigerated centrifuge.
14. RNeasy Mini Kit (Qiagen Ltd., Crawley, West Sussex, UK) containing:
    - RNeasy Mini Spin Columns.
    - Collection tubes (1.5 mL).
    - Collection tubes (2 mL).
    - Buffer RW1 (45 mL).
    - Buffer RPE (concentrated) (11 mL); *add 44 mL of pure ethanol.
    - RNase-Free Water (10 mL).
15. RNase-Free DNase Set (Qiagen, Crawley, West Sussex, UK) containing:
    - RNase-Free DNase I.
    - RNase-Free Buffer RDD.
    - RNase-Free Water.
16. Nanodrop ND-1000 UV-Vis spectrophotometer (NanoDrop Technologies Inc., Wilmington, DE, USA) or conventional spectrophotometer.
17. 2100 Agilent Bioanalyzer (Agilent Technologies, Santa Clara, CA, USA).

2.3 Library Preparation and RNA Sequencing

1. Illumina® TruSeq® Stranded mRNA Library Preparation Kit (Illumina, San Diego, California, USA).
2. Qubit 4 fluorometer (Thermo Fisher Scientific, Waltham, Massachusetts, USA).
3. Illumina NovaSeq sequencer (Illumina, San Diego, California, USA) at 2 × 150 bp paired end reads (50 M reads).

4. Nanodrop ND-1000 UV-Vis spectrophotometer.
5. DNA 1000 Nano LabChip kit on the Agilent Bioanalyzer 2100.

2.4 Differential Expression Analysis

1. Software FastQC (v 0.11.8; http://www.bioinformatics.babraham.ac.uk/projects/fastqc/).
2. BBDuk Java package https://jgi.doe.gov/data-and-tools/bbtools/.
3. Spliced Transcripts Alignment to a Reference (STAR) aligner https://github.com/alexdobin/STAR.
4. Featurecounts https://rdrr.io/bioc/Rsubread/man/featureCounts.html.
5. Ensembl (v.95) https://www.ensembl.org/index.html.
6. DeSeq2 (version 1.130.0) https://bioconductor.org/packages/release/bioc/html/DESeq2.html.
7. gProfiler2 (v. 0.2.0) package https://cran.r-project.org/web/packages/gprofiler2/index.html.
8. R package rrvgo (v.1.1.4) https://bioconductor.org/packages/release/bioc/html/rrvgo.html.
9. R package Cemitools (v1.14.0) https://www.bioconductor.org/packages/devel/bioc/vignettes/CEMiTool/inst/doc/CEMiTool.html.

3 Methods

3.1 Sample Size Calculation and Tissue Collection from the Cervix

1. A sample size calculation should be performed in advance of undertaking the study to determine the n required to detect, as significant, a true effect between treatments (*see* **Note 1**).
2. The cervix is typically recovered postmortem (animals) or post hysterectomy (women) following which they are washed on Petri dishes with Dulbecco's PBS. Spray RNaseZap solution onto the surface of the table, Petri dish, scalpel, tweezers, scissors, etc. For gene expression work the sampling should take place as soon as possible after tissue recovery.
3. Open the cervix longitudinally from the anterior (end closest to the uterus) to the posterior (end closest to the vagina) endocervical canal to avoid contamination from the vagina.
4. Thin cervical tissue sections of the inner most layer are taken from the area of interest (*see* **Note 2**).
5. Transfer the sections to individual cryogenic vials immediately (*see* **Note 3**) and snap freeze at −196 °C (liquid nitrogen) followed by long-term storage at −80 °C for RNA-sequencing analysis.

3.2 Preparing an RNAse-Free Environment

1. Before starting, clean all working surfaces and pipettes using RNaseZap and 70% ethanol.
2. Always use gloves. Change gloves when they have been in contact with potential RNase contamination.
3. Spray and wipe the probes of the homogenizer with RNaseZap and 70% ethanol.
4. Rinse each probe of the homogenizer in 20 mL of ethanol in a 50 mL tube and 20 mL of RNA-free water. Repeat this step two times.
5. Transfer each probe to an empty 50 mL conical tube to dry.

3.3 Tissue Processing and RNA Extraction

1. Half fill a small polystyrene box with liquid nitrogen to transport samples from the −80 °C freezer to the fume hood.
2. Add 350 μL of Trizol reagent to bottomed tubes (*see* **Note 4**).
3. Transfer tissue from cryogenic vial to Trizol using an autoclaved forceps sprayed with RNaseZap.
4. Attach autoclaved and cleaned probe to homogenizer and homogenize sample for 1–2 min until all sample is fully homogenized (*see* **Note 5**).
5. Incubate the homogenized sample for 5 min at room temperature to permit complete dissociation of the nucleoprotein complex.
6. Add 70 μL of chloroform (per 350 μL of Trizol; 1 in 5 dilution) for homogenization. Shake tube vigorously by hand for 15 s.
7. Incubate for 2–3 min at room temperature.
8. Centrifuge the sample at 12,000 × *g* for 15 min at 4 °C.
9. Remove the colorless upper aqueous phase by angling the tube at 45° and pipetting the solution out. Avoid drawing the interphase or organic phase layer into the pipette.
10. Add 350 μL of 70% ethanol to the lysate and mix immediately by pipetting. Do not centrifuge (*see* **Note 6**).
11. Transfer up to 700 μL of the sample to an RNeasy column placed in a 2 ml collection tube. Centrifuge for 15 s at >8000 × *g*. Discard the flow-through. If sample volume exceeds 700 μL, centrifuge successive aliquots in the same RNeasy spin column.
12. Add 700 μL of Buffer RW1 to the RNeasy spin column. Centrifuge for 15 seconds at >8000 × *g* to wash the spin column membrane. Discard the flow-through.
13. Add 10 μL of rehydrated DNase to 70 μL of RDD buffer and then add to column (*see* **Note 7**).
14. Incubate for 15 min at room temperature.

15. Wash with 350 μL of Buffer RW1. Centrifuge for 15 s at >8000 × *g* to wash the spin column membrane. Discard the collection tube and place column in a new collection tube.
16. Add 500 μL Buffer RPE to the RNeasy spin column. Centrifuge for 15 s at >8000 × *g* to wash the spin column membrane. Discard the flow-through.
17. Add 500 μL of Buffer RPE to the RNeasy spin column. Centrifuge for 2 min at >8000 × *g* to wash the spin column membrane (long centrifugation dries the membrane).
18. Optional: Place the RNeasy spin column in a new 2 mL collection tube and discard the old collection tube with flow-through. Centrifuge at full speed for 1 min (eliminates any possible carryover of Buffer RPE).
19. Place the RNeasy spin column in a new 1.5 mL collection tube. Add 30 μL of RNase-free water directly to the spin column membrane. Centrifuge for 1 min at >8000 × *g* to elute the RNA.
20. Quantify 1 μL of the RNA by using a Nanodrop Spectrophotometer or by measurement of the optical density at 260 nm after dilution (1/100) using a conventional spectrophotometer.
21. Check the RNA integrity number (RIN) by using 2100 Agilent Bioanalyzer (*see* **Note 8**).
22. Freeze RNA aliquots in the −80 °C freezer after RNA extraction.

3.4 Library Preparation and RNA Sequencing

1. Prepare RNA libraries using an mRNA Library Preparation Kit to convert mRNA into cDNA libraries for sequencing (*see* **Note 9**).
2. Allocate indexes to specific samples prior to library construction so that each sample within a pool has a unique bar code.
3. Following adapter ligation, enrich DNA fragments by performing PCR (*see* **Note 10**).
4. Individual cDNA libraries prepared from mRNA are validated through the DNA 1000 Nano LabChip kit on the Agilent Bioanalyzer 2100 ensuring that library fragment size is ~ 260 bp.
5. Assess cDNA concentration using Nanodrop Spectrophotometer ND-1000; samples with > 25 ng/μL are deemed suitable for further analysis.
6. Sequencing is performed for each sample at 2 × 150 bp paired end reads (50 M reads). Sequence all libraries on an Illumina NovaSeq sequencer by Macrogen, Inc. (Seoul, Republic of Korea) or similar platform.

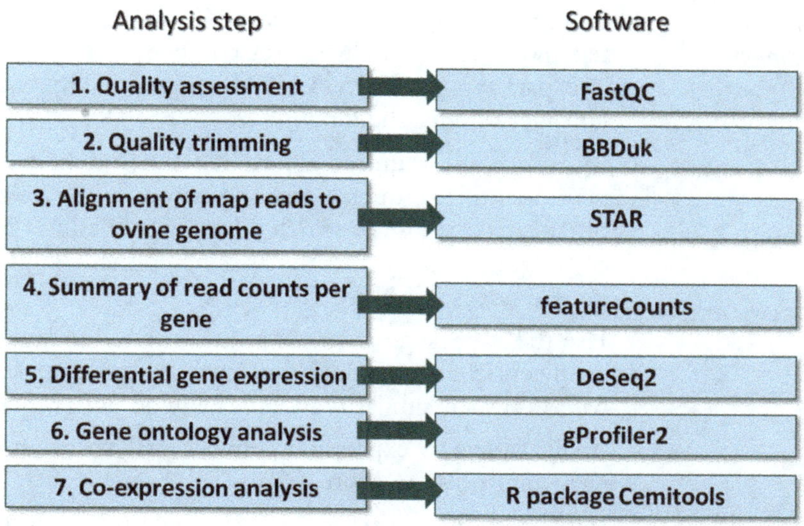

Fig. 2 Overview of data analysis pipeline and the software used in each step from quality assessment to co-expression analysis

3.5 Differential Gene Expression Analysis

1. Use the software FastQC (v 0.11.8; http://www.bioinformatics.babraham.ac.uk/projects/fastqc/) to do a quality assessment of the raw sequence data (Fig. 2).
2. Trim data by using the BBDuk Java package to remove Illumina adapter sequences and any low-quality bases (Phred score < 20) from the $3'$ end of sequence read pairs.
3. Align reads to the most recently published reference genome using the Spliced Transcripts Alignment to a Reference (STAR) aligner (*see* **Note 11**).
4. Read counts overlapping all protein coding genes in the Ensembl annotation are estimated using featureCounts software.
5. To filter out lowly expressed genes, discard the genes with less than one count per million (*see* **Note 12**).
6. Normalize the remaining gene counts using the median of ratios method as implemented in DeSeq2 to account for varying sequencing depth between samples.
7. Model transcript counts by fitting the data to a negative binomial distribution using genewise dispersion estimates and differentially will be identified with a generalized linear model likelihood ratio test.
8. Use statistical tests for multiple testing with the Benjamini-Hochberg method. Establish a $\log_2$ fold change (FC) threshold value (*see* **Note 13**).

3.6 Functional and Pathway Enrichment Analysis

1. Identify aggregated functional profiles of genes and gene clusters in the differentially expressed genes lists using the gProfiler2 package (*see* **Note 14**). This functionality is available in gprofiler2 under the function gost.
2. Introduce the required inputs for this function, a vector of gene identifiers, query, and the name of the corresponding organism which is constructed by concatenating the first letter of the genus name and the specific epithet, i.e., hsapiens for human genes (*see* **Note 15**).
3. Use the R package rrvgo to reduce the redundancy of significantly enriched gene ontology terms by grouping similar terms based on their similarity within the gene ontology hierarchy.
4. Analyze gene co-expression network analyses using the R package Cemitools (*see* **Note 16**).
5. Implement an overrepresentation analysis to identify enriched biological functions in each module.

4 Notes

1. Design of the experiment and the biological replicates to include in the study, having enough biological replicates (i.e., animals) in your study. Ensure treatments are balanced at all stages of the protocol.
2. Before doing the section, define the tissue of interest (the cervix is composed of connective tissue with two dominant cell types in both portions of the cervix consisting of stromal and epithelial cells). The size of the section should not exceed one centimeter in order to facilitate the homogenization of the tissue, if possible take biopsies with similar sizes and avoid bloody tissue as this will increase the amount of Trizol that you need to use in the RNA extraction protocol.
3. More than one section should be taken to allow technical replicates. Keep different sections in separate cryotubes.
4. Use 1 mL of Trizol per 50–100 mg tissue. Use bottomed tubes as these allow a better homogenization than using conical tubes.
5. Do not spend too much time homogenizing as it will degrade RNA.
6. The volume of lysate may be lower than the starting volume due to loss during homogenization and centrifugation.
7. Optional step as in general, DNase digestion is not required for RNA purified with RNeasy Kits since the silica-gel–membrane, spin-column technology efficiently removes the majority of the DNA without DNase treatment. However, more complete

DNA removal may be necessary for certain RNA applications that are sensitive to very small amounts of DNA.

8. RIN values between 8 and 10 are deemed to be of sufficiently high quality for subsequent RNAseq analysis.
9. This step involves isolation of poly-A tailed mRNA using poly-T oligo attached magnetic beads, reverse transcription to form double-stranded cDNA, and the ligation of indexing adaptors.
10. The number of PCR cycles is 15 cycles as recommended in the manufacturer's protocol.
11. A maximum of two mismatches with the reference genome should be allowed and only uniquely mapped read pairs should be retained for downstream analysis.
12. To filter out lowly expressed genes, genes with less than one count per million in at least five samples should be discarded from the analysis.
13. Differentially express genes with an adjusted *P* value <0.05 and a log2 FC threshold of (1.5) for further differentially expressed gene data exploration and pathway analysis.
14. Gene Ontology terms and Reactome pathways are analyzed with an enrichment threshold cutoff of $P < 0.05$.
15. The query vector can include mixed types of gene/protein identifiers, SNP rs-IDs, chromosomal intervals, or term IDs.
16. For any modules identified, a gene set enrichment analysis is carried out to indicate if each module is induced or repressed in the different variables.

Acknowledgements

This work was funded by ERANet (grant no. 16/RD/SusAn/ERA-NET). S. Fair is funded by Science Foundation Ireland (project 16/IA/4474). L. Abril-Parreño is funded by MICINN (Spain) through the program Juan de la Cierva-Formación (FJC-2021-047738-J37703).

Disclaimer The above protocol is a recommendation based on the author's experience of conducting this work in sheep cervical tissue. The authors have no vested interests to declare and researchers are advised to consider alternative platforms and products which can perform similar function.

References

1. Gipson IK, Moccia R, Spurr-Michaud S, Argüeso P, Gargiulo AR, Hill JA 3rd, Offner GD, Keutmann HT (2001) The amount of MUC5B mucin in cervical mucus peaks at mid-cycle. J Clin Endocrinol Metab 86(2): 594–600. https://doi.org/10.1210/jcem.86.2.7174
2. Ma X, Pan Q, Feng Y, Choudhury BP, Ma Q, Gagneux P, Ma F (2016) Sialylation facilitates the maturation of mammalian sperm and affects its survival in female uterus. Biol Reprod 94(6):123. https://doi.org/10.1095/biolreprod.115.137810
3. Curlin M, Bursac D (2013) Cervical mucus: from biochemical structure to clinical implications. Front Biosci (Schol Ed) 5:507–515. https://doi.org/10.2741/s386
4. Kesimer M, Makhov AM, Griffith JD, Verdugo P, Sheehan JK (2010) Unpacking a gel-forming mucin: a view of MUC5B organization after granular release. Am J Physiol Lung Cell Mol Physiol 298(1):L15–L22. https://doi.org/10.1152/ajplung.00194.2009
5. Ambort D, Johansson ME, Gustafsson JK, Nilsson HE, Ermund A, Johansson BR, Koeck PJ, Hebert H, Hansson GC (2012) Calcium and pH-dependent packing and release of the gel-forming MUC2 mucin. Proc Natl Acad Sci U S A 109(15): 5645–5650. https://doi.org/10.1073/pnas.1120269109
6. Abril-Parreño L, Morgan J, Krogenæs A, Druart X, Cormican P, Gallagher ME, Reid C, Meade K, Saldova R, Fair S (2022) Biochemical and molecular characterisation of sialylated cervical mucins in sheep. Biol Reprod 107:419. https://doi.org/10.1093/biolre/ioac077
7. Gipson IK, Blalock T, Tisdale A, Spurr-Michaud S, Allcorn S, Stavreus-Evers A, Gemzell K (2008) MUC16 is lost from the uterodome (pinopode) surface of the receptive human endometrium: in vitro evidence that MUC16 is a barrier to trophoblast adherence. Biol Reprod 78(1):134–142. https://doi.org/10.1095/biolreprod.106.058347
8. van Putten JPM, Strijbis K (2017) Transmembrane mucins: signaling receptors at the intersection of inflammation and cancer. J Innate Immun 9(3):281–299. https://doi.org/10.1159/000453594
9. Andersch-Bjorkman Y, Thomsson KA, Holmen Larsson JM, Ekerhovd E, Hansson GC (2007) Large scale identification of proteins, mucins, and their O-glycosylation in the endocervical mucus during the menstrual cycle. Mol Cell Proteom 6(4):708–716. https://doi.org/10.1074/mcp.M600439-MCP200
10. Pluta K, McGettigan PA, Reid CJ, Browne JA, Irwin JA, Tharmalingam T, Corfield A, Baird A, Loftus BJ, Evans AC, Carrington SD (2012) Molecular aspects of mucin biosynthesis and mucus formation in the bovine cervix during the periestrous period. Physiol Genomics 44(24):1165–1178. https://doi.org/10.1152/physiolgenomics.00088.2012
11. Jonckheere N, Van Seuningen I (2010) The membrane-bound mucins: from cell signalling to transcriptional regulation and expression in epithelial cancers. Biochimie 92(1):1–11. https://doi.org/10.1016/j.biochi.2009.09.018
12. Ambort D, Johansson ME, Gustafsson JK, Ermund A, Hansson GC (2012) Perspectives on mucus properties and formation – lessons from the biochemical world. Cold Spring Harb Perspect Med 2(11). https://doi.org/10.1101/cshperspect.a014159
13. Bennett EP, Mandel U, Clausen H, Gerken TA, Fritz TA, Tabak LA (2012) Control of mucin-type O-glycosylation: a classification of the polypeptide GalNAc-transferase gene family. Glycobiology 22(6):736–756. https://doi.org/10.1093/glycob/cwr182
14. Bainbridge MN, Warren RL, Hirst M, Romanuik T, Zeng T, Go A, Delaney A, Griffith M, Hickenbotham M, Magrini V, Mardis ER, Sadar MD, Siddiqui AS, Marra MA, Jones SJM (2006) Analysis of the prostate cancer cell line LNCaP transcriptome using a sequencing-by-synthesis approach. BMC Genomics 7(1):246. https://doi.org/10.1186/1471-2164-7-246
15. Stark R, Grzelak M, Hadfield J (2019) RNA sequencing: the teenage years. Nat Rev Genet 20(11):631–656. https://doi.org/10.1038/s41576-019-0150-2
16. Nayak R, Hasija Y (2021) A hitchhiker's guide to single-cell transcriptomics and data analysis pipelines. Genomics 113(2):606–619. https://doi.org/10.1016/j.ygeno.2021.01.007
17. Wu D-Y, Bittencourt D, Stallcup MR, Siegmund KD (2015) Identifying differential transcription factor binding in ChIP-seq. Front Genet 6. https://doi.org/10.3389/fgene.2015.00169

Chapter 24

Recombinant Production of Glycoengineered Mucins in HEK293-F Cells

Ling-Ting Huang, Marshall J. Colville, and Matthew Paszek

Abstract

Recombinant mucins are attractive polymeric building blocks for new biomaterials, biolubricants, and therapeutics. Advances in glycoengineered host cell systems now enable the recombinant production of mucins with tailored *O*-glycan side chains, offering new opportunities to tune the functionality of mucins and investigate the biology of specific *O*-glycan structures. Here, we provide a protocol for the scalable production of glycoengineered mucins and mucin-like glycoproteins in suspension-adapted HEK293-F cells. The protocol includes the preparation of engineered cell lines with homozygous knockout (KO) of glycosyltransferases using CRISPR/Cas9 and homology-directed repair (HDR) templates designed for efficient screening of clones. Strategies are provided for the stable introduction of mucin expression cassettes into the HEK293-F genome and the subsequent isolation of high-expressing cell populations. The high-titer production of recombinant mucins in conventional shaker flasks is described as an example production strategy using these cell lines.

Key words Mucin, Recombinant, CRISPR, Glycoengineering, HEK293-F, HEK, *O*-Glycan

1 Introduction

Mucins and mucin-like glycoproteins are characterized by their unstructured polypeptide chains that are densely grafted with *O*-GalNAc glycans linked to serine and threonine residues [1, 2]. The high hydrophilicity, low immunogenicity, and anti-adhesive properties of mucins make them attractive candidates for new biomaterials and biomedical device coatings [3, 4]. Mucins with engineered glycan structures are also attracting attention as potential therapeutics for diverse applications, including as immune modulators and antimicrobials [3, 4]. Since the mucin backbone sequence is genetically encoded, advances in custom gene synthesis can be leveraged to construct cDNAs for mucins with altered biochemical and biophysical properties [5, 6]. Furthermore, modern genome editing tools combined with greater understanding of *O*-glycan

Akihiko Kameyama (ed.), *Mucins: Methods and Protocols*, Methods in Molecular Biology, vol. 2763,
https://doi.org/10.1007/978-1-0716-3670-1_24,

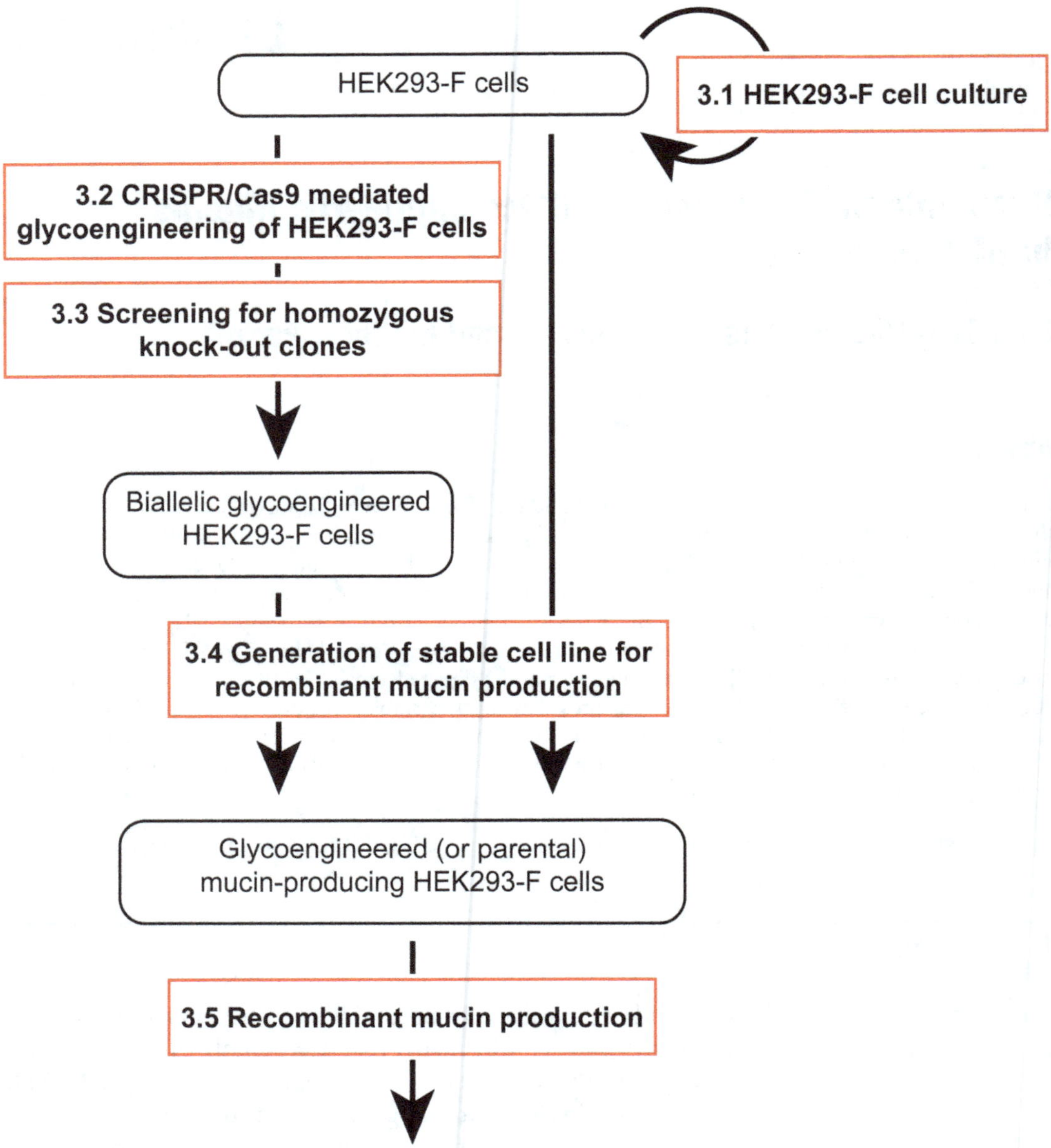

Fig. 1 Workflow of glycoengineered recombinant mucin production

biosynthetic pathways have made the engineering of *O*-glycans on recombinant mucins a practical reality [1, 7–9].

As summarized in Fig. 1, this protocol describes a complete workflow for the production of mucins with either native or engineered glycosylation patterns in HEK293-F cells, a suspension-adapted human embryonic kidney cell line. HEK293-F have emerged as a promising platform for recombinant production of mucins due to their recapitulation of human *O*-glycosylation, high productivity in serum-free media, and commercial availability, as well as the accessibility of *O*-glycosylation pathway knowledgebases that supports their glycoengineering [10, 11]. Here, glycoengineering is achieved through homozygous knockout (KO) of glycosyltransferases or other glycogenes with CRISPR/Cas9 genome editing technology. The protocol also takes advantage of modern transposon biotechnology for the stable incorporation of mucin expression cassettes into the HEK293-F genome.

1.1 CRISPR/Cas9-Mediated Cellular Glycoengineering

CRISPR has emerged as a powerful strategy for genome editing, including for the modification of genes that control glycosylation [8, 12, 13]. The CRISPR-associated protein Cas9 is an endonuclease that uses a guide RNA (gRNA) that directs Cas9 to a targeted DNA locus where it introduces a double-stranded (ds) DNA break [14]. The gRNA is a duplex of CRISPR RNA (crRNA) and trans-activating CRISPR RNA (tracrRNA). The crRNA recognizes the target DNA sequence through a complimentary 20-nucleotide (nt) spacer sequence (also referred to as a guide sequence), while the tracrRNA recruits Cas9 to form a gRNA/Cas9 ribonucleoprotein (RNP) complex (Fig. 2). Notably, a validated library of guide sequences is available for highly efficient CRISPR/Cas9 targeting of the complete glycosyltransferase genome in human cells, such as HEK293 [7, 13].

Following cleavage by Cas9, the targeted DNA is naturally repaired through one of two pathways in eukaryotic cells. Nonhomologous end joining (NHEJ) is an error-prone mechanism in which the broken ends of DNA are joined together without the use of a repair template. The alternative repair mechanism, called homology-directed repair (HDR), uses a homologous donor DNA template to repair the DNA in a precise way and allows for the introduction of new DNA sequences into the targeted locus. In this protocol, we leverage the HDR mechanism by introducing a synthetic single-stranded DNA (ssDNA) as an HDR donor oligonucleotide into the cells along with a recombinant Cas9 and gRNA duplex (Fig. 3). The ssDNA HDR donor oligonucleotide consists of a desired insert sequence flanked by 5′ and 3′ homology arms that match the sequences adjacent to the dsDNA break. Small molecules that inhibit the NHEJ repair pathway to increase HDR efficiency have been reported and are included in our protocol [15, 16].

One of the primary challenges in CRISPR/Cas9 protocols is the isolation of homozygous edited cell populations. Here we provide a protocol for generating homozygous KO of glycogenes using an HDR template that introduces an in-frame stop codon followed by a restriction recognition site into the targeted genomic locus. The incorporation of the restriction site enables simple and cost-effective screening to identify biallelic mutants by restriction enzyme digestion of an amplicon of the targeted DNA locus.

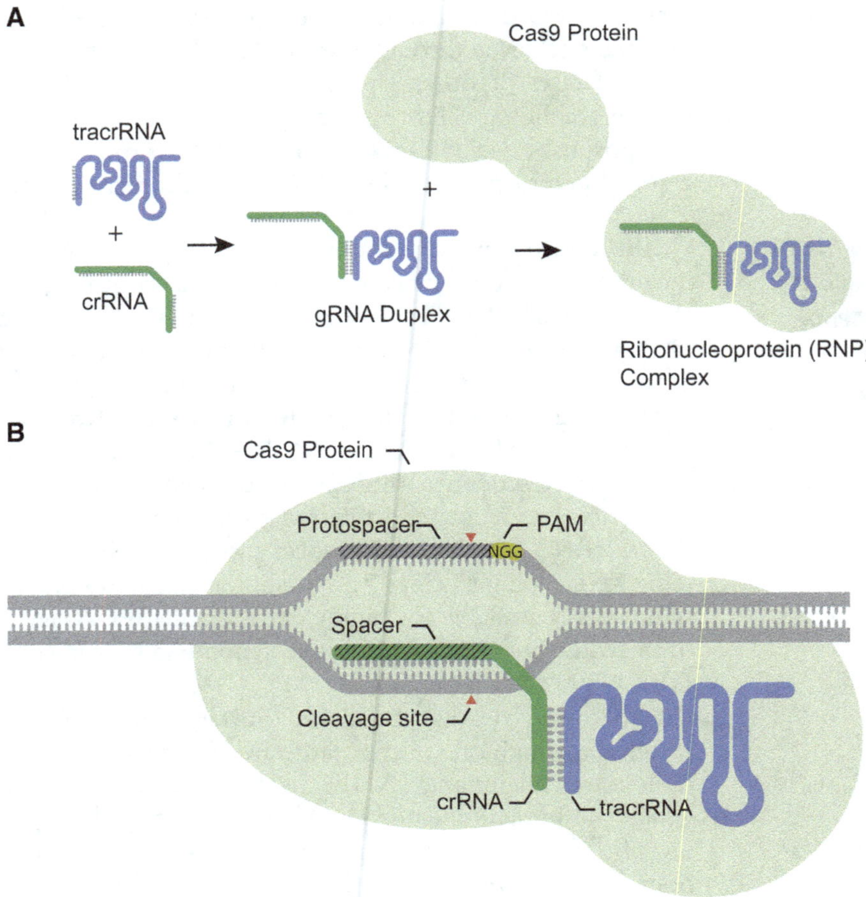

Fig. 2 CRISPR/Cas9 system. (**a**) Formation of ribonucleoprotein complex (RNP) comprised of Cas9 bound to the guide RNA (gRNA) duplex. (**b**) The 20-nucleotide (nt) spacer sequence of the gRNA directs Cas9 to a targeted DNA locus, where the Cas9 excises both DNA strands proximal to the protospacer adjacent motif (PAM: NGG for Streptococcus pyogenes (s.p.Cas9)

1.2 Generation of Mucin-Producing HEK293-F Cells Using The PiggyBac Transposon System

Stable integration of mucin cDNAs and expression cassettes into the HEK293-F cells is achieved using mobile genetic elements called transposons, which can be stably transferred from a DNA vector to the host cell genome [17]. Our protocol has adopted the moth-derived piggyBac transposon system due to its large DNA payload capacity and fidelity in transfer of more repetitive genes [18, 19]. We note that viral vectors, including retrovirus and lentivirus, are not recommended for mucins with extended tandem repeats due to the propensity for recombination of more repetitive DNA sequences with these systems [19, 20]. The piggyBac transposon system adopted in this method consists of two vectors. The first contains the transposable DNA element, which is comprised of an expression cassette for the desired mucin, as well as a second cassette for a selection marker, flanked by two terminal inverted repeats (IRs; *see* Fig. 4). The second vector encodes a hyperactive

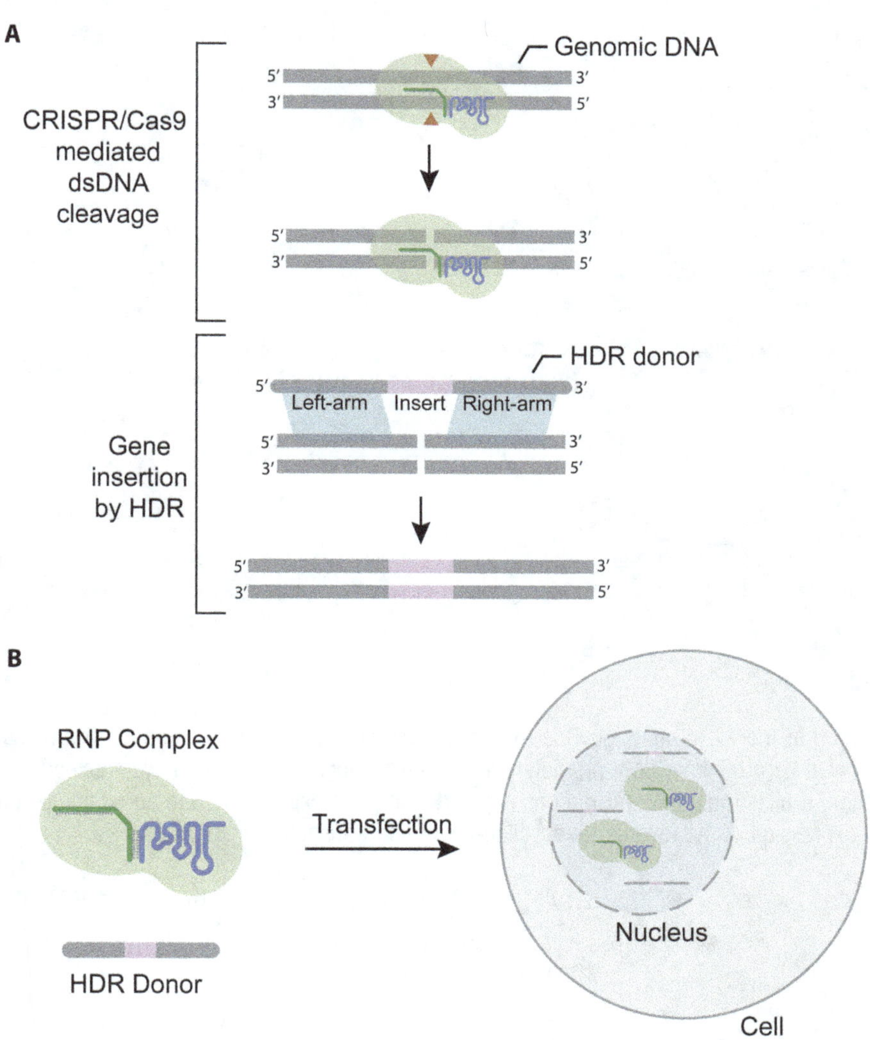

Fig. 3 Genomic DNA editing by CRISPR/Cas9 system and homology-directed repair (HDR). (**a**) HDR enables the insertion of new DNA sequences into the genome at CRISPR/Cas9 cleavage sites. In this protocol, the HDR template is designed to introduce an in-frame stop codon for gene knockout (KO) and a restriction site for screening purposes. (**b**) Recombinant Cas9 in complex with a gRNA duplex is delivered along with a synthetic HDR template into cells via transfection for genome editing

mutant of the piggyBac transposase, which catalyzes the transposon's integration into the HEK293-F genome through a "cut-and-paste" mechanism[21].

2 Materials

2.1 HEK293-F Cell Culture

All the supplies used for culturing, transferring, or containing the cells should be sterile. All the reagents used in cell culture should be tissue culture grade.

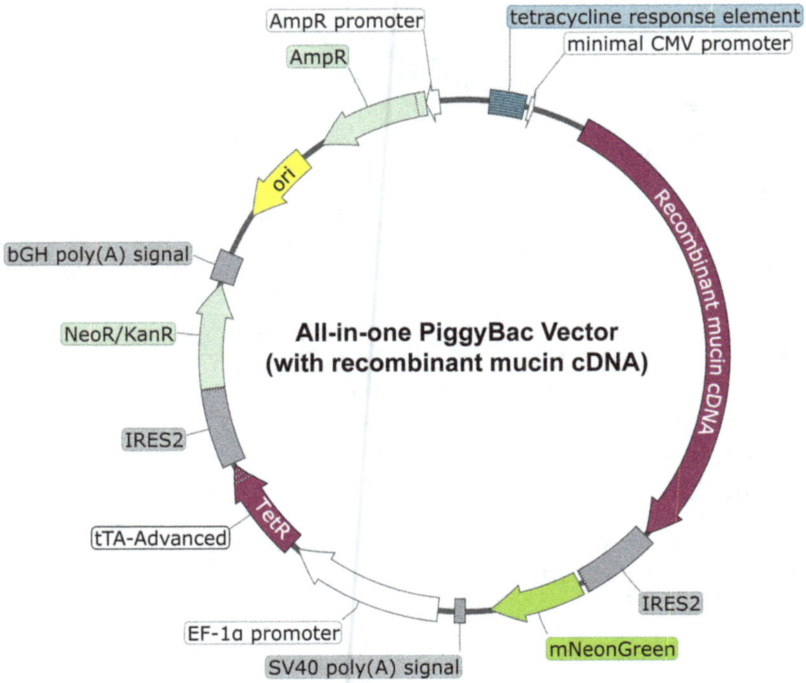

Fig. 4 DNA map of the all-in-one piggyBac expression vector. The vector contains a tetracycline response element followed by a minimal CMV promoter, recombinant mucin cDNA, a neomycin selection marker for G418 selection, a bicistronic mNeonGreen as a reporter of gene expression, and an additional cassette for constitutive expression of the reverse tetracycline-controlled transactivator (rtTA)

1. FreeStyle™ 293-F cells (Thermo Fisher, cat # R79007).
2. FreeStyle™ 293 expression medium (Thermo Fisher, cat # 12338018) (*see* **Note 1**).
3. Erlenmeyer shaker flasks with vent caps.
4. CO_2 incubator shaker operating at 37 °C supplied with 8% CO_2 in humidified air and shaking at 130 rpm.
5. Hemocytometer.
6. 0.4% trypan blue solution.
7. Water bath for warming cell culture media.
8. Light microscope for counting cells and monitoring viability.
9. Centrifuge.
10. Dimethyl sulfoxide (DMSO).
11. Cryovials (e.g., 2 mL cryogenic vial).
12. −80 °C freezer.
13. Controlled rate cell freezing device (e.g., CoolCell® freezing container).
14. Cryo-storage system with liquid nitrogen.

2.2 CRISPR/Cas9-Mediated Gene Editing

All tubes for DNA and RNA should be nuclease-free.

1. Alt-R™ s.p. Cas9 nuclease V3 (Integrated DNA Technologies, cat # 1081058).
2. Nuclease-Free Duplex Buffer (Integrated DNA Technologies, cat # 11-01-03-01).
3. Nuclease-Free IDTE buffer, pH 7.5 (Integrated DNA Technologies, cat # 11-01-02-02).
4. HDR enhancer V2 (Integrated DNA Technologies, cat # 10007910).
5. Alt-R® CRISPR-Cas9 crRNA (Integrated DNA Technologies, customized order): Resuspend the crRNA in nuclease-free IDTE buffer to make 100 μM stock solution. Store at −20°C.
6. Alt-R® CRISPR-Cas9 tracrRNA ATTO™ 550 (Integrated DNA Technologies, cat # 1075927): Resuspend the tracrRNA in nuclease-free IDTE buffer to make 100 μM stock solution. Store at −20°C.
7. Alt-R® HDR donor oligo (Integrated DNA Technologies, customized order): Resuspend the HDR donor oligo in nuclease-free IDTE buffer to make 100 μM stock solution. Store at −20°C.
8. Lipofectamine™ CRISPRMAX™ Cas9 Transfection Reagent Kit, including CRISPRMAX™ reagent and Cas9 PLUS reagent (Thermo Fisher, cat # CMAX00003).
9. Opti-MEM™ I reduced serumSerum medium (Thermo Fisher, cat # 31985062).
10. Heat block.
11. Fluorescent microscope for assessing transfection efficiency.

2.3 Clonal Isolation, Selection, and Expansion

1. 40 μm cell strainer mesh.
2. 96-, 48-, 12-, and 6-well tissue culture-treated microplates with lids.
3. T25 tissue culture flask (e.g., CELLSTAR® filter cap cell culture flasks, Greiner, cat # 690175, or the equivalent).

2.4 Screening and Detection of Mutations

1. Genomic DNA extraction kit (e.g., Quick-DNA Microprep Kit, Zymo Research, cat # D3020, or the equivalent).
2. Q5® Hot Start High-Fidelity 2× Master Mix (New England Biolabs, cat # M0494S).
3. PCR primers (e.g., custom DNA oligos, Integrated DNA Technologies, or the equivalents).
4. Nuclease-free water.

5. Thermocycler.
6. PCR cleanup kit (e.g., QIAquick PCR purification kit, QIAGEN, cat # 28104, or the equivalent).
7. Restriction enzyme: EcoRI-HF (New England Biolabs, cat # R3101S) or similar products (*see* **Note 2**).
8. 1× TAE buffer: 40 mM Tris base, 20 mM acetic acid and 1 mM EDTA.
9. Microwave.
10. Microwave-safe flask.
11. Electrophoresis-grade agarose powder.
12. 10,000× GelGreen® nucleic acid gel stain (Biotium, cat # 41005).
13. Gel tray, well comb(s), and gel tank.
14. DNA loading dye.
15. DNA molecular weight ladder.
16. Gel imaging system.
17. Power supply.

2.5 Generation of Stable Cell Line for Recombinant Mucin Production

All the reagents used for DNA transfection should be sterile and nuclease-free. All the reagents used for selection and induction should be sterile and tissue culture grade.

1. PiggyBac expression vector [5] and piggyBac hyperactive transposase vector [21] (*see* **Note 3**).
2. OptiPRO™ SFM (Thermo Fisher, cat # 12309050).
3. PEI MAX®-transfection grade linear polyethylenimine hydrochloride (MW 40,000) (Polysciences, cat # 24765): To make PEI stock solution, dissolve PEI MAX® powder in ultrapure water at 1 mg/mL. Filter through a 0.22 μm sterile filter and store at 4 °C.
4. Geneticin (G418): To make G418 stock solution, rehydrate the G418 powder with FreeStyle™ 293 expression medium to a concentration of 500 mg/mL and filter through a 0.22 μm sterile filter. Store at −20°C.
5. Doxycycline: To make doxycycline stock solution, dissolve doxycycline powder in ultrapure water at 1 mg/mL. Filter through a 0.22 μm sterile filter and store at −20 °C.
6. FACS tubes with cell strainer caps.

3 Methods

3.1 HEK293-F Cell Culture

HEK293-F cells should be handled in a biosafety cabinet (BSC) for sterile work. The cell growth medium is FreeStyle™ 293 expression medium, pre-warmed to 37 °C before use, unless otherwise noted. The cells are incubated at 37 °C in an incubator shaker supplied with 8% CO_2 in humidified air and shaking at 130 rpm, unless otherwise noted. All centrifugation steps are at room temperature unless otherwise noted.

3.1.1 Thawing Cells

1. Thaw a vial containing 1×10^7 frozen cells in a 37 °C water bath, keeping the tube in constant motion. Remove immediately upon thawing of the last visible piece of ice.
2. In a biosafety cabinet (BSC), transfer the cells immediately into a 15 mL centrifuge tube with a sterile pipette and add 5 mL of pre-warmed FreeStyle™ 293 expression medium. Gently pipette the cells up and down to resuspend and break up any cell clumps.
3. Transfer the cell suspension into a 125 mL shaker flask containing 24 mL of pre-warmed growth medium.
4. Transfer the flask to an incubator shaker and culture under agitation.

3.1.2 Measurement of Cell Density and Viability

1. In a BSC, transfer an aliquot of 10 μL of cell suspension from the shaker flask into a sterile 1.5 mL microcentrifuge tube. Return the culture flask to the incubator shaker.
2. Mix the aliquot of the cell suspension with 10 μL of 0.4% trypan blue and pipette up and down to mix.
3. Determine the cell density and viability using a hemocytometer following standard conventions.

3.1.3 Propagation of HEK293-F Suspension Cultures

1. Incubate the cells in a shaker flask under constant agitation in an incubator shaker.
2. Monitor the cell density and viability every 1–2 days.
3. Passage the cells whenever the density reaches $1–3 \times 10^6$ cells/mL. Passage by diluting the cells to $0.2–0.3 \times 10^6$ cells/mL in a volume of growth medium that is recommended for the flask size (Table 1) (*see* **Notes 4** and **5**).

3.1.4 Generation of HEK293-F Conditioned Medium

1. Seed exponentially growing cells at $0.2–0.3 \times 10^6$ viable cells/mL in 30 mL of pre-warmed growth medium in a 125 mL shaker flask.
2. Incubate the cells in the incubator shaker under constant shaking until the cell density reaches $1–1.5 \times 10^6$ viable cells/mL without diluting or exchanging the medium, typically 48 h.

Table 1
Recommended fill volume and shaking speed for different sizes of flasks

Flask size	Recommended fill volume (mL)	Recommended shaking speed (rpm) for 25-mm orbit diameter
25 mL	2–5	130
50 mL	5–15	130
125 mL	15–30	130
250 mL	30–60	130
500 mL	60–150	130
1 L	150–350	130
2 L	350–400	90–130 (lower speed if foaming)

3. In a BSC, transfer the culture into a sterile 50 mL centrifuge tube.
4. Pellet the cells by centrifuging at 3000 × *g* for 10 min.
5. In a BSC, collect the conditioned medium (supernatant) and filter it through a 0.22 μm syringe filter.
6. Aliquot the conditioned medium into sterile centrifuge tubes and store at 4 °C for up to a week.

3.1.5 Cryopreservation

1. Grow cells to 1–2 × 10^6 viable cells/mL in shaker flasks of the desired size. A 30 mL culture in a 125 mL flask is sufficient to make 3–6 cryovial stocks with 10^7 cells each.
2. Prepare freezing medium by mixing equal volumes of fresh and conditioned growth media and adding DMSO to 10%. Chill and keep the freezing medium on ice until needed.
3. Count cells and determine viability.
4. Transfer cells to a 50 mL centrifuge tube and pellet by centrifuging at 100–150 × *g* for 5 min.
5. Decant the supernatant and resuspend the cell pellet in ice-cold freezing medium to a final cell density of 1 × 10^7 viable cells/mL.
6. Transfer 1 mL of the cell suspension into each cryovial.
7. Transfer the filled cryovials to a controlled rate cell freezing device, and place in a −80 °C freezer overnight.
8. The following day, transfer the frozen cells to a cryo-storage system supplied with liquid nitrogen for long-term storage.

Table 2
Example crRNA, tracrRNA, and HDR donor sequences for CRISPR/Cas9 targeting of *C1GALT1* using the Alt-R® system

crRNA	5′-GUAAAGCAGGGCUACAUGAGGUUUUAGAGCUAUGCU-3′
tracrRNA	5′-AGCAUAGCAAGUUAAAAUAAGGCUAGUCCGUUAUCAACUUGAAAAA GUGGCACCGAGUCGGUGCUUU-3′
HDR donor	5′-CTTTGGGAGAAGATTTAAGCCTTATGTAAAGCAGGGCTACTAAGAATT CAGTGGAGGAGCAGGATATGTACTAAGCAAAGAAGCCTTGA-3′

3.2 CRISPR/Cas9-Mediated Gene Editing

The general strategy for the design of a gRNA is described. The sequences and strategy for KO of the *C1GALT1* gene, which encodes the core 1 synthase, is provided as a specific example.

3.2.1 Design of CRISPR/Cas9 gRNA for Gene KO

1. Identify a 20-nt spacer sequence that recognizes the target genomic locus. For example, from the validated library of spacer sequences for targeting human glycosyltransferases [13], we use GTAAAGCAGGGCTACATGAG for targeting *C1GALT1* (*see* **Note 6**).
2. Construct a crRNA sequence through the fusion of the 20-nt spacer sequence with the 16-nt Alt-R™ complementary tracrRNA sequence, GUUUUAGAGCUAUGCU [22]. Table 2 shows the sequences for the example Alt-R® crRNA for *C1GALT1* editing and the 67-nt Alt-R® CRISPR-Cas9 tracrRNA [22], which can be ordered as a custom Alt-R® oligo pre-labeled with ATTO™ 550 from the manufacturer (*see* **Note 7**).

3.2.2 Manual Design of the ssDNA HDR Template

1. Find the whole genomic DNA sequence for the targeted gene including its introns. Sequences can be found in the NCBI database (https://www.ncbi.nlm.nih.gov).
2. Import the DNA sequence into gene editing software, such as SnapGene, and annotate the exons and translated amino acid sequence. The exon information can be found in the NIH consensus coding sequence (CCDS) database.
3. Search and locate the protospacer sequence (crRNA binding site) and the protospacer adjacent motif (PAM) for your gRNA. The PAM sequence for *Streptococcus pyogenes* (Sp) Cas9 is 5′-NGG-3′. Note that the dsDNA cleavage site is typically 3–4 nt upstream of the PAM.
4. Edit the insertion sequence to introduce an in-frame stop codon and a unique restriction site, such as EcoRI (GAATTC) (*see* **Note 8**). Mutation sites closer to the Cas9 cleavage site usually show higher DNA editing efficiency. In the *C1GALT1* KO example, the methionine codon at the cleavage site is

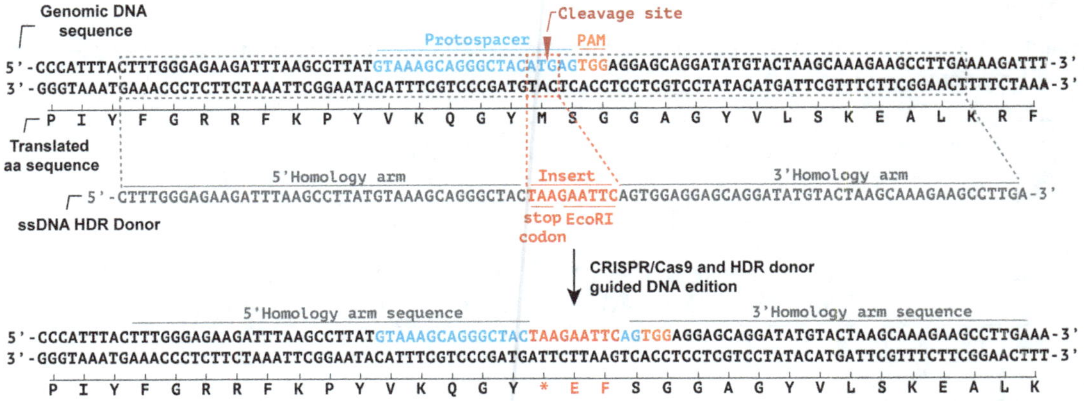

Fig. 5 Illustration of DNA editing with HDR donor template

replaced by an insertion sequence that contains an in-frame stop codon followed by an EcoRI recognition sequence (insertion sequence: TAAGAATTC) (Fig. 5).

5. The length of homology arms for the ssDNA HDR donor oligos is typically 30–60 nt on both ends of the insertion sequence [23]. In the example of the ssDNA HDR donor for *C1GALT1* KO, 40-nt homology arms have been included both upstream and downstream of the insertion sequence (Fig. 5). The 5′-homology arm has the 40-nt sequence upstream of the methionine codon and the 3′-homology arm has the 40-nt sequence downstream of the methionine codon. Table 2 shows the sequences for the example of ssDNA HDR donor.
6. The ssDNA donor oligo with Alt-R® HDR modification, which increases DNA stability, is ordered from Integrated DNA Technologies (*see* **Note 9**).

3.2.3 Software-Assisted Design of the ssDNA HDR Template

Software tools can assist with the design of the HDR donor template. The method here uses the IDT Alt-R® HDR Design Tool: (https://www.idtdna.com/site/order/designtool/index/HDRDESIGN).

1. On the design tool webpage, choose the species and enter the gene symbol for a target gene. Figure 6 illustrates an example for *C1GALT1* KO.
2. Input a validated spacer sequence [13] that targets the gene.
3. Click on the Search button to go to the page for interactive DNA editing (Fig. 6b).
4. On the DNA editing page, move the sliders on the DNA sequence to locate the mutation site of your choice. Mutation sites closer to the Cas9 cleavage site, 3–4 nt upstream of the PAM, usually show higher DNA editing efficiency.

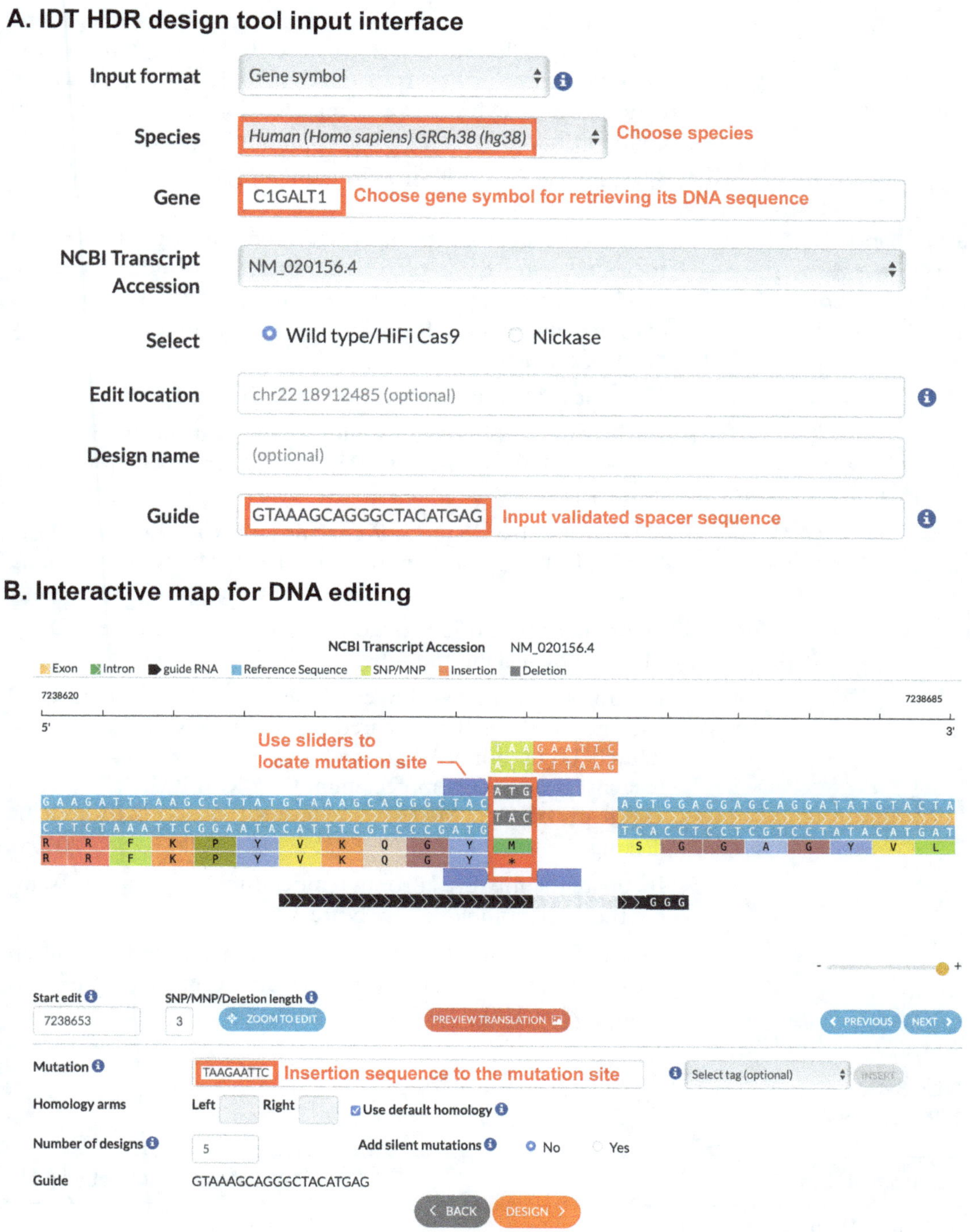

Fig. 6 IDT HDR design tool.

5. Input the desired insertion sequence that includes an in-frame stop codon followed by a restriction site for screening. In the example shown in Fig. 6 for *C1GALT1* KO, the methionine codon near the PAM site is mutated into an in-frame stop codon followed by an EcoRI recognition sequence.
6. Click on the Design button to get the HDR donor design.

3.2.4 Transfection of HEK293-F Cells with CRISPR/Cas9 Machinery and HDR Template

1. In a sterile microcentrifuge tube, combine the crRNA and tracrRNA in equimolar concentrations and dilute in Nuclease-Free Duplex Buffer to a final concentration of 5 μM of each component to form 5 μM of gRNA duplex. Mix well by pipetting up and down.
2. Heat the gRNA duplex solution at 95 °C for 5 min.
3. Remove the gRNA duplex solution from heat and allow to cool to room temperature (20–25 °C) on the bench top (*see* **Note 10**).
4. Dilute the Alt-R® S.p. Cas9 enzyme (*see* **Note 11**) to 5 μM in Opti-MEM™ in a microcentrifuge tube. Mix well by flicking the tube.
5. Dilute the HDR donor oligo to 5 μM in nuclease-free IDTE. Mix well by flicking the tube.
6. In a microcentrifuge tube, combine 5 μL of the diluted Cas9 enzyme, 5 μL of the prepared gRNA duplex, 1.7 μL of the diluted HDR donor oligo, and 12.5 μL of the Cas9 PLUS reagent (from the Lipofectamine™ CRISPRMAX™ kit) in 100.8 μL of Opti-MEM™ medium (Table 3). Mix well by flicking the tube.
7. Incubate the mixture at room temperature for 5 min to assemble the RNP complexes (*see* **Note 12**).
8. In a BSC, transfer 2×10^5 cells (for 1 transfection) from an actively growing culture into a 15 mL centrifuge tube and centrifuge at 100–150 × *g* for 5 min.

Table 3
RNP complex

Component	Amount for 1 transfection (μL)
Opti-MEM™ medium	100.8
5 μM Cas9 enzyme	5
5 μM gRNA oligos	5
5 μM HDR donor oligo	1.7
Cas9 PLUS™ reagent (from the Lipofectamine™ CRISPRMAX™ kit)	12.5
Total	125

Table 4
FreeStyle™ 293 expression medium with HDR enhancer

Component	Amount for 1 transfection (μL)
3 mM Alt-R® HDR enhancer V2	25
FreeStyle™ 293 expression medium	1475
Total	1500

Table 5
Diluted CRISPRMAX reagent

Component	Amount for 1 transfection (μL)
Opti-MEM™ medium	119
CRISPRMAX™ reagent (from the Lipofectamine™ CRISPRMAX™ kit)	6
Total	125

Table 6
Transfection complex mixture

Component	Amount for 1 transfection (μL)
RNP complex	125
Diluted CRISPRMAX™ reagent	125
Total	250

9. In a BSC, decant the supernatant and resuspend the cells into 750 μL of pre-warmed FreeStyle™ 293 expression medium.
10. Prepare 1.5 mL of fresh growth medium with 50 μM of HDR enhancer (for 1 transfection) in a 15 mL centrifuge tube (Table 4).
11. Dilute the CRISPRMAX™ reagent by adding 6 μL of CRISPRMAX™ reagent (from the Lipofectamine™ CRISPR-MAX™ kit) into 119 μL of Opti-MEM™ medium (Table 5) in a microcentrifuge tube and mix well by flicking the tube.
12. In a microcentrifuge tube, combine 125 μL of RNP complex and 125 μL of diluted CRISPRMAX™ reagent to form the transfection complex (Table 6). Mix well by flicking the tube.
13. Incubate the transfection complex mixture for 5–10 min at room temperature.

Table 7
Transfection culture mixture

Component	Amount for 1 transfection (μL)
FreeStyle™ 293 expression medium with HDR enhancer V2	1500
Transfection mixture	250
Cell suspension	750
Total	2500

14. In a 25 mL shaker flask, combine the prepared 750 μL cell suspension, 250 μL transfection complex mixture, and 1.5 mL growth medium with HDR enhancer V2 to form the transfection culture mix (Table 7).
15. Incubate the transfection culture in the incubator shaker for 24 h.
16. Twenty-four hours after transfection, count the cells and calculate the viability. Check the transfection efficiency by sampling the cells and viewing the ATTO555 signal in cells with an epi-fluorescence microscope.
17. Thirty-six to forty-eight hours after transfection, count the cells and calculate the viability.
18. In a BSC, transfer the cell suspension into a centrifuge tube and centrifuge at 100–150 × *g* for 5 min and discard the supernatant to remove the transfection complex and HDR enhancer.
19. Resuspend the cells in 1–2 mL of pre-warmed growth medium containing 25% of conditioned medium at 0.2–0.5 × 10^6 viable cells/mL. Transfer the cell suspension into a new 25 mL shaker flask. Return the shaker flask to the incubator shaker.
20. Monitor the cell density and viability every day. When the cell density reaches 1–2 × 10^6 viable cells/mL, dilute the cells to 0.2–0.5 × 10^6 viable cells/mL with pre-warmed growth medium in a 25 mL or 50 mL shaker flask. Return the shaker flask to the incubator shaker.
21. Monitor the cell density and viability every day. When the cell density reaches 1–2 × 10^6 viable cells/mL, maintain the cells according to Subheading 3.1.3 until the cells exhibit a doubling time of 20–24 h and cell viability >90% (*see* **Note 13**). Proceed to Subheading 3.2.5 to isolate biallelic KO clones.

3.2.5 Clonal Expansion of HEK293-F Cells

Before clonal selection, make sure the cells exhibit a doubling time of 20–24 h and cell viability >90%. Monoclonal cells are isolated by limiting dilution.

1. In a BSC, break up any cell clumps by passing the cells through a cell strainer mesh.

2. Count cells and calculate the viability (*see* **Note 14**).
3. Dilute cells to 5 viable cells/mL in pre-warmed FreeStyle™ 293 expression medium with 25% conditioned medium. Ten milliliters of diluted cell suspension will be needed for each 96-well plate.
4. Add 100 μL of the 5 cells/mL solution into each well of a 96-well plate (*see* **Note 15**). Return the 96-well plate to the incubator *without shaking*.
5. Culture until distinct colonies are visible using a phase contrast or brightfield microscope, typically 4–6 days (*see* **Note 16**).
6. Identify wells with single colonies for continued clonal expansion (*see* **Note 17**).
7. Grow clonal cells in each well to near confluency. To reach confluency faster, the cells of a colony can be dispersed in the well by gently pipetting the cells up and down within a BSC (*see* **Note 18**).
8. Passage cells to a 48-well plate: Detach and resuspend the cells by gently pipetting the cells up and down (*see* **Note 19**). Transfer all the cell suspension to a well of a 48-well plate and mix with 200 μL of pre-warmed growth medium containing 25% conditioned medium. Return the 48-well plate to the incubator *without shaking*.
9. When the monoclonal cells in a well of a 48-well plate become 80–90% confluent, detach and resuspend the cells by gently pipetting the cells up and down. Transfer all the cells to a well of a 12-well plate and mix with 1 mL of pre-warmed growth medium containing 25% conditioned medium. Return the 12-well plate to the incubator *without shaking*.
10. When the monoclonal cells in a well of a 12-well plate become 80–90% confluent, detach and resuspend the cells by gently pipetting the cells up and down. Collect one-third of the cell suspension in a microcentrifuge tube for genomic DNA extraction and transfer two-thirds of the cells to a well of a 6-well plate and mix with 1 mL of pre-warmed growth medium (*see* **Notes 20** and **21**). Return the 6-well plate to the incubator *without shaking*.
11. Pellet the cells collected for genomic DNA extraction by centrifuging at 500 × *g* for 5 min. Discard the supernatant and proceed to Subheading 3.3 to screen for clones with biallelic KO of the targeted gene.
12. Continue expanding the positively identified biallelic KO clones. When cells in a well of a 6-well plate become 80–90% confluent, detach and resuspend the cells by gently pipetting the cells up and down. Transfer all the cell suspension to a T25 flask and add 3 to 4 mL of FreeStyle™ 293 expression medium. Return the T25 flask to the incubator *without shaking*.

13. When the cells grow to 90% confluency in the T25 flask, detach and resuspend the cells by gently pipetting the cell culture up and down. Transfer all the culture to a 25 mL shaker flask. Return the 25 mL shaker flask to the incubator shaker at 130 rpm.
14. Monitor the cell density and viability every day.
15. When the cell density reaches 1–2 × 10^6 viable cells/mL, transfer the cells to a new 125 mL shaker flask and dilute to 0.2–0.5 × 10^6 viable cells/mL by adding 10–30 mL of pre-warmed growth medium. Return the shaker flask to the incubator shaker.
16. Follow Subheading 3.1.3 to maintain the clonal cell cultures until the cells exhibit a doubling time of 20–24 h and cell viability >90% (*see* **Note 22**).

3.3 Screening for Homozygous KO Clones

3.3.1 Polymerase Chain Reaction (PCR) Amplification of Genomic DNA

1. Design and order PCR primers that flank the target locus of the edited gene (*see* **Note 23**). In the example for *C1GALT1* KO, the forward and reverse primers are 5′-TCAAAACCTA GAGAAAAAGGCCAAACAC-3′ and 5′-GGAGGATAATAGT TGTAATTCCAGTACCAAAAC-3′.
2. Extract the genomic DNA from each clonal cell pellet (from Subheading 3.2.5 **step 11**) using the genomic DNA extraction kit.
3. For each clone, in a PCR tube or a well of a 96-well PCR plate on ice, prepare the PCR reaction mix by combining 2 μL of nuclease-free water, 1 μL of 5 μM forward PCR primer, 1 μL of 5 μM reverse PCR primer, 1 μL of extracted genomic DNA, and 5 μL of Q5® Hot Start High-Fidelity 2× Master Mix (Table 8) (*see* **Note 24**). Mix well by pipetting up and down several times.
4. Transfer the PCR tubes/plates to a thermocycler and run the PCR reaction using the sequence in Table 9.
5. Purify the PCR products using the PCR cleanup kit.

Table 8
PCR reaction mix

Component	Amount for 1 reaction (μL)
Nuclease-free water	2
Q5® hot start high-Fidelity 2× master mix	5
5 μM forward primer	1
5 μM reverse primer	1
Genomic DNA	1
Total	10

Table 9
Thermocycling sequence for amplification of targeted genomic locus. The anneal temperature and extension time for the *C1GALT1* example are shown

Step	Temp. (°C)	Time
Initial denaturation	98	30 s
30–35 cycles	98 67 (*see* **Note 35**) 72	10 s 20–30 s 15 s (*see* **Note 36**)
Final extension	72	2 m

Table 10
Digestion mix

Component	Amount for 1 reaction (μL)
Nuclease-free water	7.7
Purified PCR product (from Subheading 3.3.1 **step 5**)	1
10× buffer	1
EcoRI-HF enzyme	0.3
Total	10

3.3.2 Restriction Digestion of PCR Product

1. For each clone, in a PCR tube or a well of a 96-well plate on ice, make the digestion mix by combining 7.7 μL of nuclease-free water, 1 μL of purified PCR product, 1 μL of restriction enzyme buffer, and 0.3 μL of restriction enzyme. Table 10 shows the mixture for the *C1GALT1* KO example. Mix well by pipetting up and down several times.
2. Incubate the reaction mix at 37 °C for 1 h.

3.3.3 Analysis of Digested Product on a DNA Gel

1. Add electrophoresis-grade agarose powder to 1× TAE buffer in a microwave-safe flask to make a 0.8% agarose solution.
2. Microwave the agarose solution. Stop the microwave instantly when the solution is about to boil. Swirl the flask to mix. Repeat microwaving and swirling until the agarose is completely dissolved (*see* **Note 25**).
3. Add 10,000× GelGreen® to the agarose solution to a final 1× concentration. Thoroughly mix the solution.
4. Pour the agarose solution into a gel tray and set well comb(s) in place. Once the gel solidifies, remove the well comb(s) and place the gel into a gel tank. Fill the gel tank with 1× TAE buffer until the gel is completely covered.

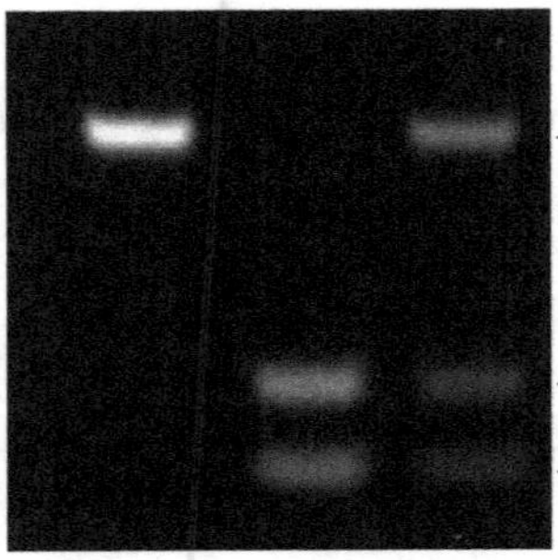

Fig. 7 DNA gel electrophoresis for validating homozygous mutation. The genomic DNAs from isolated clones were extracted, PCR amplified, and restriction digested. The representative digestion results are shown: (1) parental, (2) homozygous KO, and (3) heterozygous KO

5. For each clone, add concentrated DNA loading dye to the digested product to a final 1× concentration. Mix the digested PCR product by pipetting up and down.
6. Load the digested product into the wells of the agarose gel. Spare one well for DNA molecular weight ladder.
7. Connect the gel tank to a power supply and run the gel at 100 V until the loading dye migrates ~80% of the gel length (typically, 60–90 min).
8. Remove the gel from the gel tank and place it into a gel imaging system to image the gel.
9. Identify the biallelic KO clones from the restriction digest. The homozygotes will show complete digestion of the PCR product, whereas heterozygotes will show partial digestion (Fig. 7).
10. Confirm in-frame introduction of the stop codon by Sanger sequencing (Fig. 8): Submit purified PCR products (from Subheading 3.3.1 **step 5**) of the clones with homozygous restriction sites for Sanger sequencing (*see* **Note 26**).

3.4 Generation of Stable Cell Line for Recombinant Mucin Production in Glycoengineered or Parental HEK293-F Cells

3.4.1 Stable Introduction of Mucin Expression Cassettes into HEK293-F Cells

1. Follow Subheading 3.1.3 to grow glycoengineered or parental HEK293-F cells to a density of 1–2 × 10^6 cells/mL as a starter culture.
2. Twenty-four hours before transfection, dilute a sufficient volume of the starter culture in fresh growth medium to yield a final viable cell density of 0.5–0.6 × 10^6 cells/mL in 30 mL FreeStyle™ 293 expression medium. Transfer to a 125 mL Erlenmeyer flask and return to the incubator shaker.
3. Incubate for 24 h in the incubator shaker (*see* **Note 27**).
4. On the day of transfection, prepare two 15 mL conical tubes.

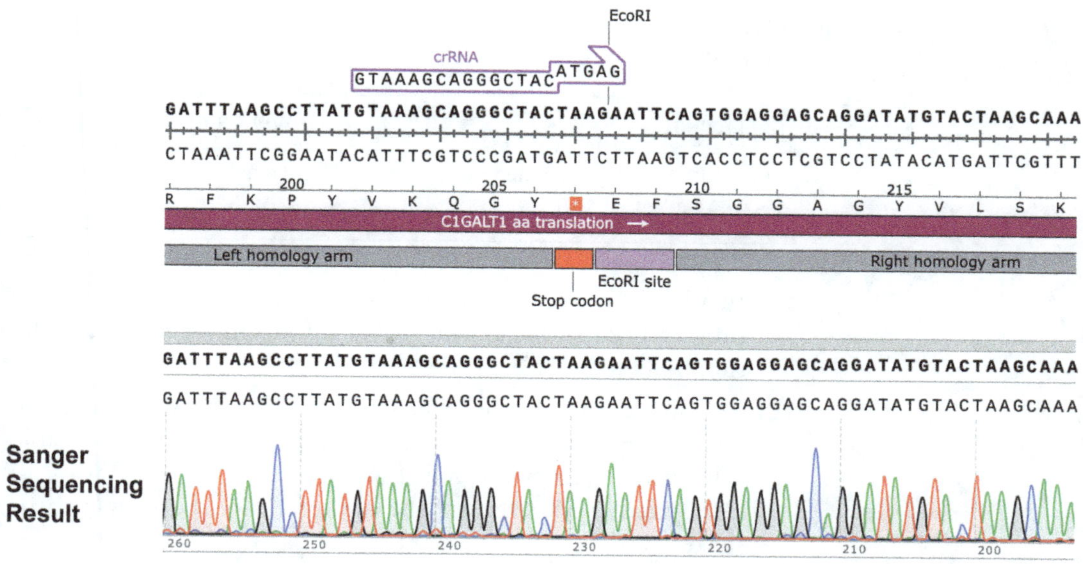

Fig. 8 Sanger sequencing result to confirm the in-frame introduction of the stop codon. The figure shows the sequencing analysis of the targeted genomic region, verifying the successful insertion of an in-frame stop codon in the gene of interest.

Table 11
DNA mix

Component	Amount for 1 transfection
OptiPRO™ Serum free medium	1.5 mL
PiggyBac expression vector	15 μg
Hyperactive transposase vector	15 μg

5. In one tube, mix 15 μg of piggyBac recombinant mucin expression plasmid and 15 μg of hyperactive transposase plasmid in 1.5 mL OptiPRO™ SFM (*see* **Note 28**) (Table 11).
6. In the other tube, mix 90 μL of PEI max stock solution (1 mg/mL) with 1.5 mL OptiPRO™ SFM to make 60 μg/mL PEI solution in OptiPRO™ SFM (Table 12).
7. Transfer the PEI solution (from **step 6**) to the DNA mix (from **step 5**).
8. Gently flick the tube to mix and incubate at room temperature for 30 min.
9. Add the PEI–DNA mixture (from **step 8**) to the cell culture (from **step 3**). Return the flask to the incubator shaker and incubate for 24 h.
10. After the 24 h, count cells and calculate the viability.

Table 12
PEI solution

Component	Amount for 1 transfection
OptiPRO™ Serum free medium	1.5 mL
PEI max stock solution (1 mg/mL)	90 μL

11. In a BSC, transfer the cell suspension into a 50 mL centrifuge tube and centrifuge at 100–150 × *g* for 5 min and discard the supernatant.
12. Resuspend the cells in FreeStyle™ 293 expression medium containing 25% of conditioned medium at 0.2–0.5 × 10^6 viable cells/mL. Transfer the cell suspension into a 125 mL or 250 mL shaker flask. Return the shaker flask to the incubator shaker.
13. After 24 h, count cells and calculate the viability.
14. In a BSC, start selection by adding geneticin (G418) to the culture to give a final concentration of 500 μg/mL. Return the shaker flask to the incubator shaker.
15. Maintain cells according to Subheading 3.1.3 in selection medium (FreeStyle™ 293 expression medium and 500 μg/mL G418) (*see* **Note 29**).
16. Continue maintaining the cells in selection medium until the viability recovers to >90%. Selection usually takes 10–14 days (*see* **Note 30**).
17. Following selection, culture and maintain cells in FreeStyle™ 293 expression medium without G418 until the cells exhibit a doubling time of 20–24 h and cell viability >90%. Follow the cryopreservation protocol in Subheading 3.1.5 to make cryo-stocks.

3.4.2 Isolation of High Mucin-Expressing Subpopulations by Fluorescence Activated Cell Sorting (FACS)

Following selection and recovery, stable HEK293-F cells can be sorted for high expression using the mNeonGreen reporter, if desired.

1. Following selection and recovery of stable mucin-expressing HEK293-F cells, seed the cells into 30 mL of FreeStyle™ 293 expression medium at 0.3 × 10^6 cells/mL. Return the cell culture to the incubator shaker and incubate for 24 h.
2. After 24 h, add 30 μL of 1 mg/mL doxycycline to the culture (1 μg/mL final concentration) in a BSC. Return the culture to the incubator shaker and incubate for 24 h.
3. After 24 h, transfer the culture to a 50 mL centrifuge tube. Centrifuge the cells at 100–150 × *g* for 5 min.

4. In a BSC, decant the supernatant and resuspend the cell pellet in pre-warmed FreeStyle™ 293 expression medium at 2–5 × 10^6 viable cells/mL (*see* **Note 31**).
5. Transfer 3 mL of the transfected cell suspension to a FACS tube by using a 1 mL micropipette to transfer and filter the cell suspension through the cell strainer mesh on top of the FACS tube (*see* **Note 32**).
6. Sort the cell subpopulation that is top 20% among the mNeonGreen-positive population (*see* **Notes 33** and **34**).
7. Centrifuge the collected cells at 100–150 × *g* for 5 min.
8. In a BSC, decant the supernatant and resuspend the cell pellet in FreeStyle™ 293 expression medium with 25% conditioned medium at 0.2–0.5 × 10^6 cells/mL in a 25 mL shaker flask. Return the shaker flask to the incubator shaker.
9. Monitor the cell density and viability every day. When the viable cell density reaches 1–2 × 10^6 cells/mL, dilute the cells to 0.2–0.5 × 10^6 viable cells/mL with pre-warmed FreeStyle™ 293 expression medium and transfer the diluted cell to a 25 mL or 50 mL shaker flask. Return the shaker flask to the incubator shaker.
10. When the viable cell density reaches 1–2 × 10^6 viable cells/mL, transfer the cells to a 125 mL shaker flask and dilute to 0.2–0.5 × 10^6 viable cells/mL with pre-warmed FreeStyle™ 293 expression medium. Return the shaker flask to the incubator shaker.
11. Follow Subheading 3.1.2 to keep culturing the cell. When the cells recover from sorting and exhibit a doubling time of 20–24 h and cell viability >90%, follow the cryopreservation protocol in Subheading 3.1.5 to make cryo-stocks.

3.5 Recombinant Mucin Production

In this section, we demonstrate the batch production of secreted mucin in 2-L shaker flasks beginning with a 30 mL starter culture.

1. Seed mucin-expressing cells into a 125-mL shaker flask containing 30 mL of pre-warmed FreeStyle™ 293 expression medium at 0.2–0.5 × 10^6 cells/mL. Return the shaker flask to the incubator shaker.
2. When the cell density reaches 1–2 × 10^6 viable cells/mL in the 125-mL flask, transfer the cells to a 500 mL shaker flask and dilute to 0.2–0.5 × 10^6 viable cells/mL by adding up to 120 mL of pre-warmed growth medium. Return the shaker flask to the incubator shaker.
3. When the cell density reaches 1–2 × 10^6 viable cells/mL in the 500 mL flask, transfer sufficient amount of cells into a 2-L shaker flask and dilute the cells with pre-warmed growth medium to a total volume of 400 mL with 0.2–0.5 × 10^6 viable cells/mL. Return the shaker flask to the incubator shaker.

4. When the cell density reaches 1–1.5 × 10^6 cells/mL in the 2-L shaker flask, add 400 μL of 1 mg/mL doxycycline stock solution to a final concentration of 1 μg/mL. Return the shaker flask to the incubator shaker and monitor the cell density and viability every day.
5. Add 400 μL of 1 mg/mL doxycycline stock solution every 48 h.
6. When the cell viability drops below 80% (typically 4–8 days), centrifuge the culture at 3000 × *g* for 15 min and collect the supernatant. The mucin-containing medium/supernatant can be aliquoted and stored at −80 °C until purification.

4 Notes

1. Unless otherwise noted, the FreeStyle™ 293 expression medium is used in cell culture routine as well as protein expression.
2. The restriction enzyme is used to verify the HDR-mediated insertion of the template DNA into the targeted locus. A restriction site that is unique in the immediate vicinity of the targeted locus should be chosen for the HDR template that introduces the in-frame stop codon and restriction site. For the *C1GALT1* examples, we use EcoRI, but you can use other restriction enzymes of your choice.
3. The quality of the plasmid preparation is critical for transfection. Use a commercially available Midi- or Maxi-prep kit that has an endotoxin removal step.
4. When passaging the cells, there is no need to centrifuge the cells unless otherwise noted.
5. To avoid clumping, cells should not be allowed to settle for extended periods during routine culture and subculture. The duration of time that the culture is kept static without shaking should be minimized.
6. Available web tools for gRNA design include Synthego Knockout Guide Design (https://design.synthego.com/#/) and CRISPOR (http://crispor.tefor.net/).
7. The gRNA can also be synthesized as a single molecule, called single-guide RNA (sgRNA), which includes a short linker nucleotide to fuse crRNA and tracrRNA. Manufacturers, such as IDT, usually offer sgRNA synthesis as the alternative to a two-part gRNA duplex.
8. The insert should be as close to the cleavage site as possible to give higher HDR efficiency. HDR-mediated insertion of the

DNA into the cleaved protospacer sequence to interrupt it is recommended to minimize Cas9 retargeting of the locus following editing.

9. The HDR ssDNA donor oligo can be either the top strand or the bottom strand. They have been shown to have similar insertion efficiency when the insertion site is in close proximity to the Cas9 cleavage site [23].
10. crRNA/tracrRNA duplexes are stable for at least 6 months at −20 °C.
11. Before use, thoroughly mix the stock Alt-R® S.p. Cas9 enzyme by inverting the tube several times, and briefly centrifuge the tube.
12. The RNP complex can be stored for up to 4 weeks at 4 °C or for up to 6 months at −80 °C.
13. Follow Subheading 3.1.3 and 3.1.5 for expansion into 125 mL flasks and cryo-preservation.
14. An accurate cell count is critical for limiting dilution to enhance the likelihood of seeding a single cell into each well of the plate.
15. For isolating single cells in a 96-well plate by limiting dilution, the final density of the resuspended cell solution should be 5 cells/mL so that the expected number of cells will be 0–1 cells in each well seeded with 100 μL of the cell solution. Thus, approximately half of the wells in a 96-well plate will be seeded with individual cells. Additional plates can be seeded to expand and screen for more clones.
16. Cells in the wells with more than one colony are not monoclonal and should be discarded.
17. The liquid evaporation in each well should be monitored. In a well-humidified incubator, the liquid level can be stable for 10–14 days. If noticeable media evaporation occurs, aspirate most of the media, taking care not to disturb the weakly adhesive cells, and add pre-warmed fresh growth media containing 25% condition.
18. Colonies from single cells grow as patches in wells. Cells might start to detach or form a second layer when they grow too densely in a patch. The cells can be dispersed in the well by gently pipetting the cells up and down.
19. Trypsin or other cell dissociation agents are not typically required to detach HEK293-F cells due to their weak adherence, which allows for mechanical resuspension through gentle pipetting.
20. After the cells have reached a stable growth rate in the 12-well plate, use of conditioned media for cell propagation is no longer typically necessary.

21. Heterozygous and nonmutant clones can be discarded.
22. Follow Subheading 3.1.3 and 3.1.5 for expansion into 125 mL flasks and cryo-preservation.
23. Using the genomic DNA sequence of the glycosyltransferase gene (including exons and introns) as the template to design the PCR primers that amplify the target locus. We typically design primers to amplify approximately a 500–1000 bp sequence with the restriction site centered in the amplicon. These guidelines provide for easy discernment of successful restriction digest of the amplicon on a gel during screening. They also allow for Sanger sequencing of the edit site in the amplicon with the same primer set.
24. Always keep the Q5® Hot Start High-Fidelity 2× Master Mix on ice or a microtube cooler.
25. The agarose solution can be superheated in the microwave and, if so, may suddenly boil after removal from the microwave. Handle with proper personal protective equipment (PPE), including eye protection and heat-resistant gloves.
26. Design a primer for Sanger sequencing. If either of the PCR amplification primers has the binding site within 100–400 bp away from the target mutation site, you can use that primer for Sanger sequencing.
27. On the day of transfection, the cell density should be 1.2–1.4 × 10^6 cells/mL. Make sure the cells exhibit a doubling time of 20–24 h and cell viability >90%.
28. For transient transfection, use 30 μg of piggyBac expression vector without hyperactive transposase vector.
29. If the viable cell density does not reach 1 × 10^6 cells/mL in 5 days, transfer the cell culture to a 50 mL centrifuge tube and centrifuge at 100–150 × *g* for 5 min. Discard the supernatant and resuspend the cell pellet in FreeStyle™ 293 expression medium containing 25% of conditioned medium and 500 μg/mL G418.
30. Mock transfected cells can be maintained in the selection medium as a control for G418 killing. After all mock transfected cells are killed, maintaining the stably transfected cells in selection medium for an additional 3–5 days will help to ensure the completion of selection.
31. Resuspension of cells at lower cell densities, such as 1 × 10^6 cells/mL, may help to reduce cell clumping if aggregation is a problem during sorting.
32. Depending on the resuspended cell density, the total cell number in a FASC tube is 3–15 × 10^6 cells. To sort more cells, repeat **step 5** to prepare more cells in additional FACS tubes.

33. In brief, on a FASC machine, gate the forward scatter- and side scatterplot for singlet live cells. Use 488-nm laser and a detector with 530-nm/30-nm filter (or similar filter) to detect mNeonGreen fluorescent signal. Use parental HEK293-F cells as a negative control to get background fluorescent. Identify mNeonGreen-positive signal on the transfected cells and gate the mNeonGreen-positive population and sort into a collection tube.
34. We recommend collecting at least 0.6 million cells from the top 20% or higher expressing population in order to sufficiently seed a 2-mL culture at a density of 0.2–0.5 × 10^6 cells/mL in a 25-mL shaker flask following sorting.
35. The annealing temperature depends on the design of your primers. Use NEB Tm calculator to calculate the annealing temperature (https://tmcalculator.neb.com/#!/main).
36. The elongation time depends on the size of the PCR product you design. For Q5® Hot Start High-Fidelity 2× Master Mix, the elongation time is 20–30 s/kilo base pair (kbp).

Acknowledgments

This work was supported by the National Institute of General Medical Sciences R01 GM138692 (M.J.P), National Science Foundation 1752226 (M.J.P.), and the Cornell Fleming graduate research fellowship (L.H.).

References

1. Varki A, Cummings RD, Esko JD, Stanley P, Hart GW, Aebi M, Mohnen D, Kinoshita T, Packer NH, Prestegard JH, Schnaar RL, Seeberger PH (2022) Essentials of glycobiology. Cold Spring Harbor Laboratory Press
2. Bansil R, Turner BS (2006) Mucin structure, aggregation, physiological functions and biomedical applications. Curr Opin Colloid Interface Sci 11:164–170. https://doi.org/10.1016/j.cocis.2005.11.001
3. Petrou G, Crouzier T (2018) Mucins as multifunctional building blocks of biomaterials. Biomater Sci 6:2282–2297. https://doi.org/10.1039/C8BM00471D
4. Wheeler KM, Cárcamo-Oyarce G, Turner BS, Dellos-Nolan S, Co JY, Lehoux S, Cummings RD, Wozniak DJ, Ribbeck K (2019) Mucin glycans attenuate the virulence of Pseudomonas aeruginosa in infection. Nat Microbiol 4:2146–2154. https://doi.org/10.1038/s41564-019-0581-8
5. Pan H, Colville MJ, Supekar NT, Azadi P, Paszek MJ (2019) Sequence-specific mucins for Glycocalyx engineering. ACS Synth Biol 8:2315–2326. https://doi.org/10.1021/acssynbio.9b00127
6. Shurer CR, Wang Y, Feeney E, Head SE, Zhang VX, Su J, Cheng Z, Stark MA, Bonassar LJ, Reesink HL, Paszek MJ (2019) Stable recombinant production of codon-scrambled lubricin and mucin in human cells. Biotechnol Bioeng 116:1292–1303. https://doi.org/10.1002/bit.26940
7. Narimatsu Y, Joshi HJ, Nason R, Van Coillie J, Karlsson R, Sun L, Ye Z, Chen Y-H, Schjoldager KT, Steentoft C, Furukawa S, Bensing BA, Sullam PM, Thompson AJ, Paulson JC, Büll C, Adema GJ, Mandel U, Hansen L, Bennett EP, Varki A, Vakhrushev SY, Yang Z, Clausen H (2019) An atlas of human glycosylation pathways enables display of the human Glycome by gene engineered cells. Mol Cell 75:394–407.

e5. https://doi.org/10.1016/j.molcel.2019.05.017

8. Nason R, Büll C, Konstantinidi A, Sun L, Ye Z, Halim A, Du W, Sørensen DM, Durbesson F, Furukawa S, Mandel U, Joshi HJ, Dworkin LA, Hansen L, David L, Iverson TM, Bensing BA, Sullam PM, Varki A, de Vries E, de Haan CAM, Vincentelli R, Henrissat B, Vakhrushev SY, Clausen H, Narimatsu Y (2021) Display of the human mucinome with defined O-glycans by gene engineered cells. Nat Commun 12: 4070. https://doi.org/10.1038/s41467-021-24366-4
9. Park S, Chin-Hun Kuo J, Reesink HL, Paszek MJ (2023) Recombinant mucin biotechnology and engineering. Adv Drug Deliv Rev 193: 114618. https://doi.org/10.1016/j.addr.2022.114618
10. Pi F, Binzel DW, Lee TJ, Li Z, Sun M, Rychahou P, Li H, Haque F, Wang S, Croce CM, Guo B, Evers BM, Guo P (2018) Nanoparticle orientation to control RNA loading and ligand display on extracellular vesicles for cancer regression. Nat Nanotechnol 13:82–89. https://doi.org/10.1038/s41565-017-0012-z
11. Portolano N, Watson PJ, Fairall L, Millard CJ, Milano CP, Song Y, Cowley SM, Schwabe JWR (2014) Recombinant protein expression for structural biology in HEK 293F suspension cells: a novel and accessible approach. J Vis Exp:e51897. https://doi.org/10.3791/51897
12. Zhu Y, Groth T, Kelkar A, Zhou Y, Neelamegham S (2020) A GlycoGene CRISPR-Cas9 lentiviral library to study lectin binding and human glycan biosynthesis pathways. Glycobiology 31:173–180. https://doi.org/10.1093/glycob/cwaa074
13. Narimatsu Y, Joshi HJ, Yang Z, Gomes C, Chen Y-H, Lorenzetti FC, Furukawa S, Schjoldager KT, Hansen L, Clausen H, Bennett EP, Wandall HH (2018) A validated gRNA library for CRISPR/Cas9 targeting of the human glycosyltransferase genome. Glycobiology 28: 295–305. https://doi.org/10.1093/glycob/cwx101
14. Doudna JA, Charpentier E (2014) The new frontier of genome engineering with CRISPR-Cas9. Science 346:1258096. https://doi.org/10.1126/science.1258096
15. Bischoff N, Wimberger S, Maresca M, Brakebusch C (2020) Improving precise CRISPR genome editing by small molecules: is there a magic potion? Cell 9:1318. https://doi.org/10.3390/cells9051318
16. Pinder J, Salsman J, Dellaire G (2015) Nuclear domain 'knock-in' screen for the evaluation and identification of small molecule enhancers of CRISPR-based genome editing. Nucleic Acids Res 43:9379–9392. https://doi.org/10.1093/nar/gkv993
17. Sandoval-Villegas N, Nurieva W, Amberger M, Ivics Z (2021) Contemporary transposon tools: a review and guide through mechanisms and applications of sleeping beauty, piggyBac and Tol2 for genome engineering. Int J Mol Sci 22:5084. https://doi.org/10.3390/ijms22105084
18. Woodard LE, Wilson MH (2015) piggyBacing models and new therapeutic strategies. Trends Biotechnol 33:525–533. https://doi.org/10.1016/j.tibtech.2015.06.009
19. Shurer CR, Colville MJ, Gupta VK, Head SE, Kai F, Lakins JN, Paszek MJ (2018) Genetically encoded toolbox for Glycocalyx engineering: tunable control of cell adhesion, survival, and cancer cell behaviors. ACS Biomater Sci Eng 4: 388–399. https://doi.org/10.1021/acsbiomaterials.7b00037
20. Henderson RA, Konitsky WM, Barratt-Boyes SM, Soares M, Robbins PD, Finn OJ (1998) Retroviral expression of MUC-1 human tumor antigen with intact repeat structure and capacity to elicit immunity in vivo. J Immunother 21:247
21. Yusa K, Zhou L, Li MA, Bradley A, Craig NL (2011) A hyperactive *piggyBac* transposase for mammalian applications. Proc Natl Acad Sci 108:1531–1536. https://doi.org/10.1073/pnas.1008322108
22. Jacobi AM, Rettig GR, Turk R, Collingwood MA, Zeiner SA, Quadros RM, Harms DW, Bonthuis PJ, Gregg C, Ohtsuka M, Gurumurthy CB, Behlke MA (2017) Simplified CRISPR tools for efficient genome editing and streamlined protocols for their delivery into mammalian cells and mouse zygotes. Methods 121–122:16–28. https://doi.org/10.1016/j.ymeth.2017.03.021
23. Schubert MS, Thommandru B, Woodley J, Turk R, Yan S, Kurgan G, McNeill MS, Rettig GR (2021) Optimized design parameters for CRISPR Cas9 and Cas12a homology-directed repair. Sci Rep 11:19482. https://doi.org/10.1038/s41598-021-98965-y

Part V

Interaction of Mucins and Other Biomolecules

Chapter 25

Analysis of the Interaction Between Mucin and Green Fluorescent Protein (GFP)-Tagged Galectin-2 Using a 96-Well Plate

Mayumi Tamura and Yoichiro Arata

Abstract

Due to a significant proportion of glycans binding to the peptide (constituting approximately 50–90% of the molecular weight), analyzing the interaction between the entire mucin molecule and its recognition protein (lectin) can be challenging. To address this, we propose a semiquantitative approach for measuring the interaction between mucin and lectin, which involves immobilizing mucin in a 96-well plate and subsequently adding lectin tagged with green fluorescent protein.

Key words Mucin, Fluorescent tag, Lectin, 96-well plate, Galectin, Green fluorescent protein

1 Introduction

Mucin is a vital component of the mucus barrier in the digestive tract and lungs [1]. It is a glycoprotein characterized by its numerous *O*-linked glycans, which are nonuniform in their structure [2]. Multiple lectins have been reported to interact with mucins [3–5]. However, quantitatively measuring and comparing the interactions between whole mucin molecules and lectins are challenging.

We have been investigating galectins, which are a type of animal lectin that recognize β-galactoside structures [6–8]. Galectin-2 (Gal-2), a member of the galectin family, is highly expressed in the stomach [9] and is thought to play a role in gastric protection. In a previous study, we reported that Gal-2 recognizes mucins from gastric mucus in a carbohydrate-dependent manner [10]. Additionally, we identified MUC5AC, one of the primary mucins in the gastric mucosa, as a potential Gal-2 ligand from mouse gastric mucus fractions and reported that Gal-2 binds to MUC5AC via its sugar chain [11].

Akihiko Kameyama (ed.), *Mucins: Methods and Protocols*, Methods in Molecular Biology, vol. 2763,
https://doi.org/10.1007/978-1-0716-3670-1_25,

Here, we introduce a method for examining the interaction between green fluorescent protein (GFP)-tagged Gal-2 and mucin derived from porcine gastric mucus. The reasons for using GFP as a label for Gal-2 include its high sensitivity, low cost as a recombinant protein tag, and the multiple possible locations for adding the tag, such as the N- or C-terminus of the protein.

To implement this method, mucin was added to a 96-well plate, immobilized, and subsequently blocked with bovine serum albumin (BSA). The GFP-tagged Gal-2 was subsequently added to the plate, and the amount of bound Gal-2 was determined by measuring the fluorescence intensity, allowing for the analysis of the interaction between mucin and Gal-2.

2 Materials

All solutions were prepared using Ultrapure water and analytical-grade reagents; all reagents were stored at 4 °C unless otherwise indicated.

Gal-2-GFP protein was produced as a recombinant protein by adding a GFP tag to the C-terminal side of Gal-2 using *Escherichia coli*.

2.1 Reagents for Preparation of Recombinant Gal-2-GFP

The Gal-2-GFP protein was prepared using the method described in Ref. [12]. However, for our preparation, we used the 2 × YT/ampicillin (Amp) medium for *E. coli* culture and asialofetuin (ASF)-immobilized columns for purification by affinity chromatography.

1. 2 × YT/Amp medium: To prepare 250 mL of 2 × YT media, use a 500 mL shaking Erlenmeyer flask with baffles. Autoclave at 120 °C for 15 min for sterilization. After cooling, add 250 μL of 125 mg/mL Amp, which has been sterilized by filtration (*see* **Note 1**).
2. 0.1 M isopropyl β-D-1-thiogalactopyranoside (IPTG): Dissolve 0.24 g of IPTG in water to a total volume of 10 mL. Sterilize by filtration using a 0.2 μm filter unit. Dispense 1 mL into each 1.5 mL sample tube and store at −20 °C until use.
3. 125 mg/mL Amp: Dissolve 1.25 g of Amp sodium salt in water to a total volume of 10 mL. Sterilize by filtration using a 0.2 μm filter unit. Dispense 0.5 mL into each 1.5 mL sample tube and store at −20 °C until use.
4. 20× EDTA-PBS (400 mM NaH_2PO_4, 3 M NaCl, 20 mM EDTA, pH 7.2): Dissolve 62.4 g of $NaH_2PO_4{\cdot}2H_2O$ and 175.3 g of NaCl in approximately 700 mL of water. Add 40 mL of 0.5 M EDTA, pH 8.0, and adjust the pH to 7.2 using NaOH (*see* **Note 2**). Using water, bring the solution to a total volume of 1 L.

5. EDTA-PBS: Dilute 20× EDTA-PBS 20-fold with water.
6. EDTA-ME-PBS: Dilute 20× EDTA-PBS 20-fold with water and add 2-mercaptoethanol, bringing the final concentration to 4 mM. Store at 4 °C until use.
7. EDTA-ME-PBS + 0.1 M lactose: Add 5 mL of 20× EDTA-PBS and 3.6 g of lactose monohydrate to water to reach a final volume of 100 mL. Add 2-mercaptoethanol to bring the final concentration to 4 mM and keep at 4 °C until use.
8. ASF-immobilized HiTrap column: Fetuin was desialylated and immobilized on a HiTrap NHS-activated column. Dissolve 50 mg of fetuin in 2.5 mL of 0.15 M NaCl and add approximately 20 μL of 6 N HCl to adjust the pH to 2. Incubate in an 80 °C water bath for 1 h. Neutralize by adding approximately 200 μL of 1 M NaOH and dialyze against 0.1 M $NaHCO_3$, 0.5 M NaCl, pH 8.3. After dialysis, add 0.1 M $NaHCO_3$, 0.5 M NaCl, pH 8.3 to reach a final volume of 5 mL and couple the desialylated fetuin to HiTrap NHS-activated columns following the manufacturer's instruction (*see* **Note 3**).
9. Centrifugal filter unit: Use an Amicon Ultra-15 with a cutoff molecular weight of 10,000 (Millipore®, Merck KGaA, Darmstadt, Germany).
10. Luria-Bertani (LB) broth/ampicillin (Amp) medium: LB containing 125 μg/mL ampicillin. Use a 500 mL shaking Erlenmeyer flask with baffles for 250 mL of LB media. Sterilize by autoclaving (at 120 °C for 15 min). After cooling, add 250 μL of 125 mg/mL Amp that has been sterilized by filtration.
11. *E. coli* expression vector pET21a: Purchased from Novagen (Merck KGaA, Germany.)
12. pAcGFP-N1 vector: Purchased from Takara Bio Inc., Siga, Japan.
13. Enzyme for PCR: PrimeSTAR Max DNA Polymerase purchased from Takara Bio Inc., Siga, Japan.
14. *E. coli* BL21(DE3) competent cells: Purchased from Merck KGaA, Germany.

2.2 Analysis of the Interaction Between Mucin and Gal-2-GFP

1. Gal-2-GFP solution: After purification by affinity chromatography and removal of lactose, dilute the Gal-2-GFP solution in EDTA-PBS to create solutions with concentrations of 0, 0.1, 0.5, 1, 2, and 3 ng/μL (*see* **Note 4**).
2. EDTA-PBS: Dilute 20× EDTA-PBS 20-fold with water, as described in **item 5** in Subheading 2.1 (*see* **Note 5**).
3. 0.25 mg/mL mucin solution: Mucin is commercially available (*see* **Note 6**). Weigh approximately 20 mg of mucin into a 15- or 50 mL conical tube, add EDTA-PBS, and dissolve the mucin

to make a concentration of 2.0 mg/mL. Dilute the solution eight-fold with EDTA-PBS.

4. 0.25 mg/mL BSA/EDTA-PBS: BSA is commercially available (*see* **Note 7**). Weigh approximately 5 mg of BSA into a 1.5 mL tube, add EDTA-PBS, and dissolve the BSA to make a concentration of 5 mg/mL. Dilute the solution 20-fold with EDTA-PBS.
5. 0.2 M lactose/EDTA-PBS: Add 0.5 mL of 20× EDTA-PBS and 0.72 g of lactose monohydrate to water to reach a final volume of 10 mL, as described in **item** 7 in Subheading 2.1.
6. 96-well white plate: Use a 96-well plate with a white cover for cell cultivation (*see* **Note 8**).
7. Microplate reader: Use a fluorescence reader with excitation at 475 nm and emission at 505 nm. We used the Spectra Max M5e microplate reader from Molecular Devices, LLC., CA, USA.

3 Methods

3.1 Preparation of Recombinant Gal-2-GFP

The preparation of Gal-2-GFP is carried out using the same methodology as described previously [12]. However, for the *E. coli* culture, 2 × YT/Amp is used, and purification is performed using ASF-immobilized columns through affinity chromatography (Fig. 1).

1. For the expression of Gal-2-GFP, use the *E. coli* expression vector pET21a, with the AcGFP1 sequence from pAcGFP1-N1 (Takara Bio Inc., Siga, Japan) serving as the GFP DNA sequence. Amplify the AcGFP DNA using PCR and insert the AcGFP DNA into the pET21a cloning vector to form GFP-pET21a. Amplify the Gal-2 DNA template via PCR, using a forward primer containing the *Nde* I recognition sequence and a reverse primer that deletes the Gal-2 termination codon and contains the *Bam*H I recognition sequence, and insert the resulting Gal-2 DNA into the GFP-pET21a vector using *Nde* I and *Bam*H I sites.
2. *E. coli* BL21(DE3) is transformed with the Gal-2-GFP-pET21a prepared in **step 1**.
3. The expression and purification of Gal-2-GFP protein is done using a method similar to that described previously [12]. Specifically, inoculate the *E. coli* culture prepared in Step 2 into two disposable 50 mL conical tubes containing 10 mL of Luria-Bertani (LB) broth/Amp medium.
4. Incubate the tubes overnight at 37 °C.

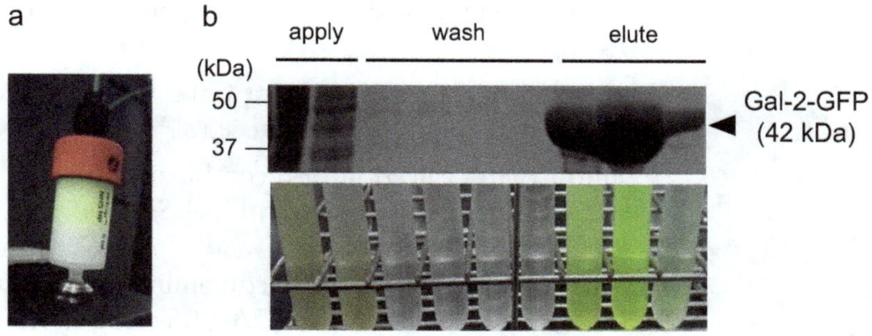

Fig. 1 Purification of recombinant Gal-2-GFP by affinity chromatography on an ASF-immobilized Sepharose column. (**a**) ASF-immobilized column before lactose elution. (**b**) SDS-PAGE and CBB staining of the eluted fractions from the column (upper figure) and the corresponding test tubes containing eluted fractions (lower figure) (*see* **Note 18**)

5. After incubation, inoculate the cultured media from the two tubes into two 500 mL shaking Erlenmeyer flasks with baffles containing 250 mL of 2× YT Amp medium. Shake the cultures at 37 °C until their optical density at 600 nm (OD_{600}) reaches between 0.6 and 0.8 (approximately 2–5 h of incubation).
6. Cool the flasks with iced water.
7. To induce protein expression, add 1 mL of 0.1 M IPTG to the culture media, resulting in a final concentration of 0.4 mM. Additionally, add 0.25 mL of 125 mg/mL Amp to the culture media. Shake the cultures at 20 °C overnight.
8. Recover the culture media into 50 mL conical tubes and centrifuge at 1580 × *g* at 4 °C for 20 min.
9. Remove the supernatant, and suspend the remaining *E. coli* pellet in 5 mL of EDTA-ME-PBS in each tube.
10. Combine all suspensions into one tube and centrifuge again at 1580 × *g* at 4 °C for 15 min.
11. Remove the supernatant.
12. Suspend the *E. coli* pellet in 10 mL of EDTA-ME-PBS and disrupt the *E. coli* cells by sonication (*see* **Note 9**) on ice. Set the duty cycle to 50% and output control to 2.5. Sonicate for 5 min and cool down by incubating on ice for 3 min.
13. Repeat this sonication-cool down process three times. After cell disruption, centrifuge the mixture at 9400 × *g* for 45 min at 4 °C. Collect the supernatant fraction and add it to an ASF-immobilized column with a bed volume of 5 mL, which should be equilibrated with 50 mL of EDTA-ME-PBS. The flow rate of the solution should be between 0.3 and 0.4 mL/min.
14. Collect the eluted fraction in different tubes (5 mL each).

15. Wash the column with 6–8 column bed volumes of EDTA-ME-PBS.
16. Elute the adsorbed recombinant Gal-2 with 20 mL of EDTA-ME-PBS containing 0.1 M lactose (*see* **Note 10**).
17. Confirm the purification of recombinant Gal-2 by subjecting each eluted fraction to conventional SDS-polyacrylamide gel electrophoresis (*see* **Note 11**) and Coomassie Brilliant Blue staining. Collect the fractions containing lactose-eluted Gal-2 and exchange the buffer to EDTA-PBS using an Amicon Ultra-15 device to remove lactose and 2-mercaptoethanol (*see* **Notes 12** and **13**).
18. Add the lactose-eluted fraction containing Gal-2 to an Amicon Ultra-15 device and centrifuge until the volume in the filter device is reduced to approximately 1 mL.
19. Add an additional 10 mL of EDTA-PBS to the filter device and recentrifuge in the Amicon Ultra-15 device.
20. Repeat this process five times to exchange the buffer to EDTA-PBS (*see* **Note 14**).

3.2 Standard Curve of Gal-2-GFP

1. Create a calibration curve to determine the amount of Gal-2-GFP bound to the wells by adding 100 μL of 0, 0.1, 0.5, 1, 2, and 3 ng/μL of Gal-2-GFP in EDTA-PBS to each well. This results in protein amounts of 0, 10, 50, 100, 200, and 300 ng per well (*see* **Note 4**).
2. Measure the fluorescence (excitation, 475 nm; emission, 505 nm) in a fluorescence microplate reader.
3. Plot a graph of the amount of Gal-2-GFP versus the corresponding fluorescence intensity to generate a calibration curve (Fig. 2).

3.3 Analysis of the Interaction Between Mucin and Gal-2-GFP Using a 96-Well Plate

1. Add 200 μL of a 0.25 mg/mL porcine mucin solution to the wells of a 96-well plate, dispensing three wells for each concentration of Gal-2-GFP (*see* **Note 15**).
2. Incubate the plate at 37 °C for 1 h to immobilize the mucin in the wells.
3. Discard the added mucin solution and wash the wells with 200 μL of EDTA-PBS. Discard the EDTA-PBS and add 200 μL of fresh EDTA-PBS to the wells (*see* **Note 16**).
4. Discard the EDTA-PBS completely and add 200 μL of a 0.25 mg/mL BSA/EDTA-PBS solution to each well.
5. Incubate the plate at 37 °C for 1 h.
6. Wash the wells in the same manner as in **step 3**.

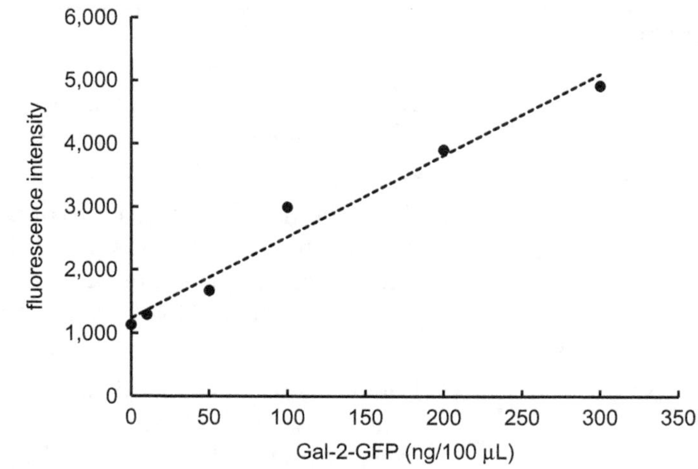

Fig. 2 Standard curve of Gal-2-GFP. The *X*-axis represents the amount of Gal-2-GFP (ng/100 μL), and the *Y*-axis represents the fluorescence intensity

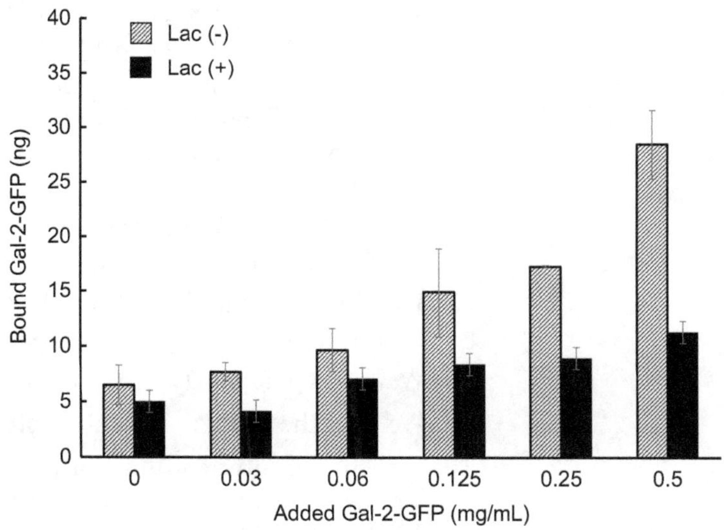

Fig. 3 Experimental results showcasing that the binding of Gal-2-GFP to mucin was inhibited by the addition of competitive sugar lactose

7. Add 50 μL of Gal-2-GFP solution at various concentrations. If lactose was added, for example, prepare the Gal-2-GFP solution in EDTA-PBS containing the lactose (*see* **Note 17**).
8. Incubate the plate at 4 °C for 1 h.
9. Wash the wells in the same manner as in **step 3**.
10. Add 100 μL of EDTA-PBS to each well to prevent drying.
11. Measure the fluorescence intensity (excitation, 475 nm; emission, 505 nm) using a fluorescence microplate reader. An example of the results is shown in Fig. 3.

4 Notes

1. Difco™ 2× YT Medium (cat. #244020) is easy to handle and would simplify the media preparation compared to weighing and dissolving media powder.
2. We recommend adding approximately 12 g of NaOH pellets first and then slowly adding NaOH solution, as using NaOH solution directly may not be sufficient for pH adjustment and could cause the total volume to exceed the targeted volume of 1 L.
3. For purification, we used the AKTA system. A commercially available Lac-immobilized column for AKTA was not available; thus, we prepared and used an ASF-immobilized column (ASF immobilized to a HiTrap™ NHS-activated HP Column, #17071701, Cytiva, Tokyo, Japan) instead.
4. Plot fluorescence intensity against the protein amount of Gal-2-GFP. In our experiment, there is linearity at least in the range of Gal-2-GFP 0–50 ng/μL (0–5000 ng/100 μL/well).
5. We used this method to examine the interaction between Gal-2 and mucin. We used PBS containing 1 mM EDTA to prevent oxidation of Gal-2 because Gal-2 loses its sugar-binding activity when oxidized.
6. We used Sigma-Aldrich® mucin from porcine stomach (Type III, partially purified powder).
7. We used globulin-free BSA (01281-97 Nacalai Tesque Inc., Kyoto, Japan).
8. We used Nunc no. 136101 plates.
9. We used the TOMY, UD-211 system for sonication.
10. If there is insufficient time for the buffer exchange procedure, we recommend leaving the eluted recombinant Gal-2 in the EDTA-ME-PBS solution containing 0.1 M lactose at 4 °C, rather than stopping the buffer exchange process. Gal-2 is quite stable in the presence of lactose and 2-mercaptoethanol and can be stored for up to 3 days at 4 °C.
11. We used a 15% gel.
12. As galectins contain cysteine residues that are important for their lectin activity, the presence of 2-mercaptoethanol in the solution is usually necessary.
13. We do not recommend buffer exchange by dialysis of a high-concentration Gal-2 solution, as this often results in precipitation of the Gal-2 protein.
14. Gal-2 in EDTA-PBS can be stored at −80 °C in aliquots for up to 12 months without losing its activity. Avoid repeated

freezing and thawing or storing at high concentrations (higher than approximately 10 mg/mL), as these could result in insoluble precipitates of Gal-2.

15. The reason for adding 200 μL of 0.25 mg/mL porcine mucin instead of smaller amounts of mucin with a higher concentration is to ensure equal and even immobilization in every well.
16. Discard EDTA-PBS by turning over the plate. For the second wash step, tap the inverted plate on a paper towel to remove excess solution.
17. In experiments in which lactose was added, we mixed equal amounts of the sugar solution and Gal-2-GFP solution and added the mixture to the wells. For example, we mixed 80 μL of 0.2 M Lac/EDTA-PBS with 80 μL of 0.5 mg/mL Gal-2-GFP/EDTA-PBS.
18. Even under visible light, purified fluorescent Gal-2-GFP can be observed.

Acknowledgements

The authors thank Mr. Dai Sato and Ms. Moeko Nakajima for technical assistance. This work was supported by research grants from the Teikyo University and Josai University.

References

1. McGuckin MA, Lindén SK, Sutton P et al (2011) Mucin dynamics and enteric pathogens. Nat Rev Microbiol 9:265–278
2. Brockhausen I, Wandall HH, ten Hagen KG et al (2022) O-GalNAc glycans. In: Varki A, Cummings RD, Esko JD et al (eds) Essentials of Glycobiology [Internet], 4th edn. Cold Spring Harbor, New York
3. Wasano K, Hirakawa Y (1997) Recombinant galectin-1 recognizes mucin and epithelial cell surface glycocalyces of gastrointestinal tract. J Histochem Cytochem 45:275–283
4. Argüeso P, Guzman-Aranguez A, Mantelli F et al (2009) Association of cell surface mucins with galectin-3 contributes to the ocular surface epithelial barrier. J Biol Chem 284:23037–23045
5. Hnisch FG, Bonar D, Schloerer N et al (2014) Human trefoil factor 2 is a lectin that binds α-GlNAc-capped mucin glycans with antibiotic activity against helicobacter pylori. J Biol Chem 289:27363–27375
6. Kasai K-I, Hirabayashi J (1996) Galectins: a family of animal lectins that decipher glycocodes. J Biochem 119:1–8
7. Cooper DNW (2002) Galectinomics: finding themes in complexity. Biochim Biophys Acta Gen Subj 1572:209–231
8. Leffler H, Carlsson S, Hedlund M et al (2002) Introduction to galectins. Glycoconj J 19:433–440
9. Nio-Kobayashi J, Takahashi-Iwanaga H, Iwanaga T (2009) Immunohistochemical localization of six galectin subtypes in the mouse digestive tract. J Histochem Cytochem 57: 41–50
10. Tamura M, Sato D, Nakajima M et al (2017) Identification of galectin-2–mucin interaction and possible formation of a high molecular weight lattice. Biol Pharm Bull 40:1789–1795
11. Tamura M, Tanaka T, Fujii N et al (2020) Potential interaction between galectin-2 and MUC5AC in mouse gastric mucus. Biol Pharm Bull 43:356–360
12. Tamura M, Arata Y (2020) Expression, S-nitrosylation, and measurement of S-nitrosylation ratio of recombinant galectin-2. In: Hirabayashi J (ed) Lectin purification and analysis. Springer Nature, New York

Chapter 26

Solution NMR Analysis of *O*-Glycopeptide–Antibody Interaction

Ryoka Kokubu, Shiho Ohno, Noriyoshi Manabe, and Yoshiki Yamaguchi

Abstract

O-Linked glycans potentially play a functional role in cellular recognition events. Recent structural analyses suggest that *O*-glycosylation can be a specific signal for a lectin receptor which recognizes both the *O*-glycan and the adjacent polypeptide region. Further, certain antibodies specifically bind to the *O*-glycosylated peptide. There is growing interest in the mechanism by which *O*-glycans on proteins are specifically recognized by lectins and antibodies. The recognition system may be common to many *O*-glycosylated proteins; however, there is limited 3D structural information on the dual recognition of glycan and protein. This chapter describes a solution NMR analysis of the interaction between MUC1 *O*-glycopeptide and anti-MUC1 antibody MY.1E12.

Key words Antibody, Epitope mapping, *O*-Glycopeptide, MUC1, NMR, Titration, Signal assignment

1 Introduction

Biosynthesis of *O*-linked glycan (mucin-type, GalNAc-type glycan) starts with GalNAc addition to Ser/Thr of the polypeptide chain by polypeptide GalNAc transferases. It is simpler than the biosynthesis of *N*-linked glycan in which a lipid-linked oligosaccharide precursor ($Glc_3Man_9GlcNAc_2$) is attached to the Asn-X-Ser/Thr consensus sequence of the nascent polypeptide. Unlike that of *N*-glycosylation, the consensus sequence for GalNAc addition is not known. It is therefore difficult to predict the site of *O*-glycosylation. Changes in *O*-glycosylation elaborations on proteins are found in cancer cells, which may induce antigenicity of the peptide [1]. Currently, information is rather limited on the structure–function relationships of *O*-glycans.

Mucin is heavily glycosylated with *O*-glycans and the thick layer of sugars with intercalated water molecules is thought to act as a protective barrier at the cell surface. In general, *O*-glycan chains are

Akihiko Kameyama (ed.), *Mucins: Methods and Protocols*, Methods in Molecular Biology, vol. 2763,
https://doi.org/10.1007/978-1-0716-3670-1_26,

shorter than those of *N*-glycan. Because of the role of *O*-glycans in mucin persuaded many researchers, the abundant *O*-glycan presented on secreted proteins does not take part in specific recognition by lectin receptors. Recent structural analyses, however, suggest that *O*-glycosylation can be a specific recognition signal for a lectin receptor which recognizes both the *O*-glycan and the adjacent polypeptide region [2–4]. It has also been reported that certain antibodies can enhance the antigen-binding specificity through recognition of both *O*-glycan and the peptide. The anti-human podoplanin antibody, LpMab-3, binds to an epitope that includes both peptide and disialyl core 1 *O*-glycan at Thr76 [5]. MY.1E12 is a unique anti-MUC1 antibody which binds to *O*-glycan on a specific Thr residue of MUC1 [6]. For a more detailed view of structure–function relationships of *O*-glycans, we need to understand how lectins and antibodies specifically recognize *O*-glycosylated peptides and proteins. X-Ray crystallographic analysis is the method of choice for atomic-level structure, but it does need crystals. Solution NMR analysis is an alternative and complementary approach which reveals modes of interaction under more physiological conditions. This chapter describes solution NMR analysis of the interaction between MUC1 *O*-glycopeptide and anti-MUC1 antibody MY.1E12, focusing on our NMR signal assignment strategy of the *O*-glycopeptide and on NMR titration using the antibody [7].

2 Materials

1. *O*-Glycopeptide sample of interest (0.5–1.0 mg).
2. Antibody sample of interest (1.0–5.0 mg).
3. 20 mM sodium phosphate buffer (pH 6.8, $H_2O/D_2O = 9:1$): 20 mM sodium phosphate buffer (pH 6.8, H_2O) is prepared and the buffer is lyophilized. Then the H_2O/D_2O mixture ($H_2O:D_2O = 9:1$) is added. Alternatively, 0.1 mL of D_2O is added to 0.9 mL of 22 mM sodium phosphate buffer (pH 6.8, H_2O).
4. Deuterium oxide (D_2O).
5. 4,4-Dimethyl-4-silapentane-1-sulfonic acid (DSS).
6. Disposable plastic tube 1.5 mL.
7. 5 mm NMR sample tube.
8. Pasteur pipettes with a long tip for NMR tubes.
9. Parafilm.

3 Methods

Recommended ^{1}H resonance frequency of the NMR spectroscopy is 600 MHz (corresponds to 14.1 Tesla) or higher to avoid signal overlapping.

3.1 Sample Preparation for NMR Study

1. Prepare 0.5–1.0 mg of *O*-glycopeptide sample with high purity (>95%) (*see* **Note 1**).
2. In a disposable plastic tube, dissolve the glycopeptide in 500–600 μL of 20 mM sodium phosphate buffer (pH 6.8, $H_2O/D_2O = 9{:}1$) (*see* **Note 2**).
3. Add a small amount of DSS (a chemical shift standard for ^{1}H).
4. Transfer the *O*-glycopeptide solution to an NMR tube using a Pasteur pipette with a long tip.
5. Cap and tightly seal the top of the NMR tube with Parafilm.
6. Similarly, prepare the antibody sample using 1.0–5.0 mg of antibody using the same buffer (*see* **Note 3**).

3.2 NMR Signal Assignment of O-Glycopeptide

1. Set the probe temperature of the NMR magnet to 278 K and transfer the NMR tube with a spinner turbine into the NMR magnet (*see* **Note 4**).
2. After tuning and shimming, collect a one-dimensional ^{1}H-NMR spectrum with proper suppression of the H_2O signal, such as with WATERGATE (WATER suppression by GrAdient-Tailored Excitation) [8] (*see* **Note 5**).
3. Identify the DSS methyl signal and set it to 0 ppm.
4. To obtain intra-residue connectivity, collect a set of 2D NMR spectra including double-quantum filtered correlation spectroscopy (DQF-COSY), homonuclear Hartmann–Hahn spectroscopy (HOHAHA)/total correlation spectroscopy (TOCSY) with suppression of water signal (*see* **Note 6**).
5. To obtain inter-residue connectivity, collect 2D nuclear Overhauser effect spectroscopy (NOESY). The three 2D NMR spectra are shown in Fig. 1.
6. (Option) Prepare the sample in D_2O buffer and collect the NMR spectra when the anomeric signals overlap the water signal.

3.3 NMR Titration Study

1. Collect a 1D ^{1}H-NMR spectrum of the antibody solution without glycopeptide at 278 K with a good signal-to-noise ratio.
2. After measuring the ^{1}H-NMR, collect the antibody solution from the NMR tube using a Pasteur pipette with a long tip, and transfer it to a disposable plastic tube (1.5 mL).

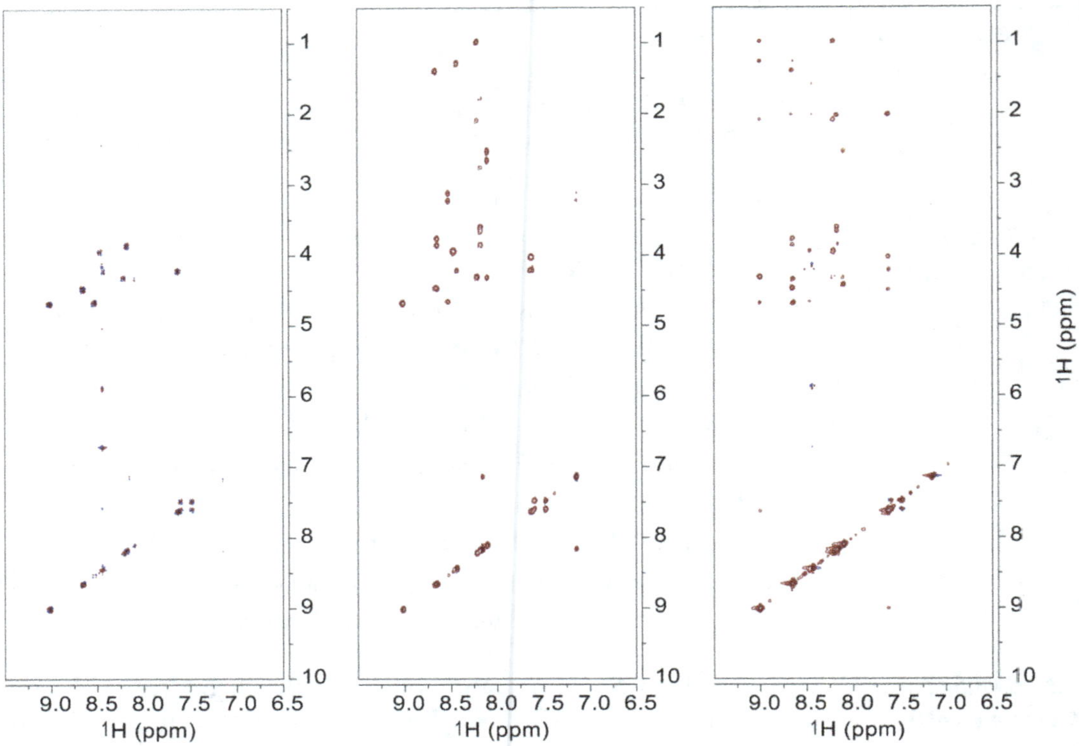

Fig. 1 2D NMR spectra (NH-aliphatic region) of MUC1 *O*-glycopeptide (7AA, AHGVTSAPD) modified with Neu5Acα(2–3)Galβ(1–3)GalNAc to Thr8. DQF-COSY (left), HOHAHA (middle), and NOESY (right) spectra are shown. NMR spectra were measured at ^{1}H observation frequency of 600 MHz, and the probe temperature was set to 278 K. ^{1}H chemical shifts were referenced to internal DSS. Further, 1.6 mg of MUC-1 glycopeptide was dissolved in 600 μL of 20 mM sodium phosphate buffer, pH 6.8 ($H_2O/D_2O = 9:1$)

3. Add the glycopeptide solution to the antibody solution at a molar ratio of 1:0.2, 1:0.4, 1:0.6, 1:0.8, 1:1, etc (antibody binding site/glycopeptide) (*see* **Note 7**).
4. Gently mix the solution using the Pasteur pipette and transfer it to the NMR tube.
5. Collect a 1D ^{1}H-NMR spectrum and compare the NMR spectrum with that of the glycopeptide-free antibody solution.
6. Repeat **steps 2–5** with increasing concentrations of antibody and observe line broadening and/or chemical shift changes of glycopeptide signals (Fig. 2). This result will be used for identification of the epitope (*see* **Note 8**).

4 Notes

1. The required amount of sample (mg) is dependent on the purpose of the study.

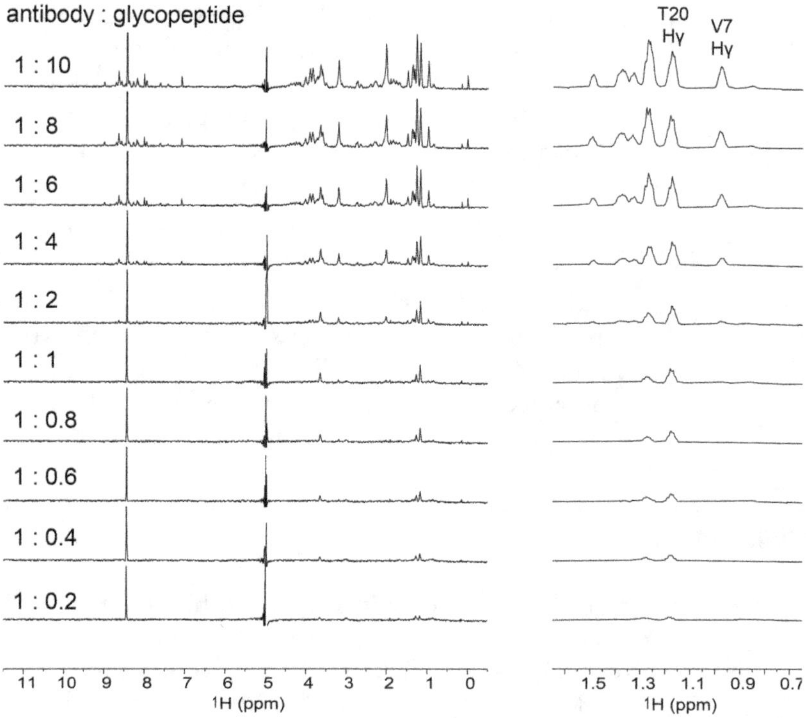

Fig. 2 NMR titration experiments using MUC1 *O*-glycopeptide (20AA, APPAHGVTSAPDTRPAPGST) modified with Neu5Acα(2–3)Galβ(1–3)GalNAc to Thr8. **(left)** 1D ^{1}H-NMR spectra (full region) of anti-MUC1 antibody MY.1E12 with increasing amount of MUC1 *O*-glycopeptide. Molar ratios (antibody binding site/glycopeptide) are indicated for each spectrum. **(right)** Expanded 1D ^{1}H-NMR spectra showing methyl signals originating from T20 and V7

2. The volume is typically 500–600 μL for a 5 mm tube, and it depends on the NMR probe you use. Phosphate buffer is used to maintain pH. Instead of D_2O, H_2O is used as a solvent to detect the N-H amide proton. The amide proton is important for sequential assignment of the peptide backbone.
3. Buffer composition and pH must be the same for both the glycopeptide and the antibody solutions. This is necessary to observe the changes originating from antibody–glycopeptide interaction.
4. A lower probe temperature, e.g., 278 K, is better for producing a sharp NH signal. Higher temperature increases the exchange rate of NH to solvent water and hence induces line broadening. Therefore, it is important to find and set the optimal temperature.
5. Intensity of the signals close to the water signal (e.g., anomeric proton) will be attenuated when the WATERGATE scheme is applied.

6. The mixing time of HOHAHA/TOCSY and NOESY needs to be properly chosen. Typically, the mixing time of HOHAHA will be 20–100 ms and that of the NOESY 40–150 ms. Instead of DQF-COSY to provide antiphase signals, CLIP-COSY (clean in-phase COSY) can be applied and this then provides in-phase signals [9].
7. The antibody/glycopeptide ratio will be 1:0.2, 1:0.4, 1:0.6, 1:0.8, 1:1, etc.
8. Other NMR techniques can be applied for epitope mapping and conformational analysis, such as STD-NMR (saturation transfer difference-NMR) [10], waterLODGY (water–ligand observed via gradient spectroscopy) [11, 12], and TR-NOE (transferred NOE) [13] experiments.

Acknowledgements

We thank Izuru Nagashima, Yayoi Yoshimura, Hiroki Shimizu, and Yasunori Chiba (AIST) for the chemoenzymatic synthesis of *O*-glycopeptides and Makoto Tsuiji (Hoshi University), Miki Noji, Kaori Denda-Nagai, and Tatsuro Irimura (Juntendo University) for the sequencing and preparation of anti-MUC1 antibody. We also thank Tomoyuki Matsuki, Shinichi Sato, Yuka Takahashi, Hirohide Kuratani (Tohoku Medical and Pharmaceutical University), Seizo Koshiba, Jin Inoue, and Kumiko Sugai (Tohoku Medical Megabank Organization) for their support in NMR measurements.

References

1. Hanisch FG (2001) O-glycosylation of the mucin type. Biol Chem 382(2):143–149. https://doi.org/10.1515/bc.2001.022
2. Nagae M et al (2014) A platform of C-type lectin-like receptor CLEC-2 for binding O-glycosylated podoplanin and nonglycosylated rhodocytin. Structure 22(12):1711–1721. https://doi.org/10.1016/j.str.2014.09.009
3. Kuroki K et al (2014) Structural basis for simultaneous recognition of an O-glycan and its attached peptide of mucin family by immune receptor PILRalpha. Proc Natl Acad Sci U S A 111(24):8877–8882. https://doi.org/10.1073/pnas.1324105111
4. Somers WS, Tang J, Shaw GD, Camphausen RT (2000) Insights into the molecular basis of leukocyte tethering and rolling revealed by structures of P- and E-selectin bound to SLe(X) and PSGL-1. Cell 103(3):467–479. https://doi.org/10.1016/s0092-8674(00)00138-0
5. Ogasawara S et al (2020) Crystal structure of an anti-podoplanin antibody bound to a disialylated O-linked glycopeptide. Biochem Biophys Res Commun 533(1):57–63. https://doi.org/10.1016/j.bbrc.2020.08.103
6. Yoshimura Y et al (2019) Products of Chemoenzymatic synthesis representing MUC1 tandem repeat unit with T-, ST- or STn-antigen revealed distinct specificities of anti-MUC1 antibodies. Sci Rep 9(1):16641. https://doi.org/10.1038/s41598-019-53052-1
7. Kokubu R et al (2022) O-glycan-dependent interaction between MUC1 Glycopeptide and MY.1E12 antibody by NMR, molecular dynamics and docking simulations. Int J Mol Sci 23(14). https://doi.org/10.3390/ijms23147855
8. Piotto M, Saudek V, Sklenar V (1992) Gradient-tailored excitation for single-quantum NMR spectroscopy of aqueous

solutions. J Biomol NMR 2(6):661–665. https://doi.org/10.1007/BF02192855

9. Koos MR et al (2016) CLIP-COSY: a clean in-phase experiment for the rapid acquisition of COSY-type correlations. Angew Chem Int Ed Engl 55(27):7655–7659. https://doi.org/10.1002/anie.201510938
10. Mayer M, Meyer B (1999) Characterization of ligand binding by saturation transfer difference NMR spectroscopy. Angew Chem Int Ed Engl 38(12):1784–1788. https://doi.org/10.1002/(SICI)1521-3773(19990614)38:12
11. Dalvit C et al (2001) WaterLOGSY as a method for primary NMR screening: practical aspects and range of applicability. J Biomol NMR 21(4):349–359. https://doi.org/10.1023/a:1013302231549
12. Dalvit C et al (2000) Identification of compounds with binding affinity to proteins via magnetization transfer from bulk water. J Biomol NMR 18(1):65–68. https://doi.org/10.1023/a:1008354229396
13. Post CB (2003) Exchange-transferred NOE spectroscopy and bound ligand structure determination. Curr Opin Struct Biol 13(5):581–588. https://doi.org/10.1016/j.sbi.2003.09.012

Part VI

Mucin and Microorganism

Chapter 27

Cultivation of Microorganisms in Media Supplemented with Mucin Glycoproteins

Hiromi Takada, Takane Katayama, and Toshihiko Katoh

Abstract

To examine the mucin-utilizing capacity of bacterial isolates from fecal samples, an in vitro cultivation method using mucins as a carbon source should be considered. This chapter describes a practical method for cultivating bacteria in media containing mucin glycoproteins; for this cultivation method, several factors are considered due to the physical nature of mucin glycoproteins.

Key words Mucin-utilizing bacteria, Mucin glycoprotein, *Bifidobacterium bifidum*, Growth curve, Colony-forming unit

1 Introduction

The mucosal surface provides an ecosystem in which mucin glycoproteins critically contribute to the interplay between the host and microbes. Recent studies examining the intestinal mucosal microbiome have revealed that some commensal gut microbes, such as species in the genera *Akkermansia, Bacteroides,* and *Bifidobacterium*, can utilize mucin *O*-glycans, including those from the gel-forming mucin MUC2, as nutrients; further, the corresponding abundance of these commensal gut microbes significantly affects human health [1–3]. In addition, many pathogenic bacteria exploit secretory mucinases (glycoprotein proteases) to degrade mucin glycoproteins and penetrate the mucus barrier [4, 5]. To examine the mucin-utilization ability of each bacterial isolate with respect to the corresponding symbiotic or pathogenic effects on host physiology, the effective cultivation of target bacterial isolates in mucin-supplemented medium is a primary and indispensable method.

When conducting in vitro microbial cultivation in mucin-containing media, the solubility of mucin samples may be problematic. Although it depends on the mucin sample preparation, highly *O*-glycosylated mucin glycoproteins are generally soluble in an

Akihiko Kameyama (ed.), *Mucins: Methods and Protocols*, Methods in Molecular Biology, vol. 2763,
https://doi.org/10.1007/978-1-0716-3670-1_27,

aqueous solution at a lower concentration without any chaotropic agent or detergent; however, these mucins do not completely dissolve at higher concentrations (>0.2%, *w*/*v*). Mucin solubility also decreases when it forms a large polymer in which they covalently connect with each other via cysteine bridges; this solubility may also decrease when glycans are removed and/or trimmed to reduce hydrophilicity. Most cultivation methods involve the addition of a high concentration of mucins to the medium as a carbon source, making it difficult to monitor bacterial growth with turbidity (optical density at 600–660 nm). To avoid this issue, determination of colony-forming units (CFUs) and/or quantitative polymerase chain reaction (qPCR) analyses are frequently used to evaluate bacterial growth [6]. CFU determination and qPCR analysis are effective methods for examining the growth of bacterial species during co-cultivation with other microbes [7, 8]. Nonetheless, some studies utilize *O*-glycans that are chemically released from mucin glycoproteins, instead of intact mucin glycoproteins, to circumvent the problem regarding insolubility; however, some microbes may exhibit different behaviors when using *O*-glycans compared to intact mucin glycoproteins.

This chapter describes an experimental method used to evaluate the growth of a bacterium in medium containing mucin glycoproteins. ***Bifidobacterium bifidum***, a probiotic in vivo mucin *O*-glycan utilizing gut microbe, are used as a model bacterium in this method [9].

2 Materials

Prepare all solutions using reagents of analytical grade. Thaw the mucin solution and ensure that the medium components remain under anoxic conditions until they are ready for use.

1. Gifu anaerobic medium (GAM) liquid medium/agar plate: Dissolve 59 g of GAM broth in 1 L of deionized water, according to the manufacturer's instructions. To prepare the liquid medium, take aliquots in glass test tubes (3–5 mL/tube), autoclave the media at 115 °C for 15 min, and then cool rapidly by placing in ice water (*see* **Note 1**). To produce agar plates, add agar powder (a final concentration of 1.5%) to the GAM medium, autoclave, and cool as described above. Pour the agar mixture aseptically into the plastic plates before solidifying. Leave the liquid media and agar plates for at least 12 h in an anaerobic jar. Store the plates at 4 °C.
2. Modified de Man-Rogosa-Sharpe (mMRS) medium: Glucose-free 2× MRS broth is prepared. The components of 2× mMRS (−) are shown in Table 1. After adjusting the pH to 6.5 using HCl, autoclave the broth at 121 °C for 15 min, and then cool

Table 1
Modified 2× MRS(−)

Component	Concentration (g/L)
Proteose peptone no.3 (Bacto)	20
Beef extract (Bacto)	20
Yeast extract (Bacto)	10
Polysorbate 80	2
Triammonium citrate	4
Sodium acetate	10
Dipotassium phosphate	4
Magnesium sulfate heptahydrate	0.082
Manganese sulfate heptahydrate	0.032
L-ascorbic acid sodium salt[a]	6.8
Cysteine HCl[a]	0.4
Adjust pH to 6.5 with HCl and autoclave	

[a]Add these compounds after autoclaving

rapidly using tap water. After cooling, add a filter-sterilized solution containing 0.34% (*w*/*v*) sodium ascorbate and 0.02% (*w*/*v*) cysteine–HCl. Maintain the medium for at least 12 h in an anaerobic jar. Store the medium at 4 °C (*see* **Note 2**).

3. 2× mucin solution: Resuspend porcine gastric mucin (PGM; type III, partially purified; Sigma-Aldrich, St. Louis, MO, USA) in Ultrapure water to a concentration two times higher % (*w*/*v*) than the final concentration (*see* **Note 3**). Pasteurize the mucin solution by incubating at 70 °C for 30 min (*see* **Note 4**). Store the samples in a freezer (−30 °C) until use.
4. Freeze-stock *B. bifidum* cells.

3 Methods

Conduct all procedures on a clean bench.

3.1 Pre-culture of B. bifidum

1. Streak the freeze-stocked *B. bifidum* cells on a GAM plate using a sterile picker (*see* **Note 5**).
2. Incubate the plate at 37 °C under anoxic conditions to form bacterial colonies.
3. Inoculate GAM liquid medium with a single colony of the *B. bifidum* bacterium and incubate at 37 °C under anoxic conditions.

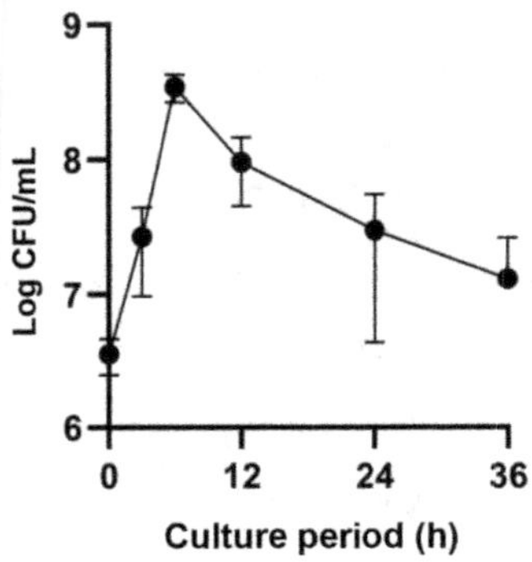

Fig. 1 Growth curve of *Bifidobacterium bifidum* JCM 1254 on a porcine gastric mucin (PGM)-containing modified de Man-Rogosa-Sharpe (MRS) medium
B. bifidum JCM1254 was used to inoculate modified MRS medium supplemented with 0.5% PGM as a carbon source. The initial OD_{600} value was 0.01. The cells were cultured at 37 °C in an anaerobic chamber. Aliquots of the culture were taken at the indicated time points and spread on GAM agar plates to determine colony-forming units. Data are presented as mean ± standard deviation of three independent experiments

3.2 Preparation of the Mucin Medium and Cultivation

1. Mix aseptically equal volumes of 2× mMRS(−) and 2× mucin solution and place under anoxic conditions.
2. Collect pre-cultured bacterial cells by centrifugation and wash twice with 1× mMRS (−) to avoid carryover of the GAM broth.
3. Resuspend the cells in an appropriate volume of 1× mMRS(−) and measure the OD_{600} value. Dilute the suspension when necessary.
4. Add the cells to the mucin medium at an OD_{600} value of 0.01–0.05 (*see* **Note 6**) and cultivate at 37 °C under anoxic conditions.
5. Collect a small amount of culture periodically, dilute appropriately with GAM liquid medium, and spread on GAM agar plates.
6. Incubate the GAM agar plates at 37 °C under anoxic conditions and count the colonies to obtain CFU (Fig. 1).

4 Notes

1. After autoclaving, cool the medium rapidly to prevent the air (oxygen) from dissolving the liquid.
2. Other media that support the growth of the bacterium are also available; however, these should not contain any known sugar sources such as glucose or starch.

3. Dialyze the PGM extensively against Ultrapure water and lyophilize prior to use. At a concentration > 1% (10 mg/mL), mucin may remain insoluble; therefore, preparation of 2× mucin solution in separate tubes is necessary. Therefore, when cultivation is carried out in triplicate, 2× mucin solution is prepared in three tubes.
4. Sterilization by autoclaving can be performed; however, this causes the release of sialic acid from the glycan.
5. For obligate anaerobes, bacterial streaking needs to be conducted quickly as exposure to air (oxygen) kills these bacteria.
6. The volume of the bacterial suspension added should be less than 4% of the medium volume.

Acknowledgements

This work was partly supported by a grant-in-aid from the JSPS Research Fellowship (21J15883 to HT), the Institution for Fermentation, Osaka (to ToK), and the Kyoto University Foundation (2022 to ToK).

References

1. Cani PD, Depommier C, Derrien M et al (2022) *Akkermansia muciniphila*: paradigm for next-generation beneficial microorganisms. Nat Rev Gastroenterol Hepatol 19:625–637
2. Hayase E, Hayase T, Jamal MA et al (2022) Mucus-degrading *Bacteroides* link carbapenems to aggravated graft-versus-host disease. Cell 185:3705–3719.e14
3. Glover JS, Ticer TD, Engevik MA (2022) Characterizing the mucin-degrading capacity of the human gut microbiota. Sci Rep 12:8456
4. Malaker SA, Pedram K, Ferracane MJ et al (2019) The mucin-selective protease StcE enables molecular and functional analysis of human cancer-associated mucins. Proc Natl Acad Sci U S A 116:7278–7287
5. Noach I, Ficko-Blean E, Pluvinage B et al (2017) Recognition of protein-linked glycans as a determinant of peptidase activity. Proc Natl Acad Sci U S A 114:E679–E688
6. Takada H, Katoh T, Katayama T (2020) Sialylated *O*-glycans from hen egg white ovomucin are decomposed by mucin-degrading gut microbes. J Appl Glycosci 67:31–39
7. Katoh T, Ojima MN, Sakanaka M et al (2020) Enzymatic adaptation of *Bifidobacterium bifidum* to host glycans, viewed from glycoside hydrolyases and carbohydrate-binding modules. Microorganisms 8:481
8. Ojima MN, Jiang L, Arzamasov AA et al (2022) Priority effects shape the structure of infant-type *Bifidobacterium* communities on human milk oligosaccharides. ISME J 16:2265–2279
9. Katoh T, Yamada C, Wallace MD et al (2023) A bacterial sulfoglycosidase highlights mucin *O*-glycan breakdown in the gut ecosystem. Nat Chem Biol 19:778–789

Chapter 28

Bacterial Enzyme Assay for Mucin Glycan Degradation

Toshihiko Katoh and Hisashi Ashida

Abstract

Bacterial sialidase and sulfoglycosidase may act on the acidic modifications of mucin *O*-glycans, producing sialic acid and 6-sulfated *N*-acetylglucosamine, respectively. Assays for these enzymes, using mucin as a substrate, are enabled by the detection and/or quantification of the free monosaccharides that are released by these enzymes. This chapter describes enzyme reactions with mucin, detection by thin-layer chromatography of sialic acid, and quantification of 6-sulfated *N*-acetylglucosamine by liquid chromatography–tandem mass spectrometry.

Key words Mucin, *O*-Glycan, Sialidase, Sulfoglycosidase, *Bifidobacterium bifidum*, TLC, LC-MS/MS

1 Introduction

Some intestinal bacteria exhibit the ability to degrade mucins to disrupt the mucus barrier or utilize mucin as a carbon and nitrogen source. These "mucin degraders" typically possess an array of mucin-degrading enzymes including glycoside hydrolases (GHs), sulfatases, and proteases. Many classes of GHs have been categorized in the Carbohydrate-Active enZyme (CAZyme) database [1]; these intestinal bacteria have been found to target glycosidic linkages found in the mucin *O*-glycan structures, reflecting the structural diversity of mucin oligosaccharides. Mucin molecules are covered in proline/threonine/serin-rich domains (PTS domains) with densely attached *O*-glycans, which account for more than 50% of its molecular weight [2]. Biosynthesis of *O*-glycans is initiated by the ppGalNAc-transferase-mediated addition of an *N*-acetylgalactosamine (GalNAc) residue, via an α-linkage, to the hydroxyl group of serine/threonine residues. This α-linked GalNAc is normally elongated by galactose (Gal) and/or *N*-acetylglucosamine (GlcNAc) residues. Further modifications of *O*-glycan, which drastically increase structural complexity, include fucosylation and/or acidic modifications such as sialylation and sulfation. The frequency

Akihiko Kameyama (ed.), *Mucins: Methods and Protocols*, Methods in Molecular Biology, vol. 2763,
https://doi.org/10.1007/978-1-0716-3670-1_28,

of acidic modifications increases gradually from the proximal colon to the rectum [3] and these modifications are inversely proportional to the burden of intestinal bacteria; therefore, acidic modifications may confer resistance to bacterial mucus degradation. Nonetheless, the removal of terminal sialic acid residues and sulfated residues is assumed to be a key reaction for mucin degradation by mucin degraders. Sialylation of *N*-acetylneuraminic acid (Neu5Ac) and/or *N*-glycolylneuraminic acid (Neu5Gc) on mammalian mucin oligosaccharides predominantly occurs as terminal α2,3- and/or α2,6-linked residues; additionally, the cleavage of sialic acid residues, to release Neu5Ac and Neu5Gc, is usually conducted by exo-sialidases in bacteria. Intramolecular *trans*-sialidases (IT-sialidases) also act on α2,3-linked Neu5Ac to produce free 2,7-anhydro-Neu5Ac [4]. Thin-layer chromatography (TLC) has previously been used to discriminate between free Neu5Ac and 2,7-anhydro-Neu5Ac [5]. Additionally, free Neu5Ac can be quantified by many colorimetric methods and specific methods using sialic acid aldolase [6]. Sulfation primarily occurs at the 3-, 4-, and 6-positions of terminal Gal and the 6-position of internal/terminal GlcNAc [7, 8]. At present, sulfatases [8, 9] and 6-SO_3-β-D-*N*-acetylglucosaminidases (also known as sulfoglycosidases) [10–12] have been determined to be involved in the degradation of sulfated glycans by gut microbes.

Among the mucin-degrading bacteria, *Bifidobacterium bifidum*, a probiotic species, possesses several mucin-targeting GHs, including GH33 sialidases, SiaBb1 [13] and SiaBb2 [14], alongside a GH20 sulfoglycosidase, BbhII, [11, 12]; these enzymes cleave mucin to produce sialic acid and 6-sulfated GlcNAc (GlcNAc-6S), respectively. These enzymes have a signal peptide, a class-E sortase-dependent [LIV][SA]XTG cell-anchoring motif [15], and a membrane-spanning region at the C-terminus; thus, it is assumed that these are cell wall–anchored extracellular enzyme. Here, we describe an enzymatic reaction with mucin using recombinant enzymes of sialidase and sulfoglycosidase from *B. bifidum*, the qualitative detection of the released sialic acid by TLC, and quantification of free GlcNAc-6S by liquid chromatography–tandem mass spectrometry (LC-MS/MS).

2 Materials

Prepare all solutions using Ultrapure water and analytical grade reagents.

2.1 Enzyme Reaction

1. Porcine gastric mucin (PGM; type III, partially purified; Sigma-Aldrich, St. Louis, MO, USA) solution: Resuspend the PGM powder in Ultrapure water to obtain a concentration of

1% (w/v) (10 mg/mL) (*see* **Note 1**). Store the mucin solution in a freezer.

2. Enzyme preparation: Prepare SiaBb2 and BbhII enzyme solutions (*see* **Note 2**) and store at 4 °C.
3. Reaction buffer: 0.5 M sodium acetate buffer, pH 6.0 (stock solution). Store the buffer at room temperature (25 °C) or at 4 °C.
4. Incubator (heating block or water bath, set at 37 °C).

2.2 TLC Components

1. TLC Silica gel 60 aluminum plates (Merck Millipore, Billerica, MA, USA).
2. Development solvent: 1-butanol/acetic acid/water (2/1/1, *v/v/v*) (*see* **Note 3**).
3. Developing glass chamber.
4. Hair dryer.
5. Diphenylamine–aniline–phosphoric acid reagent: Dissolve diphenylamine (1 g) in acetone (100 mL) and then add aniline (1 mL) and phosphoric acid (10 mL) (*see* **Note 4**). Store the reagent at 4 °C.
6. Glass sprayer.
7. Heating device (a household toaster can be used).
8. Neu5Ac standard solution (1–5 mM): Dissolve in Ultrapure water and store in a freezer.
9. Imager (scanner).
10. Draft chamber.

2.3 Quantification of GlcNAc-6S by LC-MS/MS

1. Acetone.
2. Centrifugal concentrator.
3. Syringe-driven filter unit (0.45 μm pore membrane filter).
4. A liquid chromatography mass spectrometer equipped with a heated electrospray ionization probe (*see* **Note 5**).
5. Eluents: 10 mM ammonium bicarbonate buffer (pH 10) (eluent A) and acetonitrile (eluent B); filtrate through a 0.45 μm pore membrane filter and degas.
6. Hypercarb™ column (2.1 × 100 mm, particle size: 3 μm; Thermo Fisher Scientific, Waltham, MA, USA).
7. GlcNAc-6S (44001, Sigma-Aldrich) standard solution: Dissolve in Ultrapure water and store in a freezer.
8. Microcentrifuge.

3 Methods

Conduct all procedures at room temperature unless otherwise indicated.

3.1 Enzyme Reaction with Mucin

1. Constitute the mixture with 20 μL of 1% mucin solution (0.5% final concentration), 4 μL of 0.5 M sodium acetate buffer (50 mM final concentration), and 12 μL of Ultrapure water. The mixture was prewarmed to 37 °C.
2. Add 4 μL of the enzyme solution (0.01–0.1 μM final concentration) and incubate the mixture with at 37 °C.
3. After an appropriate period, stop the reaction by heating at 90 °C for 5 min (*see* **Note 6**).

3.2 Detection of Released N-Acetylneuraminic Acid by TLC

1. Add the development solvent into a developing glass chamber; then, the chamber was filled with vaporized gas (*see* **Note 7**).
2. Prepare an aluminum TLC plate (*see* **Note 8**) and spot the Neu5Ac standard and the samples along this plate (*see* **Note 9**). Dry the spots using a hair dryer.
3. Place the plate in the chamber containing the development solvent. Allow the TLC plate to expand to approximately 90% of its top edge.
4. Remove the plate from the developing chamber and dry it with a hair dryer.
5. Then, spray the diphenylamine–aniline–phosphoric acid reagent onto the plate in a draft chamber (*see* **Note 10**).
6. Place the plate in a preheated heating device (toaster) until the color is visible, as shown in Fig. 1. Then, remove the plate and leave to cool to room temperature. Save the image data (*see* **Note 11**).

3.3 Measurement of GlcNAc-6S by LC-MS/MS

3.3.1 Preparation of the Samples

1. Add four times the volume of ice-cold acetone to the enzyme reaction mixture; allow the tubes to stand in a freezer at −30 °C for 1 h to precipitate mucin proteins (*see* **Note 12**).
2. Centrifuge the sample at 20,000 × *g* for 10 min to pellet the protein and collect the supernatants in new tubes.
3. Evaporate the supernatants using a centrifugal concentrator until dry.
4. Adjust the dried samples with ultrapure water, typically by the addition of approximately 100 μL of water. Filter the sample using 0.45 μm pore membrane filter device (*see* **Note 13**).
5. Add an equal volume of equilibration buffer (eluent A/eluent B = 95:5) to the sample. Repeat the **steps** (**1–5**) for GlcNAc-6S standard solutions.

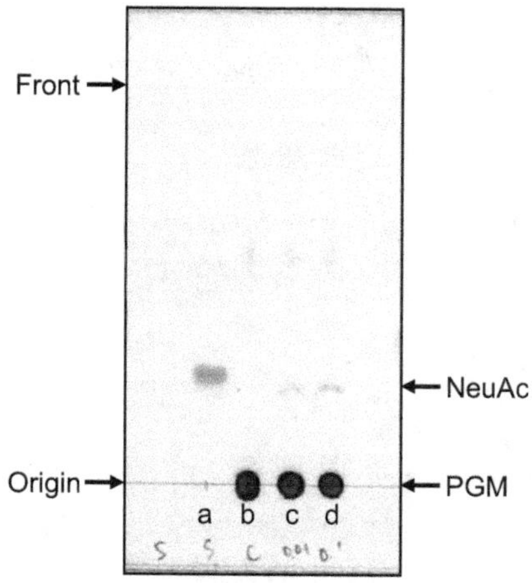

Fig. 1 Thin-layer chromatography (TLC) of recombinant sialidase (SiaBb2)-treated porcine gastric mucin (PGM). PGM (0.5%) was incubated with 0.01 and 0.1 μM SiaBb2 at 37 °C for 1 h. Lanes: (**a**), 2 mM Neu5Ac (1 μL); (**b**), PGM no-enzyme control (3 μL); (**c**), PGM treated with 0.01 μM SiaBb2 (3 μL); and (**d**), PGM treated with 0.1 μM SiaBb2 (3 μL). The TLC plate was developed in *1*-butanol/acetic acid/water (2/1/1, v/v/v). Diphenylamine–aniline–phosphoric acid reagent was used for staining. The R_f value of Neu5Ac is approximately 0.30

3.3.2 LC-MS/MS Analysis

1. Equilibrate the Hypercarb column with eluent A/eluent B (95/5) at a flow rate of 0.2 mL/min using a LC instrument. Maintain the temperature of the column oven at 45 °C.
2. Run a gradient program. For one run, the LC gradient program is set to maintain 5% B for 1 min, linearly increase eluent B up to 30% over 6 min, maintain 30% B for 1 min, return to 5% B within 6 s, and maintain 5% B for 3 min.
3. Perform a multiple reaction monitoring (MRM) with negative electrospray ionization (ESI) to detect GlcNAc-6S. Monitor the precursor ion (m/z 300.10) and CID product ion (m/z 139.05) to quantify GlcNAc-6S (*see* **Note 14**). The MS parameters used are listed in Table 1.
4. Calculate the concentration of GlcNAc-6S in the sample by creating a calibration curve using standard solutions with known concentrations (Fig. 2).

4 Notes

1. Dialyze the PGM extensively against ultrapure water and lyophilize prior to use. At a concentration of 10 mg/mL, mucin may remain insoluble, but can be used regardless.
2. Enzymes can be prepared as hexahistidine-tagged recombinant proteins that are expressed in *Escherichia coli* and purified using Ni^{2+}-affinity chromatography. Sequences for the signal peptide

Table 1
LC-MS/MS parameters for detection of GlcNAc-6S

Flow rate	0.2 mL/min
Temp. of column oven	45 °C
Sample injection volume	10 μL
Interface	Negative ESI
Interface voltage	3.0 kV
Interface temp.	300 °C
DL temp.	250 °C
Nebulizer gas	3.0 L/min
Drying gas	10.0 L/min
Heating gas	10.0 L/min

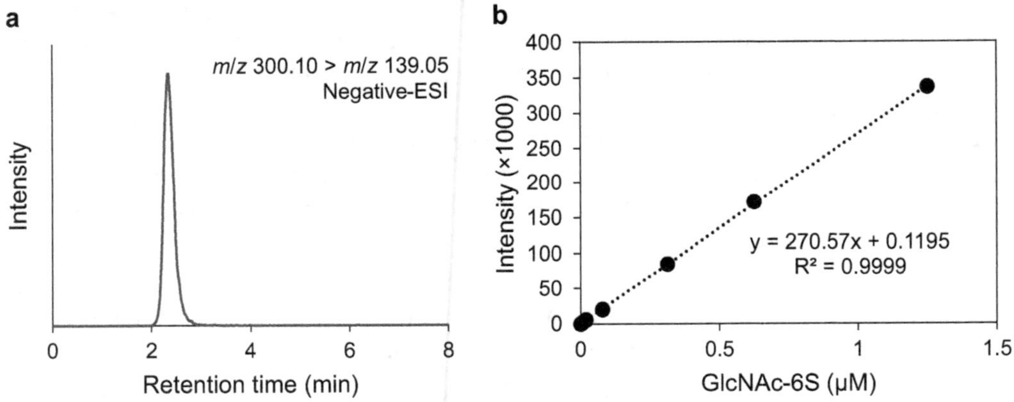

Fig. 2 Liquid chromatography–tandem mass spectrometry (LC-MS/MS) analysis for free GlcNAc-6S. (**a**) Elution pattern of GlcNAc-6S with detection by multiple reaction monitoring (MRM) at m/z 300.10 > m/z 139.05. (**b**) Standard curve of GlcNAc-6S (4.9 nM to 1.25 μM)

and membrane-spanning region should be eliminated from the construct to produce soluble enzymes. Crude enzyme solutions can be used but there is a possibility of background staining.

3. As much of this development solvent should be prepared before use because changes in composition due to volatilization will affect development.
4. This reagent should be prepared by adding all reagents to acetone in order; otherwise, it is difficult for these reagents to dissolve.
5. We use a LCMS-8045, Shimadzu, Kyoto, Japan.

6. Note that excessive heat treatment may liberate sialic acid from mucin.
7. Add the development solvent at a depth of approximately 0.5 cm so that the spot point of the sample is not soaked when the TLC plate is placed. Because organic solvents, such as acetic acid, volatilize, development is performed in a draft chamber.
8. Keep TLC plates in a dry space (e.g., desiccator) and handle with tweezers. If five samples are analyzed, cut a TLC plate to 4 cm wide and 6–7 cm high. Draw a line with a pencil (not a ballpoint pen) at a height of 1–1.5 cm from the bottom of the plate; mark a 0.5 cm-wide spot with a pencil and leave a margin of at least 1 cm at both ends (Fig. 1).
9. Spotting 1 μL of 1–5 mM monosaccharide should be sufficient to stain sugars.
10. Spray the reagent evenly from a fixed distance in a simple covered box that is placed in a draft chamber.
11. Diphenylamine–aniline–phosphoric acid reagent staining gives sialic acid a purplish-red color and GlcNAc-6S a brown color. Otherwise, fucose stains greenish-blue, Gal and glucose stain blackish-blue, GlcNAc and GalNAc stain reddish-brown, and fructose stains light brown.
12. Remove mucin from the sample by precipitation to avoid its introduction into the subsequent LC.
13. Note that some sample loss may occur due to filtering.
14. The collision energy is set to 24.0 eV in this experiment; however, this setting should be determined on an instrument basis. Quantitativity is confirmed for other CID fragments, e.g., m/z 300.10 > m/z 199.05 and m/z 300.10 > m/z 97.05. The retention time of GlcNAc-6S is 2.3 min. GalNAc-6S also produced the same fragment, but its retention time is later than that of GlcNAc-6S.

Acknowledgements

This work was partly supported by a grant-in-aid from the Institution for Fermentation, Osaka (to TK), and the Kyoto University Foundation (2022 to TK).

References

1. Drula E, Garron ML, Dogan S et al (2022) The carbohydrate-active enzyme database: functions and literature. Nucleic Acids Res 50: D571–D577
2. Hansson GC (2020) Mucins and the microbiome. Annu Rev Biochem 89:769–793
3. Robbe C, Capon C, Coddeville B et al (2004) Structural diversity and specific distribution of

O-glycans in normal human mucins along the intestinal tract. Biochem J 384:307–316

4. Tailford LE, Owen CD, Walshaw J et al (2015) Discovery of intramolecular trans-sialidases in human gut microbiota suggests novel mechanisms of mucosal adaptation. Nat Commun 6: 7624
5. Chou MY, Li SC, Li YT (1996) Cloning and expression of sialidase L, a NeuAcα2→3Gal-specific sialidase from the leech, *Macrobdella decora*. J Biol Chem 271:19219–19224
6. Cheeseman J, Kuhnle G, Spencer DIR et al (2021) Assays for the identification and quantification of sialic acids: challenges, opportunities and future perspectives. Bioorg Med Chem 30:115882
7. Holmén Larsson JM, Thomsson KA, Rodríguez-Piñeiro AM et al (2013) Studies of mucus in mouse stomach, small intestine, and colon. III. Gastrointestinal Muc5ac and Muc2 mucin *O*-glycan patterns reveal a regiospecific distribution. Am J Physiol Gastrointest Liver Physiol 305:G357–G363
8. Luis AS, Jin C, Pereira GV et al (2021) A single sulfatase is required to access colonic mucin by a gut bacterium. Nature 598:332–337
9. Praharaj AB, Dehury B, Mahapatra N et al (2018) Molecular dynamics insights into the structure, function, and substrate binding mechanism of mucin desulfating sulfatase of gut microbe *Bacteroides fragilis*. J Cell Biochem 119:3618–3631
10. Rho J, Wright DP, Christie DL et al (2005) A novel mechanism for desulfation of mucin: identification and cloning of a mucin-desulfating glycosidase (sulfoglycosidase) from *Prevotella* strain RS2. J Bacteriol 187:1543–1551
11. Katoh T, Maeshibu T, Kikkawa K et al (2017) Identification and characterization of a sulfoglycosidase from *Bifidobacterium bifidum* implicated in mucin glycan utilization. Biosci Biotechnol Biochem 81:2018–2027
12. Katoh T, Yamada C, Wallace MD et al (2023) A bacterial sulfoglycosidase highlights mucin *O*-glycan breakdown in the gut ecosystem. Nat Chem Biol 19:778–789
13. Ashida H, Tanigawa K, Kiyohara M et al (2018) Bifunctional properties and characterization of a novel sialidase with esterase activity from *Bifidobacterium bifidum*. Biosci Biotechnol Biochem 82:2030–2039
14. Kiyohara M, Tanigawa K, Chaiwangsri T et al (2011) An exo-α-sialidase from bifidobacteria involved in the degradation of sialyloligosaccharides in human milk and intestinal glycoconjugates. Glycobiology 21:437–447
15. Ishikawa E, Yamada T, Yamaji K et al (2021) Critical roles of a housekeeping sortase of probiotic *Bifidobacterium bifidum* in bacterium-host cell crosstalk. iScience 24:103363

Chapter 29

Measurement of Mucinase Activity

Hiroki Tanabe, Tatsuya Morita, and Naomichi Nishimura

Abstract

Mucinase consists of some proteases, glycosidases, sulfatases, and sialidases. It is not practical to measure individual enzyme activities when measuring mucinase activity. In this method, mucinase activity is measured using porcine gastric mucin as a substrate and feces as an enzyme source. This description includes fecal pellet preparation, reaction procedure of mucinase, measurement of reducing sugars liberated during the procedure, and determination of nitrogen content in the fecal preparations.

Key words Mucinase, Porcine gastric mucin, Feces, Rat

1 Introduction

Mucin is an important component that lubricates the luminal surface and protects it from various invaders [1]. Therefore, measuring the amount of mucin in the gastrointestinal tract is important to determine the sufficiency of mucosal defense [2]. The amount of mucin in the gastrointestinal tract is estimated as the difference between the amount of mucin secreted by goblet cells and the amount of mucin degraded by intestinal bacteria. Even when the amount of mucin in the gastrointestinal tract is the same, there are cases in which the amount of secreted and degraded mucin varies widely [3]. Therefore, it is important to measure not only the amount of mucin in the gastrointestinal tract, but also the activity of mucinase in order to learn more about the factors that cause changes in mucin levels and to develop countermeasures. Mucinase cleaves sugars from the tips of the sugar covering the surface of mucin (Fig. 1). Mucinase consists of some sialidases, glycosidases, proteases, and sulfatases [4]. Sialidases cleave the sialic acids from the mucin sugar chain tips, first. Glycosidases cleave sugars from mucin oligosaccharides. These enzymes act in conjunction with each other to degrade the mucin oligosaccharides [5]. Proteases cleave non-glycosylated regions in mucin molecule and final

Akihiko Kameyama (ed.), *Mucins: Methods and Protocols*, Methods in Molecular Biology, vol. 2763,
https://doi.org/10.1007/978-1-0716-3670-1_29,

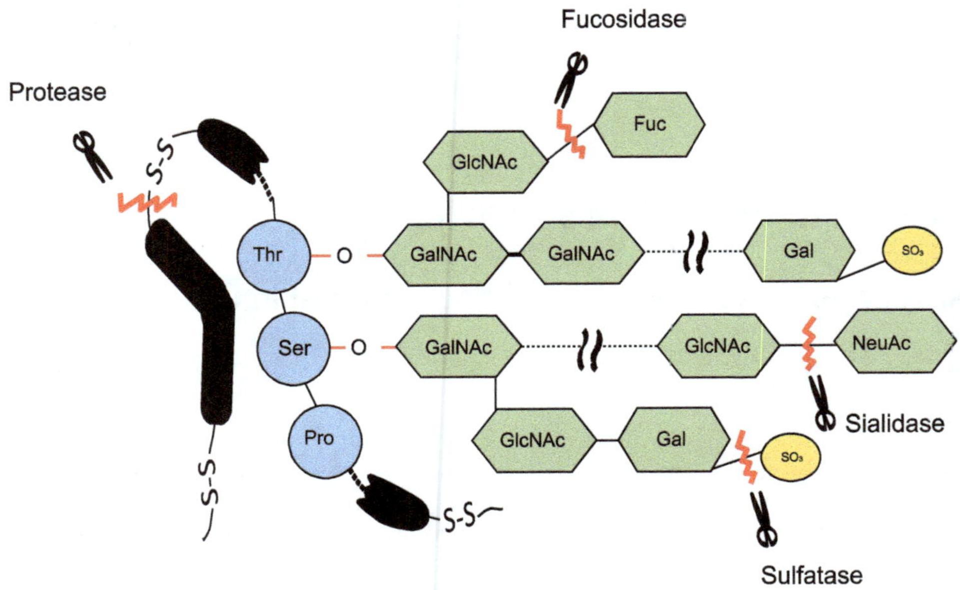

Fig. 1 Schematic diagram of the chemical structure of mucin. Corresponding enzymes act stepwise from the tip of the mucin sugar chains, and finally, the core protein is degraded by the protease

disruption of exposed protein core after deglycosylation by other enzymes [6]. Sulfatases cleave terminal sulfate residues from mucin oligosaccharides [7]. Thus, it is not practical to measure individual enzyme activities when measuring mucinase activity. Therefore, a method was established to measure mucinase activity using porcine gastric mucin as a substrate and feces as an enzyme source [8]. The method consists of fecal pellet preparation, reaction procedure of mucinase, measurement of reducing sugars liberated during the procedure, and determination of nitrogen content in the fecal preparations. Mucinase activity is expressed as nmol glucose-liberated·min^{-1}·mg^{-1} nitrogen. This method eliminates the cumbersome process of isolating the enzyme and has a low risk of the loss of enzyme activity. Here we describe a simple but reproducible and robust method for measuring mucinase activity in rat feces. This method well reflects the changes in mucinase activity when colonic fermentation is modified by the ingestion of nondigestible carbohydrates [3].

2 Materials

Prepare all solutions using Ultrapure water and analytical grade reagents. Prepare and store all reagents at room temperature (unless indicated otherwise).

2.1 Fecal Pellet Preparation

1. Sodium phosphate buffer (10 mmol/L, pH 7.5): Dissolve 2.410 g of disodium hydrogenphosphate-12 hydrate and 0.521 g of sodium dihydrogenphosphate-2 hydrate in 800 mL purified water, and make the total volume 1 L with purified water.
2. Polypropylene tube of 50 mL volume.
3. Polytron homogenizer with generator shaft of OD, 12 mmφ.
4. Plastic cup, large enough to keep 50 mL polypropylene tube on ice.
5. Fecal pellets.

2.2 Fecal Mucinase Assay

1. Porcine gastric mucin solution (0.5%, w/v): Add 250 mg of porcine gastric mucin (PGM, type II, M2378, Sigma-Aldrich, St. Louis, MO, USA) in 50 mL beaker containing 40 mL of distilled water (*see* **Note** 1). Add 0.1 mol/L sodium hydroxide solution and adjust pH to 7.2. Then, stand overnight at 4 °C. Bring to room temperature while stirring with a stirrer. Add 0.1 mol/L sodium hydroxide solution and adjust pH to 7.5. Transfer the solution to a measuring flask and fill up to 50 mL. Finally, filter the solution and save the filtrate as substrate solution.
2. Constant temperature bath.
3. Gas stove.
4. 10 mL glass test tube.
5. Boiling water bath.
6. Ice-water bath.

2.3 Reducing Sugar Measurement

1. Solution A: Add 7.5 g of copper (II) sulfate-5 hydrate in a 50 mL beaker containing 40 mL of distilled water. Transfer the solution to a measuring flask and fill up to 50 mL.
2. Solution B: Add 12.5 g of sodium carbonate, 12.5 g of potassium sodium tartrate, 10 g of sodium hydrogen carbonate, and 100 g of sodium sulfate in a 500 mL beaker containing 350 mL of distilled water. Transfer the solution to a measuring flask and fill up to 500 mL.
3. Copper regent: Mix solution A and solution B at a 1:25 ratio (v/v), freshly prepared before use.
4. Nelson's reagent: Add 12.5 g of ammonium molybdate-4 hydrate in a 500 mL beaker containing 400 mL of distilled water. Gently place 21 g of concentrated sulfuric acid to the beaker. Then, add 1.5 g of disodium hydrogenarsenate-7 hydrate, previously dissolved in 25 mL of distilled water. Transfer the solution to a measuring flask and fill up to 500 mL.

Incubate at 37 °C for 24 h under light-shielded conditions, and then store at 4 °C.

5. Glass test tube.
6. Boiling water bath.
7. Ice bath.
8. 15 mL polypropylene tube.
9. Centrifuge.
10. Spectrophotometer.
11. Glucose solution for standard curve (25–200 μg/mL)

2.4 Nitrogen Determination

1. 200 mL Kjeldahl flask.
2. Concentrated sulfuric acid.
3. Catalyzer: Mix potassium sulfate and copper sulfate pentahydrate at a 9:1 ratio (w/w), grind, and mix well in a mortar.
4. Boiling stones.
5. Electric heater.
6. 4. Kjeldahl distillation apparatus.
7. Sodium hydroxide aqueous solution (30%, w/w).
8. Indicator: Mixing 0.2% methyl red (w/v) dissolved in 95% ethanol with 0.1% methylene blue (w/v) dissolved in 95% ethanol at a 1:1 ratio (v/v) makes the pH indicator turn red on acid, green on alkaline, and clear and colorless on neutral.
9. Boric acid aqueous solution (3%, w/v) (*see* **Note 2**).
10. Volumetric burette of 10 mL.
11. Sulfuric acid aqueous solution (0.01 mol/L) (*see* **Note 3**).

3 Methods

Carry out all procedures at room temperature unless specified otherwise.

3.1 Fecal Pellet Preparation

1. Weigh approximately 0.2 g of fresh fecal pellets and place in 50 mL polypropylene tube on an ice bucket.
2. Add 100 times (w/v) of 10 mmol/L sodium phosphate buffer (pH 7.5) to the fecal pellets (approx. 20 mL in total).
3. Stand on ice for 20 min to soften the fecal pellets.
4. Homogenize for 30 s (*see* **Note 4**). The fecal homogenates should be stored on ice and used for the measurement of mucinase activity within 3 h. The fecal homogenates are also used for nitrogen determination, in which case they may be frozen and stored.

3.2 Fecal Mucinase Reaction

It is necessary to have a sample blank and substrate blank in addition to a test sample. Use glass test tubes with high thermal conductivity. Never use plastic test tubes.

1. To 10 mL glass test tubes, add 0.9 mL of fecal homogenate for the test sample and the sample blank, or 0.9 mL of 10 mmol/L sodium phosphate buffer for the substrate blank, respectively, and preincubate at 30 °C for 2 min.
2. Add 0.1 mL 0.5% porcine gastric mucin for the test sample and the substrate blank, or 0.1 mL of 10 mmol/L sodium phosphate buffer for the sample blank, respectively, and incubate at 30 °C for 20 min.
3. After reaction, place the glass test tubes in boiling water bath and stand for 1 min to stop the enzymatic reaction.
4. Place the glass test tubes in ice-water bath for cooling.

3.3 Reducing Sugar Measurement

Following mucinase reaction, the amount of reducing sugar liberated in the glass test tubes is measured by the Somogyi–Nelson method. Glucose is used as the standard in the range of 25–200 μg/mL for creation of a standard curve.

1. Add 1 mL of the copper regent to the glass test tubes (test sample, sample blank, substrate blank, and standard solution).
2. Place the glass test tubes in boiling water bath for 10 min, and then immediately place the glass test tubes in an ice bath immediately (*see* **Note 5**).
3. Add 1 mL of Nelson's reagent to the glass test tubes and stand for 20 min (molybdenum blue coloration is stable for 5 h).
4. After vortexing the glass test tubes, collect 0.45 mL each and transfer to 15 mL polypropylene tubes, respectively.
5. Add 3.3 mL of distilled water to the polypropylene tubes and mix well.
6. Centrifuge at 2300 × *g* for 10 min, and measure the absorbance of the supernatant at 660 nm in a spectrophotometer.
7. Subtract the absorbance of the sample blank and substrate blank from the absorbance of the test sample, and fit the value to the standard curve. The reducing sugar concentration/mL test sample (μg/mL) is corrected by dividing by 0.9.

3.4 Nitrogen Determination

Fecal homogenate is heated in concentrated sulfuric acid to convert the nitrogen contained in it into ammonium sulfate (Kjeldahl digestion). Following the micro-Kjeldahl method [9], the solution containing ammonium sulfate is alkalized and heated, and then the amount of ammonia generated by distilling it is determined by titration.

3.4.1 Kjeldahl Digestion

1. Place 15 mL of fecal homogenate in a 200 mL Kjeldahl flask.
2. Add a small amount of catalyzer (approx. 300–500 mg) and a few boiling stones.
3. Place 15 mL of concentrated sulfuric acid in the flask.
4. Heat the flask until the solution boils, allowing the reaction to proceed. When the liquid becomes clear (typically, takes a few h), stop heating and allow it to come to room temperature.
5. Carefully add 30 mL of distilled water to the solution in the flask while chilling, and then transfer to a measuring flask and fill up to 50 mL with distilled water (i.e., sample for distillation).

3.4.2 Kjeldahl Distillation and Titration (Fig. 2)

1. Prepare a 100 mL Erlenmeyer flask containing 20 mL of 3% boric acid solution for trapping ammonia vapor. Add 2 drops of indicator (red coloration at this point), and then attach to the distillation apparatus.

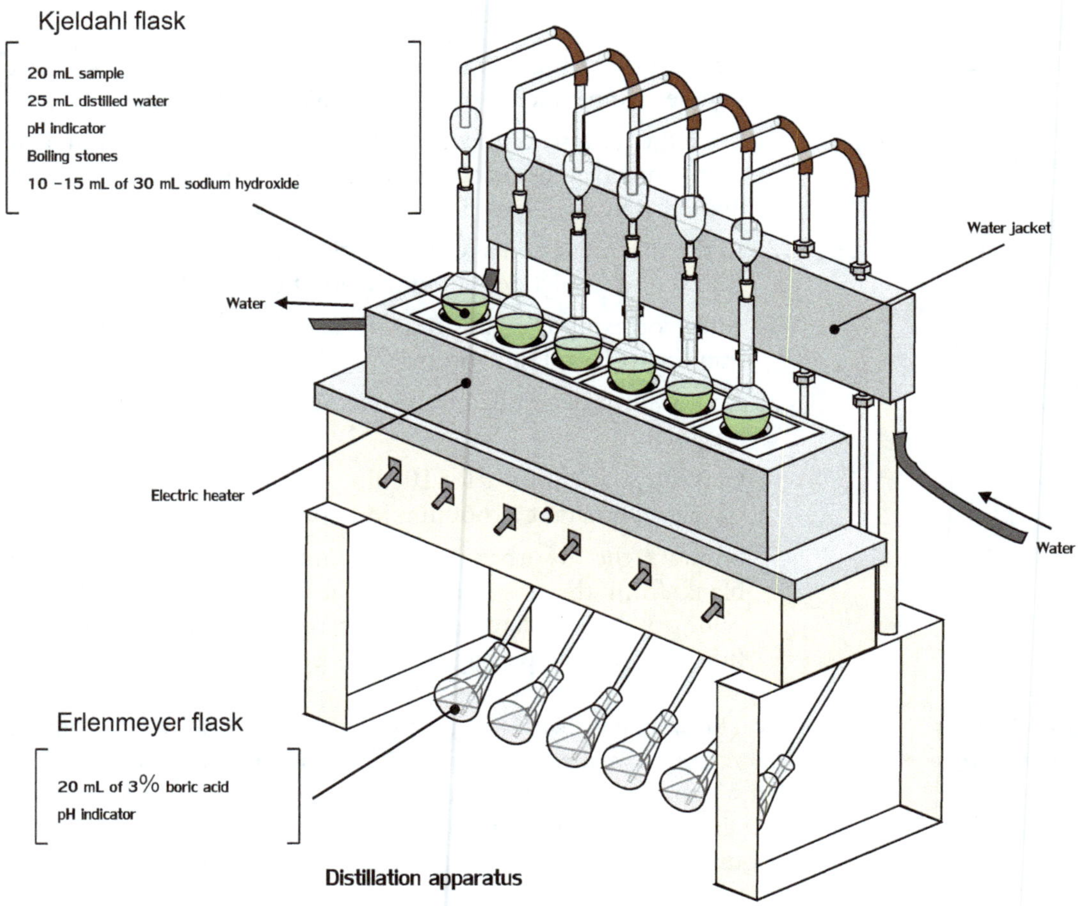

Fig. 2 Outline of the distillation apparatus. Instead of this, an ordinary steam distillation system can be used

2. To a 100 mL Kjeldahl flask for distillation, add exactly 20 mL of sample (**step 5** in Subheading 3.4.1) with a whole pipette.
3. Add 25 mL of distilled water, a few boiling stones, and 2 drops of pH indicator to the Kjeldahl flask.
4. Add 30% sodium hydroxide solution until the color of the indicator turns green (approx. 10–15 mL), and immediately attach the Kjeldahl flask to the distillation apparatus (*see* **Note 6**).
5. Heat the Kjeldahl flask and begin distillation. Heat until the remaining distillate volume is approximately 25 mL. At this point, the solution in the Erlenmeyer flask turns green, indicating alkalinity.
6. Using a biuret, titrate the solution in the Erlenmeyer flask with 0.01 mol/L of sulfuric acid solution until neutralized. Typically, the titration value is around 2 mL.
7. Nitrogen content of fecal homogenate (mg/mL) $= 0.01 \times 2 \times 14 \times \frac{\text{Titration volume (mL)}}{1000\ (\text{mL})} \times 1000 \times \frac{50\ (\text{mL})}{20\ (\text{mL})} \times \frac{1}{15\ (\text{mL})}$.

3.5 Calculation of Mucinase Activity

Specific activity (nmol·min^{-1}·mg^{-1} N) = the reducing sugar concentration (μg/mL) $\times \frac{1\ (\text{mL})}{0.9\ (\text{mL})} \times \frac{1}{180\ (\text{g glucose/mol})} \times \frac{1}{20\ (\text{min})} \times 1000 \times \frac{1}{\text{Nitrogen content of fecal homogenate (mg/mL)}}$

Normally, mucinase activity in rats fed a purified diet (AIN-76 or 93 diet) is found in the range of 20–50 nmol reducing sugar·min^{-1}·mg^{-1} nitrogen, whereas activity is about a half to one third of that in rats fed a non-purified diet.

4 Notes

1. The measurement of fecal mucinase activity is based on the method reported by Shiau et al. [8] using Sigma-Aldrich porcine gastric mucin as a substrate. There are two types of porcine gastric mucins available from Sigma-Aldrich: type II (M2378, Sigma-Aldrich) and type III (M1778, Sigma-Aldrich). Type II mucins are crude and must be purified before use. Since a method for measuring mucinase activity using type II mucins (M2378, Sigma-Aldrich) has already been established [3], it is recommended that type II mucins are used. It may be possible to use type III mucins, which are partially purified from type II mucins, but this should be considered in preliminary experiments if necessary.
2. As the solubility of boric acid in cold water is low, heat the solution as needed.
3. As it is used for titration, it must be accurately prepared and standardized.

4. To avoid mucinase denaturation, homogenize at a non-foaming speed while chilling the tube in a cup of ice.
5. A quick cooling is important to prevent a reverse oxidation of Cu_2O to CuO.
6. Note that ammonia vapor is produced as soon as the solution becomes alkaline.

References

1. Neutra MR, Forstner JG (1987) Gastrointestinal mucus: synthesis, secretion, and function. In: Johnson LR (ed) Physiology of the gastrointestinal tract, vol 2. Raven Press, New York, pp 975–1009
2. Morita T, Tanabe H, Takahashi K, Sugiyama K (2004) Ingestion of resistant starch protects endotoxin influx from the intestinal tract and reduces D-galactosamine-induced liver injury in rats. J Gastroenterol Hepatol 19(3):303–313. https://doi.org/10.1111/j.1440-1746.2003.03208.x
3. Komura M, Fukuta T, Genda T, Hino S, Aoe S, Kawagishi H, Morita T (2014) A short-term ingestion of fructo-oligosaccharides increases immunoglobulin a and mucin concentrations in the rat cecum, but the effects are attenuated with the prolonged ingestion. Biosci Biotechnol Biochem 78(9):1592–1602. https://doi.org/10.1080/09168451.2014.925782
4. Wiggins R, Hicks SJ, Soothill PW, Millar MR, Corfield AP (2001) Mucinases and sialidases: their role in the pathogenesis of sexually transmitted infections in the female genital tract. Sex Transm Infect 77(6):402–408. https://doi.org/10.1136/sti.77.6.402
5. Stewart-Tull DE, Ollar RA, Scobie TS (1986) Studies on the vibrio cholerae mucinase complex. I. Enzymic activities associated with the complex. J Med Microbiol 22(4):325–333. https://doi.org/10.1099/00222615-22-4-325
6. Dwarakanath AD, Campbell BJ, Tsai HH, Sunderland D, Hart CA, Rhodes JM (1995) Faecal mucinase activity assessed in inflammatory bowel disease using 14C threonine labelled mucin substrate. Gut 37(1):58–62. https://doi.org/10.1136/gut.37.1.58
7. Robertson AM, Wright DP (1997) Bacterial glycosulphatases and sulphomucin degradation. Can J Gastroenterol 11(4):361–366. https://doi.org/10.1155/1997/642360
8. Shiau SY, Chang GW (1983) Effects of dietary fiber on fecal mucinase and beta-glucuronidase activity in rats. J Nutr 113(1):138–144. https://doi.org/10.1093/jn/113.1.138
9. Miller L, Houghton JA (1945) The microKjeldahl determination of the nitrogen content of amino acids and proteins. J Biol Chem 169:373–383

Chapter 30

Adhesion Inhibition Assay for *Helicobacter pylori* to Mucin by *Lactobacillus*

Keita Nishiyama and Takao Mukai

Abstract

The ability of *Lactobacillus* to adhere to mucin is a parameter for evaluating the effectiveness of probiotics. In particular, a competitive inhibition assay of pathogenic bacteria using mucin-adherent lactobacilli is useful for identifying *Lactobacillus* strains capable of preventing mucus from being colonized by pathogens. Here, we describe an adhesion inhibition assay method for *Helicobacter pylori* to porcine gastric mucin by *Limosilactobacillus reuteri*.

Key words Adhesion, *Helicobacter*, *Lactobacillus*, Mucin, Probiotics

1 Introduction

Mucin is secreted by goblet cells throughout the human gastrointestinal tract, from the stomach to the colon. It forms a mucin gel layer that ranged in thickness from 50 to 800 μm, with that in the large intestine being especially thick [1]. This layer acts as a physical barrier that protects the intestinal epithelium from contact with digestive contents and prevents the bacterial invasion and infection of host cells [2]. The mucin layer also provides a scaffold for symbiotic and pathogenic bacteria in the gut.

Lactobacillus is predominant in the lower part of the small intestine and has been considered to provide a variety of beneficial functions in humans. Several *Lactobacillus* strains have been reported to prevent infection by intestinal pathogens, such as *Campylobacter jejuni*, *Escherichia coli*, *Salmonella* Typhimurium, *Enterococcus faecalis*, and *Helicobacter pylori*. The following properties are believed to provide these beneficial effects: (i) production of inhibitory compounds, such as short-chain fatty acids and bacteriocins, (ii) competition for nutrients, (iii) immune system stimulation, and (iv) competitive exclusion [3]. In particular, although the mechanism of competitive exclusion is well known in terms of the

Akihiko Kameyama (ed.), *Mucins: Methods and Protocols*, Methods in Molecular Biology, vol. 2763,
https://doi.org/10.1007/978-1-0716-3670-1_30,

properties of competition for adhesion sites, because the relevant experiments are relatively uncomplicated, there are few publications in which the corresponding methods are described in detail. We previously established a method for evaluating the adhesion of *Lactobacillus* to mucins using microtiter plates or diagnostic slides [4]. One strategy for reproducibly evaluating the adherence of lactobacilli is to use highly purified mucins obtained via gel-filtration chromatography combined with density-gradient ultracentrifugation [5].

Helicobacter pylori is a gram-negative, spiral-shaped microaerophilic bacterium that causes antral gastritis and peptic ulcers and is associated with stomach cancer. *Limosilactobacillus reuteri* JCM1081 exhibits specific adhesion to the sulfated sugar moieties of mucin and glycolipids [6, 7], which serve as receptor molecules for *H. pylori* colonization [8, 9]. Meanwhile, elongation factor–Tu (EF–Tu), an adhesion factor of *L. reuteri* JCM1081, inhibits the adhesion of *H. pylori* to porcine gastric mucin (PGM) [10]. Here, we describe an adhesion inhibition assay method for *H. pylori* to PGM by *Lactobacillus* based on *H. pylori* SS-1 and *L. reuteri* JCM 1081 as examples.

2 Materials

2.1 Bacterial Culture

1. *L. reuteri* JCM 1081 (Japan Collection of Microorganisms, Tsukuba).
2. *H. pylori* SS-1 (clinical isolates).
3. De Man, Rogosa, and Sharpe (MRS) (pH = 6.5) broth and agar plates: Autoclave at 121 °C for 15 min prior to use.
4. Brucella broth and agar plates: After sterilization by autoclaving at 121 °C for 15 min, supplement with 10% fetal bovine serum.
5. AnaeroPack and Micro-AnaeroPack (Mitsubishi Gas Chemical, Japan).
6. Dulbecco's Modified Eagle's Medium (DMEM): Adjust to pH 5.0 or 7.4 with HCl (*see* **Note 1**).
7. Spectrophotometer.

2.2 Microtiter Plate and Mucin

1. 96-well polystyrene microplate (CELLSTAR, Greiner Bio-One, Austria) (*see* **Note 2**).
2. Moisture box: Place a wipe moistened with water in a sealed container.
3. Mucin solution: Suspend 10 mg purified PGM in 10 mL of 50 mM Tris–HCl (pH 9.0), and gently rotate for 12 h at 4 °C. Centrifuge the solution at 16,000 × *g* for 10 min at 4 °C. Use the supernatant as the PGM solution. Store at 4 °C (*see* **Note 3**).

2.3 Bacterial Inhibition Assay

1. Phosphate-buffered saline (PBS): 10 mM phosphate buffer containing 150 mM NaCl. Adjust to pH 7.4 and sterilize by autoclaving at 121 °C for 15 min.
2. Acetate-buffered saline (ABS): 10 mM acetate buffer containing 150 mM NaCl. Adjust to pH 5.0 and sterilize by autoclaving at 121 °C for 15 min.
3. Blocking buffer: PBS or ABS containing 2% (w/v) bovine serum albumin (BSA).
4. Recovery buffer: PBS (pH 7.4) containing 0.05% (v/v) Triton-X100.
5. Orbital shaker.

2.4 Quantification of H. pylori

1. PCR grade water.
2. Primers: HP-FOR (5′- TTATCGGTAAAGACACCAGAAA - 3′), HP-REV (5′-ATCACAGCGCATGTCTTC-3′) (*see* Ref. 11).
3. Standard DNA for quantification of *H. pylori*: *H. pylori* SS-1 *ureC* gene fragment inserted into the pGEM-T plasmid (pGEM-T::*ureC*) (*see* **Note 4**).
4. Quantitative PCR (qPCR) SYBR® Green Master Mix (*see* **Note 5**).
5. Real-Time PCR system.
6. Spectrophotometer.

3 Methods

3.1 Bacterial Culture

1. *L. reuteri* JCM1081 is anaerobically cultured in MRS broth (5 mL) at 37 °C until the exponential phase (OD_{600} = 1.2 to 1.5). The cells are harvested by centrifugation (4000 × *g*, 5 min, 4 °C) and resuspended in DMEM (pH 5.0 or 7.4), which is adjusted to OD_{600} = 1.0 (approximate bacterial cell number is 1.0×10^8 colony-forming unit [CFU]/mL). Samples are stored on ice until further use in the assay (*see* **Note 6**).
2. *H. pylori* SS-1 is micro-anaerobically cultured in Brucella broth (5 mL) at 37 °C until the exponential phase (OD_{600} = 0.6 to 0.8). The cells are harvested by centrifugation (4000 × *g*, 5 min, 4 °C) and resuspended in DMEM (pH 5.0 or 7.4), which is adjusted to OD_{600} = 0.5 (approximate bacterial cell number is 5×10^7 CFU/mL). Samples are stored on ice until further use.

3.2 Immobilization of PGM on Microtiter Plate

1. 100 μL of PGM solution is added to each well of a microtiter plate, and the plate is kept at 4 °C for 12 h in a moisture box.
2. PGM solution is removed, and the plate gently washed twice with 200 μL of PBS.
3. The remaining protein-binding sites in the wells are blocked by adding 100 μL of blocking buffer, and the plate is incubated for 1 h at room temperature in the moisture box.
4. The plate is gently washed twice with 200 μL of ABS or PBS.

3.3 Bacterial Inhibition Assay

1. 100 μL of *L. reuteri* JCM 1081 is added to each well, and the plate incubated for 1 h at 37 °C in the moisture box.
2. To remove non-adherent bacteria, the plates are gently washed twice with 200 μL of ABS or PBS.
3. 100 μL of *H. pylori* SS-1 is added to each well, and the plate incubated for 3 h at 37 °C in the moisture box.
4. To remove non-adherent bacteria, the plates are gently washed twice with 200 μL of ABS or PBS.
5. 100 μL of recovery buffer is then added to each well, and the plate incubated for 10 min at 37 °C. Subsequently, the adhered bacterial cells can be harvested by vigorous pipetting.

3.4 Quantification of H. pylori by qPCR

1. A 100 μL bacterial suspension is diluted ten-fold with water, and the 10^2- to 10^5-fold dilutions (2 μL) are used as templates for qPCR.
2. Standard bacterial DNA (pGEM-T::*ureC*) is prepared at a linear concentration with a DNA copy number range of 10^3–10^9.
3. The qPCR solution (20 μL reaction volume) comprises the SYBR® Green Master Mix with 300 nM of HP-FOR and HP-REV primer pairs and 2 μL of standard DNA or bacterial suspension (unknown samples).
4. PCR is performed according to the SYBR® Green Master Mix manufacturer's instructions, and the annealing temperature is set at 54 °C.
5. The bacterial cells (calculated as copy numbers) are quantified using the standard curve-fitting method (*see* **Note 7**).
6. The ratio (%) of adhesion of *H. pylori* to PGM (Fig. 1) is determined as follows: 100 × (number of *H. pylori* in the *L. reuteri* added wells/number of *H. pylori* in the *L. reuteri* non-added wells) (*see* **Note 8**).

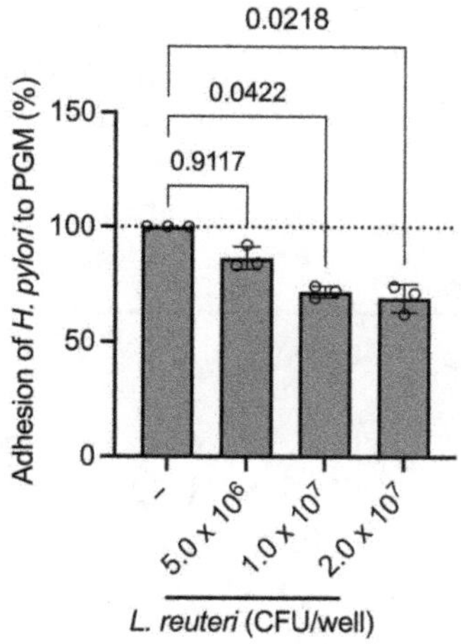

Fig. 1 Inhibition of *Helicobacter pylori* SS-1 adhesion to porcine gastric mucin by *Limosilactobacillus reuteri* JCM1081 (pH 5.0). Statistical significance was determined using the Kruskal–Wallis test with a post hoc test. The exact *P* values are indicated in the figure

4 Notes

1. The adherence of *L. reuteri* and *H. pylori* to PGM is strongly influenced by pH. Thus, two different conditions, pH = 5.0 and 7.4, should be examined. In general, the adherence of *L. reuteri* to PGM was enhanced at pH = 5.0 [7].
2. The microtiter plate is suitable for the immobilization of PGM.
3. For a detailed information of the purification of the PGM, see Ref. [4].
4. The DNA (pGEM-T::*ureC*) for standard curves is prepared by the following procedure according to [11]. *H. pylori* SS-1 DNA used as a template for PCR (50 μL reaction) using Ex-taq polymerase (TaKaRa Bio Inc., Japan). PCR amplification of the *ureC* gene fragment of *H. pylori* SS-1 is performed with the primer pair HP-FOR-out (5′-TCTGTCTGATTCG CTTTTCTG-3′) and HP-REV-out (5′-AAGCTCGCT AAAAACGACC-3′) and PCR products were purified by agarose gel electrophoresis. The PCR product is subsequently cloned using pGEM®-T Vector Systems (Promega Corp., USA).
5. The SYBR® Green Master Mix instructions of each manufacturer were followed.
6. The pH of all buffers used in the subsequent steps should be the same as that of the bacteria-suspended DMEM.
7. All reactions should be performed in duplicate.

8. The inhibition ratio of *H. pylori* SS-1 adhesion to PGM by *L. reuteri* JCM1081 is generally approximately 20% (Fig. 1). To validate these results, the number of *L. reuteri* JCM1081 cells should be halved or doubled to confirm the concentration-dependent inhibition.

Acknowledgments

This study was supported by a Grant-in-Aid for Scientific Research (KAKENHI 22 K05974; 23H02358) from the Japan Society for the Promotion of Science (JSPS).

References

1. Atuma C, Strugala V, Allen A et al (2001) The adherent gastrointestinal mucus gel layer: thickness and physical state *in vivo*. Am J Physiol Gastrointest Liver Physiol 280:G922–G929
2. McGuckin MA, Lindén SK, Sutton P et al (2011) Mucin dynamics and enteric pathogens. Nat Rev Microbiol 9:265–278
3. Dempsey E, Corr SC (2022) *Lactobacillus* spp. for gastrointestinal health: current and future perspectives. Front Immunol 13:840245
4. Nishiyama K, Mukai T (2019) Adhesion of *Lactobacillus* to intestinal mucin. Methods Mol Biol 1887:159–166
5. Nishiyama K, Kawanabe A, Miyauchi H et al (2014) Evaluation of bifidobacterial adhesion to acidic sugar chains of porcine colonic mucins. Biosci Biotechnol Biochem 78:1444–1451
6. Mukai T, Asasaka T, Sato E et al (2002) Inhibition of binding of *Helicobacter pylori* to the glycolipid receptors by probiotic *Lactobacillus reuteri*. FEMS Immunol Med Microbiol 32: 105–110
7. Nishiyama K, Ochiai A, Tsubokawa D et al (2013) Identification and characterization of sulfated carbohydrate-binding protein from *Lactobacillus reuteri*. PLoS One 8:e83703
8. Huesca M, Borgia S, Hoffman P et al (1996) Acidic pH changes receptor binding specificity of *Helicobacter pylori*: a binary adhesion model in which surface heat shock (stress) proteins mediate sulfatide recognition in gastric colonization. Infect Immun 64:2643–2648
9. Huesca M, Goodwin A, Bhagwansingh A et al (1998) Characterization of an acidic-pH-inducible stress protein (hsp70), a putative sulfatide binding adhesin, from *Helicobacter pylori*. Infect Immun 66:4061–4067
10. Nishiyama K, Kagamitani T, Yamamoto Y et al (2017) The elongation factor Tu from *Lactobacillus reuteri* inhibits the adhesion of *Helicobacter pylori* to porcine gastric mucin. Milk Sci 66:17–26
11. He Q, Wang JP, Osato M et al (2002) Real-time quantitative PCR for detection of *Helicobacter pylori*. J Clin Microbiol 40:3720–3728

Part VII

Imaging and MD Simulation of Mucins

Chapter 31

Imaging of Mucin Networks with Atomic Force Microscopy

Jerome Carpenter and Mehmet Kesimer

Abstract

Mucin networks serve as the structural scaffold of mucus and play a significant role in determining its biophysical properties. Thus, characterizing the organization, macromolecular structure, and interactions within these networks is a key step in understanding the parameters that govern mucus functionality in both health and disease. Atomic force microscopy (AFM) is uniquely suited to study mucin networks; AFM can clearly resolve nanometer-sized features, does not require fixation or metallization, and can be performed in air or aqueous solutions. In this chapter we describe protocols to image mucin networks using AFM. First, we describe two protocols to enrich and isolate mucin samples in preparation for AFM imaging. Next, we detail a protocol to deposit the samples onto a mica substrate. Finally, we give general tips to optimize and troubleshoot AFM imaging of mucin networks.

Key words Atomic force microscopy, Mucin, Mucin networks, Topography

1 Introduction

Gel-forming mucins form the structural backbone of mucus, and thus, the network they form, and their interactions with other proteins, plays a critical role in determining the biophysical properties of mucus gels. Atomic force microscopy (AFM) has proven to be an excellent tool to explore several aspects of mucins and mucin networks. Studies have used AFM to characterize the glycosylation of mucins [1, 2], as well as the biophysical properties of individual mucin molecules [3]. Beyond individual molecules, AFM imaging has also been used to study differences in the macromolecular structure of mucin networks, which has provided valuable insight into potential branching seen in MUC2 networks [4] and differences between mucin networks comprised of MUC5AC or MUC5B [5]. Finally, AFM has also been used to look at differences in mucin networks with and without interacting proteins [6], in the presence of polyphenols [7], and as a function of pH [8]. Successfully using AFM to study mucin and mucin networks relies on good sample preparation. Sample preparation determines the aspects of

Akihiko Kameyama (ed.), *Mucins: Methods and Protocols*, Methods in Molecular Biology, vol. 2763,
https://doi.org/10.1007/978-1-0716-3670-1_31,

the system that can be studied and the imaging quality. Here, we will describe two protocols to enrich samples for mucin networks with and without their interacting proteins. Next, we will describe prepping the sample for AFM imaging. Finally, we will broadly describe generalized tips for imaging mucin networks in AFM.

2 Materials

Prepare all solutions using Ultrapure water and analytical grade reagents. Prepare and store all reagents at room temperature (unless indicated otherwise).

2.1 Preparing the Mucin Solution

1. Mucin/mucus solution to enrich (*see* **Note 1**).
2. Phosphate-buffered saline (PBS).
3. Low-speed centrifuge (500 × *g*).

   Optional materials for a dissociative gradient.
4. Guanidine hydrochloride (GuHCl) 8 M buffer: Prepare an 8 M guanidine HCl buffer (pH 8.0) – 8 M guanidine hydrochloride, 0.1 M Tris–HCl, 5 mM ethylenediaminetetraacetic acid. Prepare using Ultrapure water and adjust the pH to 8 using hydrochloric acid. Filter the final buffer twice through a 0.45 μm nitrocellulose membrane before using.
5. Guanidine hydrochloride 4 M buffer: The same preparation as above but diluted 1:1 with PBS.

2.1.1 Purifying Mucin Networks with Cushion Ultracentrifugation

1. Cesium chloride (CsCl) cushion: Prepare at least 1 mL of a cesium chloride/PBS solution. Add cesium chloride until the solution reaches a final density of 1.35 g/mL. (For the dissociative prep, use 4 M GuHCl instead of PBS.)
2. Fourteen-milliliter ultracentrifuge tube (Ultra-Clear centrifuge tube, 344060, Beckman Coulter, Ca, USA) (*see* **Note 2**).
3. Swing out rotor compatible with 14 mL ultracentrifuge tubes and capable of 150,000 × *g* (SW40Ti, Beckman Coulter, Ca, USA).
4. Ultracentrifuge (L8–70, Beckman Coulter, Ca, USA).
5. Plastic syringe.
6. Four-inch Luer Lock syringe needle.
7. G25 Hi-trap desalting column (Cytiva, Sweden).

2.1.2 Purifying Mucin Networks with Gel Permeation Chromatography (GPC)

1. Gel permeation chromatography system (*see* **Note 3**).
2. Light scattering buffer: Prepare light scattering buffer consisting of 0.2 mM sodium chloride, 1 mM ethylenediaminetetraacetic acid (EDTA), and 0.05% sodium azide. Prepare using

Ultrapure water and adjust the pH to 7 using hydrochloric acid. Filter the solution twice using a 0.45 μm nitrocellulose filter, and then degas the solution before using as a buffer (*see* **Note 4**).

3. Ten-milliliter chromatography column packed with Sepharose CL-2B media.
4. Two-hundred-microliter loop.

2.2 Depositing the Sample onto Mica

All washing steps should be performed using Ultrapure water.

1. Steel specimen disc (#16218, Ted Pella, Ca, USA).
2. Super glue (*see* **Note 5**).
3. Mica disc, 9.9 mm (#50, Ted Pella, Ca, USA) (*see* **Note 6**).
4. Scotch tape.
5. Flat-tipped forceps.
6. Nickel chloride buffer: Prepare a 10 mM $NiCl_2$ buffer by dissolving nickel chloride into water. After the nickel chloride has dissolved, filter the solution using a 0.2 μm filter (*see* **Note 7**).
7. Air-spray gun (Nitro-1, Ipolymer.com) (*see* **Note 8**).

2.3 Imaging the Sample with AFM/Image Optimization

1. UHD Arrow AFM tip (Nanoworld, Neuchâtel) (*see* **Note 9**).
2. AFM (Cypher, Oxford Instruments) (*see* **Note 10**).

3 Methods

Carry out all procedures at room temperature unless specified otherwise.

3.1 Preparing the Mucin Solution

Depending on the experimental needs, mucin networks can be prepared in either an associative (without guanidinium chloride) or dissociative (with guanidinium chloride) manner. Under associative conditions the mucins will stay in their native form and proteins that interact with the mucins will still be present. Under dissociative conditions, most of the noncovalently bound proteins will be removed from the sample (*see* Fig. 1). The following methods will be written for an associative preparation, but any changes needed for a dissociative preparation will be noted in parentheses.

In the interest of flexibility, two methods for purifying mucin networks are presented. One method uses ultracentrifugation, while the other uses gel permeation chromatography (GPC). Either method can be used individually, or centrifugation (Subheading 3.1.1) followed by gel permeation chromatography (Subheading 3.1.2) can be used to further enrich the sample. Before

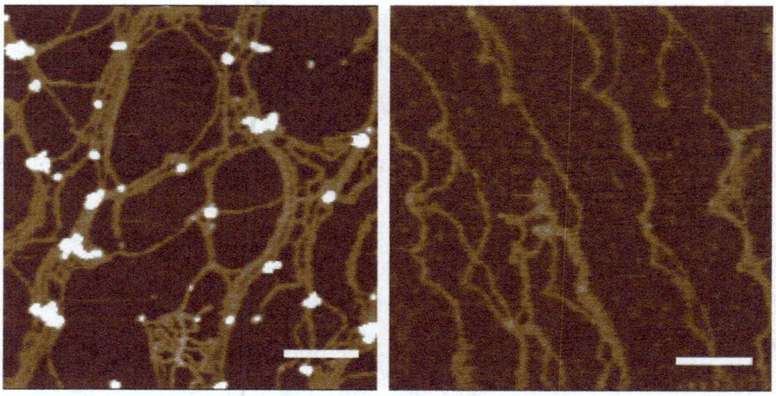

Fig. 1 AFM images demonstrating the differences between associative preparations (left) and dissociative preparations (right) of mucin samples. The associative preparation preserves more of the original structure and includes more globular proteins (white dots). The dissociative preparation removes most of the non-mucin proteins and disrupts the structure (Scale bar = 200 nm)

either method, however, the following steps should be performed to prepare the sample.

1. Centrifuge the sample at a low speed (500 × *g*) for 10 min to pellet any cells or other large debris.
2. Collect the supernatant and discard the pellet (if performing a dissociative preparation, mix the supernatant 1:1 with 8 M GuHCl).

3.1.1 Purifying Mucin Networks with Cushion Ultracentrifugation

1. Add 1 mL of a 1.35 g/mL solution of cesium chloride/PBS (cesium chloride/4 M GuHCl for a dissociative prep) to the bottom of a 14 mL ultracentrifuge tube.
2. Tilt the tube, and then slowly and carefully add 11 mL of PBS (4 M GuHCl for a dissociative prep) on top of the CsCl layer (*see* **Note 11**). There should be a clear boundary visible between the CsCl cushion and the overlying buffer due to the differences in refractive indices.
3. Add 1 mL of the sample prepared in Subheading 3.1 to the top of the tube. The sample will diffuse into the buffer prior to centrifugation (*see* Fig. 2a).
4. Load the tube along with a blank tube of the same weight to balance it into a swingout rotor and run at ~145,500 × *g* for 2 h.
5. Once the centrifugation is complete, carefully remove the tube from the rotor, and place it in a stand so that the bottom of the tube is visible.
6. After centrifugation, the mucin will be in the cesium chloride cushion region (*see* Fig. 2b). Using a syringe and a long needle,

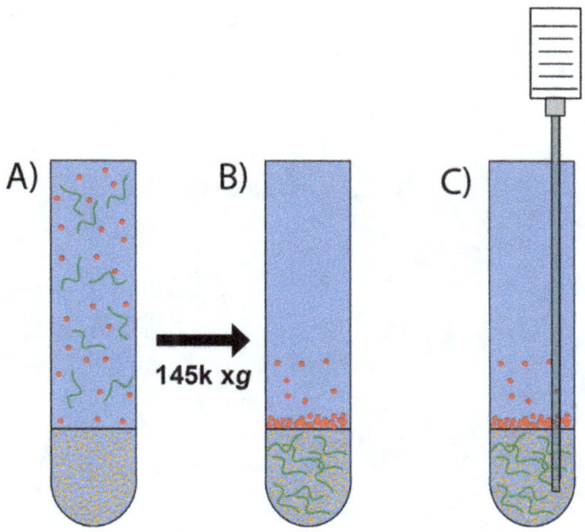

Fig. 2 A cartoon depicting cushion ultracentrifugation. (**a**) A sample consisting of proteins (red dots) and mucins (green lines) is loaded on top of a denser cesium chloride solution (orange dots). (**b**) After centrifuging at 145,500 × *g* for 2 h, the mucins penetrate the cesium chloride cushion, while other proteins form an interface at the boundary. (**c**) Insert the syringe beyond the protein interface into the cesium chloride cushion to retrieve the enriched mucins

carefully insert the needle into the tube so that it goes into the liquid, beyond the cushion interface and towards the bottom of the tube (*see* Fig. 2c and **Note 12**).

7. Collect the bottom ~0.9 mL of liquid (if using both the ultracentrifugation and GPC protocols, then skip to Subheading 3.1.2 and use the liquid collected here as the sample to be injected into the GPC) (*see* **Note 13**).
8. Equilibrate a Hi-Trap G25 desalting column with PBS (*see* **Note 14**).
9. Apply the sample collected in **step 7** into the desalting column.
10. Fill a syringe with PBS.
11. Attach the syringe to the desalting column and elute 0.5 mL.
12. Elute an additional 2 mL but collect the eluent in a microcentrifuge tube and use it as the sample to deposit onto mica in Subheading 3.2.

3.1.2 Purifying Mucin Networks with Gel Permeation Chromatography (GPC)

1. Configure the GPC system with the following parameters: light scattering buffer as the running buffer, a 10 mL CL-2B column, and a 200 μL loop. The flow rate should be compatible with the CL-2B column (*see* **Note 15**).
2. Establish a baseline in the GPC system.
3. Flush the loop with clean light scattering buffer.
4. Load the sample into the loop.

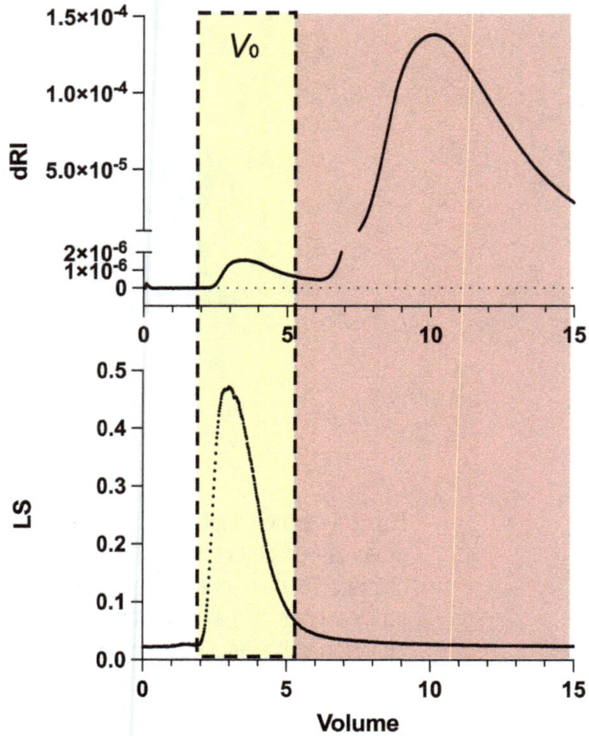

Fig. 3 Representative chromatograms showing the differential refractive index (*dRI, top*) and light scattering (*LS, bottom*) signals of a mucin sample running through a Sepharose CL-2B size-exclusion column. Mucin and anything larger than the column cutoff value will elute in the void volume, V_0, depicted in yellow. The void volume has a comparatively lower dRI, and by extension lower concentration, than the included fraction (pink), but has a much larger molecular weight. A UV chromatogram will look similar to the dRI signal, but the void intensity will be lower due to mucin's poor UV absorbance compared to other proteins

5. Start the run and inject the sample.
6. Collect a portion of the void fraction in a microcentrifuge tube (*see* Fig. 3 and **Note 16**). This will be the sample deposited onto the mica substrate in the next section.

3.2 Depositing the Mucin Solution onto Mica

1. Place a small amount of super glue onto the middle of a steel disc (*see* **Note 17**).
2. Position the mica over the super glue, and gently press it down against the disc.
3. Using a wipe or paper towel, press firmly against the mica to secure it to the steel disc.
4. Press scotch tape against the top surface of the mica.

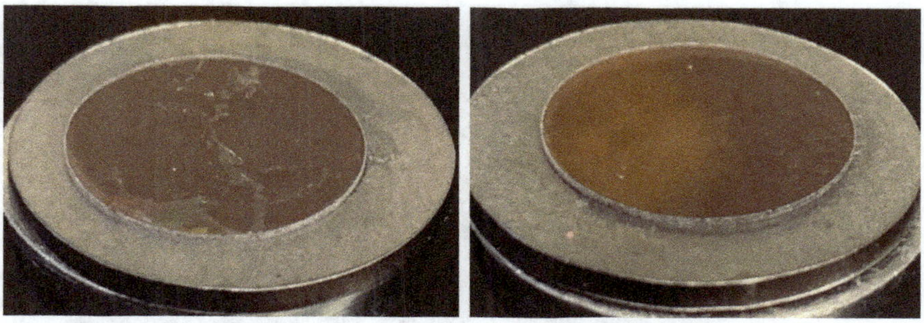

Fig. 4 Image showing a fractured layer of mica (left) vs. a smooth layer of mica (right)

5. Quickly rip away the tape to cleave the mica. If the top surface of the mica on the disc is not uniform (by visual inspection; *see* Fig. 4), repeat **steps 4** and **5** until the top surface is uniform.
6. Pipette 50 μL of $NiCl_2$ onto the mica substrate and let it remain for 1 min (*see* **Note 18**).
7. Rinse off the $NiCl_2$ by pipetting deionized water onto and off the surface (*see* **Note 19**).
8. Pipette 20 μL of sample onto the mica substrate and let it remain for 3 min (*see* **Note 20**).
9. Rinse off the excess sample by pipetting deionized water onto and off the surface.
10. Dry the sample by holding the steel disc with forceps and blowing air perpendicular to the mica (*see* **Note 21**).

3.3 Imaging the Sample with the AFM/Image Optimization

1. Load the sample into the AFM.
2. Load a cantilever into the AFM.
3. Set the focal plane for the tip and for the sample surface.
4. Align the laser onto the back of the cantilever.
5. Configure the scope for tapping mode: Perform a frequency sweep and determine the cantilever's resonance, and set the resonant frequency at the peak (*see* **Note 22**).
6. Adjust the drive amplitude until the amplitude is 1 V (*see* **Note 23**).
7. Set the setpoint to 600 mV, the integral gain to 90, and approach the surface.
8. Do one final frequency sweep, and double-check the resonance peak and amplitude values.
9. Set the phase to 90° and center it.
10. Engage the sample and begin scanning. See Fig. 5 for representative images of a mucin network with poor and good tracking (*see* **Notes 24–26**).

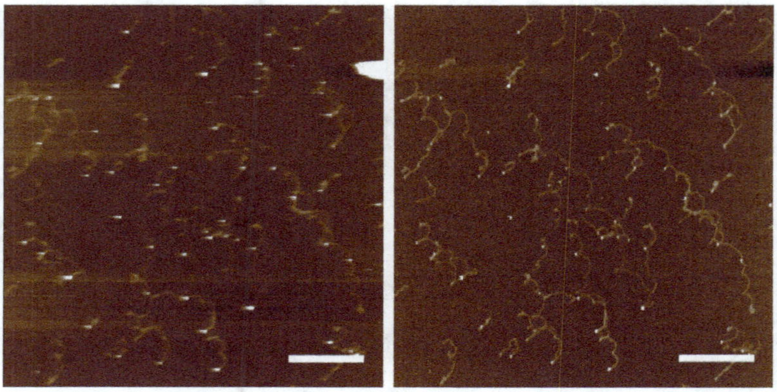

Fig. 5 Mucin sample imaged with poor tracking (left) and good tracking (right) (scale bar = 200 nm)

4 Notes

1. To image mucin networks aim for a concentration of ~100 μg/mL. The solution can be diluted more to image individual mucins or increased for a denser network.
2. The exact volume of the tube is not important, but this size is convenient for retrieving the cushion and the typical volume of material used. Adjusting the size will require changing centrifugation parameters.
3. For the sake of flexibility, the type of gel permeation chromatography system is not important. We typically use a DRI-MALS system, which gives us concentration and molecular weight. UV is fine, although not optimal for measuring mucin, it will at least give a sense of the void fraction's location. This protocol can also be performed without a sensor, but it is important to be familiar enough with the CL-2B column to only collect the void volume.
4. The light scattering buffer contains a chelator (EDTA) to reduce aggregation, as well as sodium azide to prevent any bacterial growth. Degassed PBS can be used as a substitute for the light scattering buffer.
5. Liquid super glue is preferable to gel. Using liquid, it is easier to wet the full area of the mica disc without any gaps.
6. The mica disc diameter can be varied, but you may want to adjust the volume of sample pipetted onto the mica. The most important thing is to stay consistent so that you can tweak parameters for ideal sample preparation.
7. The nickel chloride solution can be stored and used later, but if using an older solution make sure to vigorously mix and refilter before using.

8. Any air source can be used to dry, but the key is to not use something that may leave chemicals or residues.
9. In our lab we have had success using Arrow-UHFAuD tips, but other probes will work for imaging mucins. Most vendors will categorize tips by application, and in general tips for DNA imaging will also work with mucin networks.
10. In our lab we use Asylum's Cypher AFM, but other AFMs optimized for high-resolution imaging of small samples will work.
11. A cleaner interface between the cesium chloride cushion and overlying layer will give better separation, so it is important to minimize disturbances to the cushion. For the first few milliliters added, it is best to add dropwise with the tube tilted 70–80°.
12. Maintaining the interface is important, it should be visible by eye (and oftentimes after centrifugation, proteins will be present at the interface.)
13. There will be plenty of material, so it is better to stop short of the interface instead of pulling more material and "contaminating" the sample with other proteins.
14. Other desalting techniques (dialysis tubing, buffer replacement spin columns) can be used instead of a G25. The key is to replace the cesium chloride and/or guanidine with sodium chloride at physiological levels (150 mM). Excess salt will make the formation of salt crystals more likely during the drying steps.
15. If the sample is too dilute, a larger loop volume can be used. The column size can also be adjusted, but a larger column may require more material.
16. If using some sort of detector, it is important to know the delay between the detector and where the eluent can be collected.
17. When gluing the mica onto the steel disc, you want the super glue to completely fill the space between the mica and the steel, i.e., you want to minimize any air gaps. The best way to do this is to put a small drop of super glue in the middle of the steel disc, then position the mica over the super glue, and drop the mica onto the glue. Let the glue wet the area between the mica and steel disc, and then gently move the mica around to fill in any spots missing super glue. Having too much super glue (spilling out from underneath the mica) should also be avoided.
18. Once the mica is cleaved and the nickel chloride has been added, keep the mica wet until the final step. Throughout the wash steps, always try to keep 20–30 μL of fluid on top of the mica until the final drying step.

19. For these rinsing/washing steps, it is best to treat them as a gentle buffer exchange. The goal is to gently remove excess material, not to vigorously pipette water to detach material. One way to do this is to slowly pipette ~100 μL of water on one side of the sample, and then gently remove it from the other side of the sample. Doing this 3–5 times is typically sufficient to remove excess material. Also, try to keep all the liquid on the mica; do not add so much fluid that it spills onto the steel disc.
20. For pipetting the sample, pipette into the middle of the mica, you can also mix with the pipette. Regardless, there will be heterogeneity across the sample.
21. Drying the sample by blowing air perpendicularly is an attempt to reduce directionality from drying. There may be experiments where directionality could be useful, and this can be achieved by drying the sample from an angle.
22. While contact mode can be used, typically tapping, noncontact, or another variation will give better images for mucins and mucin networks.
23. With our AFM and tip configuration, an initial 1 V free oscillation with a set point of around 600 mV is typically a good starting point. With a good (non-sticky) sample, slowly bringing the amplitude and set point down to about 100–200 mV can result in good imaging of mucin features and details.
24. When imaging mucins with AFM, there is a balancing act between minimizing the force of the tip to get optimal imaging and making sure that there is enough force that the tip does not get "stuck" on highly sticky/charged regions of the sample. If you notice the tip taking a long time to return to the surface after encountering a tall feature, you may want to try increasing the force, or increasing the gain to improve the tracking of the surface. Additionally, a slightly stiffer cantilever may help with "sticking."
25. Despite your best efforts, there will always be sample heterogeneity. Sometimes there will be variations in concentration, so it is worth moving to other regions of the sample (it is a good idea to do this regardless to get a sense of the sample.) Sometimes there will be regions of the sample where you will run into technical challenges with imaging. Before changing the sample or switching tips, it is worth moving to another region to see if the problems persist in different areas.
26. For scan area, I would recommend starting with a scan area somewhere between 2 and 5 μm. Mucins can be difficult to see if you use a much larger scan area before your imaging parameters are optimized.

Acknowledgements

We would like to thank members of the Kesimer lab as well as the Marsico Lung Institute for help and support during the development of this protocol. The protocols presented here were developed for work supported by the NIH (P01HL164320-01) and Cystic Fibrosis Foundation (CFFCARPEN19I0).

References

1. Round AN, Berry N, McMaster TJ et al (2004) Glycopolymer charge density determines conformation in human ocular mucin gene products: an atomic force microscope study. J Struct Biol 145(3):246–253. https://doi.org/10.1016/j.jsb.2003.10.029
2. Carpenter J, Kesimer M (2021) Membrane-bound mucins of the airway mucosal surfaces are densely decorated with keratan sulfate: revisiting their role in the Lung's innate defense. Glycobiology 31(4):436–443. https://doi.org/10.1093/glycob/cwaa089
3. Round AN, Berry N, McMaster TJ et al (2002) Heterogeneity and persistence length in human ocular mucins. Biophys J 83(3):1661–1670. https://doi.org/10.1016/S0006-3495(02)73934-9
4. Round AN, Rigby NM, García-de la Torre A et al (2012) Lamellar structures of MUC2-rich mucin: a potential role in governing the barrier and lubricating functions of intestinal mucus. Biomacromolecules 13(10):3253–3261. https://doi.org/10.1021/bm301024x
5. Carpenter J, Wang Y, Gupta R et al (2021) Assembly and organization of the N-terminal region of mucin MUC5AC: indications for structural and functional distinction from MUC5B. Proc Natl Acad Sci U S A 118(39):e2104490118. https://doi.org/10.1073/pnas.2104490118
6. Radicioni G, Cao R, Carpenter J et al (2016) The innate immune properties of airway mucosal surfaces are regulated by dynamic interactions between mucins and interacting proteins: the mucin interactome. Mucosal Immunol 9(6):1442–1454. https://doi.org/10.1038/mi.2016.27
7. Davies HS, Pudney PD, Georgiades P et al (2014) Reorganisation of the salivary mucin network by dietary components: insights from green tea polyphenols. PLoS One 9(9):e108372. https://doi.org/10.1371/journal.pone.0108372
8. Hong Z, Chasan B, Bansil R et al (2005) Atomic force microscopy reveals aggregation of gastric mucin at low pH. Biomacromolecules 6(6):3458–3466. https://doi.org/10.1021/bm0505843

Chapter 32

Molecular Dynamics Simulation and Docking of MUC1 *O*-Glycopeptide

Ryoka Kokubu, Shiho Ohno, Noriyoshi Manabe, and Yoshiki Yamaguchi

Abstract

Advances in computer performance and computational simulations allow increasing sophistication in applications in biological systems. Molecular dynamics (MD) simulations are especially suitable for studying conformation, dynamics, and interaction of flexible biomolecules such as free glycans and glycopeptides. Computer simulations are best performed when the scope and limitations in performance have been thoroughly assessed. Proper outputs are obtained only under suitable parameter settings, and results need to be properly validated. In this chapter, we will introduce an example of molecular dynamics simulations of MUC1 *O*-glycopeptide and its docking to anti-MUC1 antibody Fv fragment.

Key words MD simulation, MUC1, Antibody, Docking simulation, Glycopeptide, Homology modeling

1 Introduction

Structural biology aims to understand the function of biomolecules from their 3D structures. Atomic-level images of biomolecules have largely been obtained through experimental approaches, such as X-ray crystallography, nuclear magnetic resonance (NMR), and cryo-electron microscopy (cryo-EM). Each method has advantages and disadvantages and needs to be chosen through consideration of the nature of the target biomolecule. Solution NMR spectroscopy has proven to be especially superior to other methods for analyzing flexible biomolecules of small to medium size, such as intrinsically disordered proteins [1], glycans [2], and glycopeptides [3]. Conformational analysis of proteins with NMR mainly relies on ^{1}H-^{1}H NOE and the spin–spin coupling constant. In principle, the solution conformations of glycans and glycopeptides should, likewise, lend themselves to NMR analysis. However, the problem is that ^{1}H-^{1}H NOEs between distant regions are seldom observed in extended glycans, and this can make it difficult to define the

Akihiko Kameyama (ed.), *Mucins: Methods and Protocols*, Methods in Molecular Biology, vol. 2763,
https://doi.org/10.1007/978-1-0716-3670-1_32,

conformation. To overcome this issue, researchers are developing additional observables to provide distance information in NMR analysis [4, 5]. A complementary approach can also be taken, namely, molecular dynamics (MD) simulation, which is the topic of this chapter.

MD simulations are now widely applied to many biological systems, keeping pace with the continuous advances in computer performance. However, a thorough understanding of the scope as well as the limitations of MD simulations is essential. Appropriate parameter settings including the force field need to be ascertained prior to simulation. Incorrect settings can profoundly affect the outcome. Even when settings have been carefully determined, the outputs need to be statistically validated. There are now several reports on MD simulations of glycans and glycoproteins, and these provide views of the dynamics of glycans and glycoconjugates [6–9]. In this chapter, we present an example of MD simulation of a MUC1 *O*-glycosylated peptide and show a docking model of the glycopeptide in complex with an anti-MUC1 antibody Fv fragment [10].

2 Hardware and Software

2.1 Computer for Running Discovery Studio

OS: Windows 10 Pro.

CPU: Intel(R)Xeon(R) Gold 5120.

Memory: 192 GB.

GPU: Intel(R) NVIDIA GeForce RTX 3060.

2.2 Software

BIOVIA Discovery Studio 2021 (Dassault Systèmes).

The PyMOL Molecular Graphics System (Schrodinger, LLC) (https://www.schrodinger.com/products/pymol).

3 Methods

3.1 Molecular Dynamics Simulation of MUC1 O-Glycopeptide

1. (Option) Coordinates of free *O*-glycan are prepared using Carbohydrate Builder in GLYCAM (https://glycam.org/). For example, the sequence of the glycan is entered as DNeup5Aca2-3DGalpb1-3DGalpNAca1-OH. Then the corresponding PDB file is downloaded.
2. Coordinates of the MUC1 peptide are prepared using the Build and Edit Protein tool in Discovery Studio 2021. The conformation of the peptide is set to "Extended", and the coordinate file is downloaded as PDB format (Fig. 1a).
3. *O*-Glycan is attached to the MUC1 peptide using Glycoprotein Builder in GLYCAM (https://glycam.org/). The PDB file of

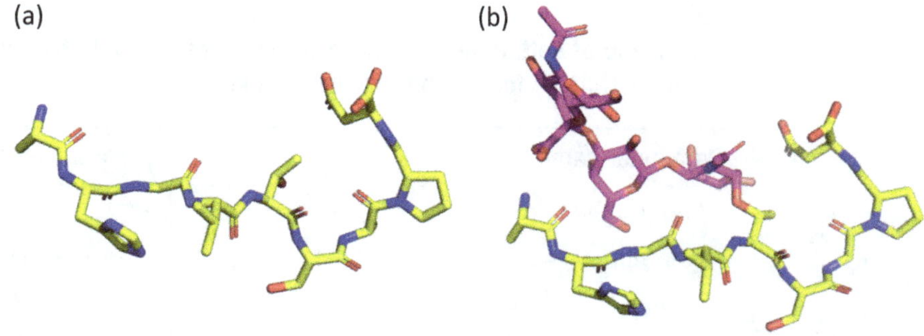

Fig. 1 Initial coordinates for MD simulation. These coordinates were prepared using Discovery Studio and GLYCAM. (a) Non-glycosylated MUC1 peptide (AHGVTSAPD) and (b) *O*-glycosylated MUC1 peptide. *O*-Glycan (pink) and peptide are shown in stick representation. This figure was prepared using PyMOL software

Table 1
An example of setting in the Solvation tool in the Discovery Studio

Parameter name	Parameter value
Solvation model	Explicit periodic boundary
Cell shape	Orthorhombic
Minimum distance From boundary	7 Å
Cation type	Sodium
Anion type	Chloride

the MUC1 peptide is loaded in GLYCAM and *O*-glycan is attached onto the peptide. The sequence of the *O*-glycan is entered as DNeup5Aca2-3DGalpb1-3DGalpNAca1-OH. It should be noted that the reducing terminal D-GalNAc is α-linked to a Thr/Ser side chain. Then the coordinate file (PDB file) is downloaded (Fig. 1b).

4. PDB file of the MUC1 *O*-glycopeptide is loaded into Discovery Studio as the initial coordinate of the MD simulation (*see* **Note 1**).
5. CHARMm is assigned as the force field in the Discovery Studio. Orthorhombic cell shape is used in explicit periodic boundary solvation model. The minimum distance from the periodic boundary is set to 7.0 Å.
6. Water molecules and ions are placed around the MUC1 glycopeptide using the Solvation tool in Discovery Studio (Table 1). Moreover, 439–1045 water molecules are explicitly placed and TIP3 is used as the force field template (*see* **Note 2**).
7. For minimization and heating, the Standard Dynamics Cascade tool is applied to the coordinate file including waters and ions

Table 2
An example of setting in the Standard Dynamics Cascade tool and Dynamics (NAMD) tool in the Discovery Studio

Parameter name	Parameter value
Minimization	
Algorithm	Steepest descent
Max steps	1000 steps
Root mean square (RMS) gradient	1.0 kcal/(mol*Å)
Minimization2	
Algorithm	Adopted basis NR
Max steps	2000 steps
RMS gradient	0.01 kcal/(mol*Å)
Heating	
Simulation time	20 ps
Time step	2 fs
Initial temperature	50 K
Target temperature	310 K
NAMD	
Simulation time	10,000 ps
Ensemble	*nPT*

(Table 1). Minimization is performed in two steps. The first step eliminated distortion of the entire structure with the steepest descent algorithm. In the second step, minimization is performed with adopted basis Newton–Raphson (NR). Heating is carried out at 310 K.

8. After equilibration, the Nanoscale Molecular Dynamics (NAMD) program is carried out under an *nPT* ensemble (amount of substance [*n*], pressure [*P*], and temperature [*T*] are conserved) with the time step of 2 fs (Table 2). The simulation time is set to 10 ns (*see* **Note 3**).

3.2 Homology Modeling of Fv Domain

1. The 3D coordinates of the antibody Fv fragment are generated with a homology modeling technique using Discovery Studio (*see* **Note 4**).
2. Amino acid sequences of V_H and V_L are separately written as text files in a FASTA format. The FASTA files are loaded into Discovery Studio.

3. Complementarity-determining regions (CDRs) are then identified using the Annotate Sequence tool in Discovery Studio. The IMGT scheme was selected as our choice but other schemes (Chothia, Kabat and Honegger) can also be used.
4. The Identify Framework Templates tool is used to search for candidate templates using BLAST in the Discovery Studio.
5. Modeling is performed with the Model Antibody Framework tool. The Model Antibody Loop was subsequently performed to rebuild the CDRs.
6. The model structure is verified by Ramachandran plot and other programs.

3.3 Docking Simulations of MUC1 Peptides–Antibody

1. Docking simulation of the antibody–glycopeptide complex is performed using ZDOCK in Discovery Studio. Structures of MUC1 glycopeptides for docking are derived from those at the end of the MD simulation.
2. (Option) For docking of alternative ligand structures, several MUC1 conformers are selected from the MD trajectory.
3. Before the docking simulations, active sites are set for both glycopeptide and antibody Fv fragment. On the antibody side, all the CDRs are defined as active site and the other sites are defined as blocking sites (*see* **Note 5**).
4. The docked poses are obtained with scores (*see* **Note 6**). The coordinate file can be downloaded as a PDB file (Fig. 2).
5. The contact surface area between ligand and antibody is calculated using Accessible Surface Area using Solvent Accessibility tool in Discovery Studio (*see* **Note 7**).

4 Notes

1. Many other platforms are available for MD simulation.
2. The number of water molecules depends on the size of the ligand.
3. There are several programs to perform homology modeling such as SWISS-MODEL (https://swissmodel.expasy.org/). AI-based AlphaFold [11] can be used as an alternative to obtain a 3D model of the Fv fragment.
4. The simulation time must be carefully chosen. A short simulation time may not cover a slow exchange system. However, a longer simulation time, while superior in terms of statistical verification, may involve the use of supercomputer resources.
5. Active sites of the ligand and proteins can be defined using experimental data. It is highly recommended that, in addition,

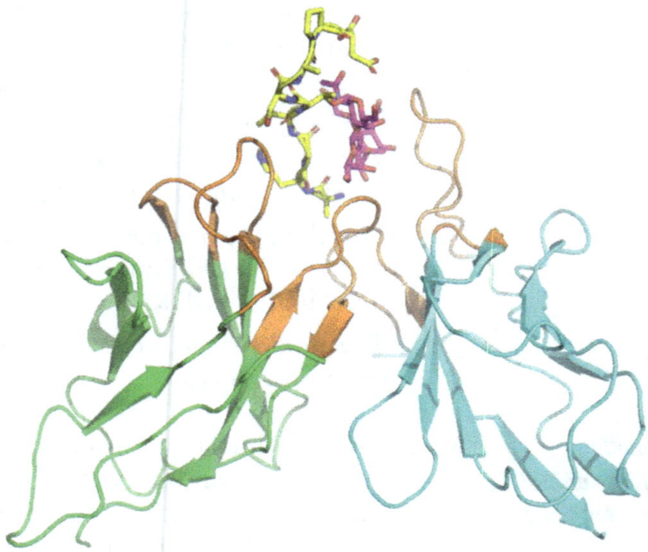

Fig. 2 One of the docking poses (MUC1 glycopeptide-MY.1E12 Fv fragment) obtained from ZDOCK program. Fv is presented in the ribbon model, and the MUC1 glycopeptide is shown in stick representation. V_H is shown in green, V_L in sky blue, CDR regions in orange, MUC1 *O*-glycan in pink, and MUC1 peptide in yellow. This figure was prepared using PyMOL software

experimental data on epitope and paratope, such as from mutagenesis studies of antibody fragment and/or epitope mapping by STD-NMR and NMR titration, are incorporated.

6. The docked poses need to be carefully validated from many aspects.
7. Accessible surface area can be also analyzed in other programs including AREAIMOL in CCP4i (https://www.ccp4.ac.uk/), UCSF Chimera (https://www.cgl.ucsf.edu/chimera/), and PyMOL.

Acknowledgements

We thank Re Suyong (NIBIOHN) for his kind assistance and support in performing MD simulations. Some results were obtained using the supercomputing resources at Cyberscience Center, Tohoku University.

References

1. Dyson HJ, Wright PE (2019) Perspective: the essential role of NMR in the discovery and characterization of intrinsically disordered proteins. J Biomol NMR 73(12):651–659. https://doi.org/10.1007/s10858-019-00280-2
2. Widmalm G (2013) A perspective on the primary and three-dimensional structures of carbohydrates. Carbohydr Res 378:123–132.

https://doi.org/10.1016/j.carres.2013.02.005

3. Barchi JJ Jr (2013) Mucin-type glycopeptide structure in solution: past, present, and future. Biopolymers 99(10):713–723. https://doi.org/10.1002/bip.22313
4. Battistel MD, Azurmendi HF, Yu B et al (2014) NMR of glycans: shedding new light on old problems. Prog Nucl Magn Reson Spectrosc 79:48–68. https://doi.org/10.1016/j.pnmrs.2014.01.001
5. Valverde P, Quintana JI, Santos JI et al (2019) Novel NMR avenues to explore the conformation and interactions of glycans. ACS Omega 4(9):13618–13630. https://doi.org/10.1021/acsomega.9b01901
6. Woods RJ (2018) Predicting the structures of glycans, glycoproteins, and their complexes. Chem Rev 118(17):8005–8024. https://doi.org/10.1021/acs.chemrev.8b00032
7. Re S, Watabe S, Nishima W et al (2018) Characterization of conformational ensembles of protonated N-glycans in the gas-phase. Sci Rep 8(1):1644. https://doi.org/10.1038/s41598-018-20012-0
8. Yamaguchi Y, Nishima W, Re S et al (2012) Confident identification of isomeric *N*-glycan structures by combined ion mobility mass spectrometry and hydrophilic interaction liquid chromatography. Rapid Commun Mass Spectrom 26(24):2877–2884. https://doi.org/10.1002/rcm.6412
9. Grant OC, Montgomery D, Ito K et al (2020) Analysis of the SARS-CoV-2 spike protein glycan shield reveals implications for immune recognition. Sci Rep 10(1):14991. https://doi.org/10.1038/s41598-020-71748-7
10. Kokubu R, Ohno S, Kuratani H et al (2022) O-glycan-dependent interaction between MUC1 glycopeptide and MY.1E12 antibody by NMR, molecular dynamics and docking simulations. Int J Mol Sci 23(14):7855. https://doi.org/10.3390/ijms23147855
11. Jumper J, Evans R, Pritzel A et al (2021) Highly accurate protein structure prediction with AlphaFold. Nature 596(7873):583–589. https://doi.org/10.1038/s41586-021-03819-2

Part VIII

Mucin-Hydrogel and Physicochemical Properties

Chapter 33

Fabrication and Characterization of Mucin Nanoparticles for Drug Delivery Applications

Ceren Kimna, Theresa M. Lutz, and Oliver Lieleg

Abstract

Mucin glycoproteins are ideal biomacromolecules for drug delivery applications since they naturally offer a plethora of different functional groups that can engage in specific and unspecific binding interactions with cargo molecules. However, to fabricate drug carrier objects from mucins, suitable stabilization mechanisms have to be implemented into the nanoparticle preparation procedure that allow for drug release profiles that match the requirements of the selected cargo molecule and its particular mode of action. Here, we describe two different methods to prepare crosslinked mucin nanoparticles that can release their cargo either on-demand or in a sustained manner. This method chapter includes a description of the preparation and characterization of mucin nanoparticles (stabilized either with synthetic DNA strands or with covalent crosslinks generated by free radical polymerization), as well as protocols to quantify the release of a model drug from those nanoparticles.

Key words Glycoprotein, MUC5AC, Antibiotic, DNA strand displacement, Methacrylation

1 Introduction

Physiologically, mucins are present on all wet epithelia including the corneal surface, the gastrointestinal tract, and the female reproductive system [1–3]. Mucins constitute a group of large glycoproteins with molecular weights of up to a few MDa (the protein backbone alone has a molecular weight of >0.5 MDa), and they exhibit a complex architecture: The polypeptide backbone is decorated with a high density of glycan motifs [4], and only the non-glycosylated termini of mucins are (partially) folded. Moreover, mucins comprise hydrophobic and hydrophilic domains as well as both anionic and cationic regions. Mucins can trap or repel pathogens such as viruses or bacteria [5, 6], they are well-hydrated, and they can adsorb to various surfaces. Owing to the latter two properties, mucin solutions can serve as excellent lubricants and protect epithelial cells from stress-induced damage

Akihiko Kameyama (ed.), *Mucins: Methods and Protocols*, Methods in Molecular Biology, vol. 2763,
https://doi.org/10.1007/978-1-0716-3670-1_33,

[7, 8]. To develop functional biomaterials using mucins as key components, the glycoproteins have to be carefully purified; damage inflicted to their structural motifs has been shown to strongly affect their important properties, e.g., their lubrication and gel-forming abilities [9]. And indeed, purified mucins have been used in several research projects as main components of materials that were developed for a wide variety of biomedical applications including selective filters [10], lubricating solutions [11, 12], and antiviral [13] and antibacterial agents [14]. Importantly, mucin solutions and surface-immobilized mucins do not lose their functionalities when they are sterilized with suitable methods [15, 16], and they can be used as anti-(bio)fouling coatings on medical devices including contact lenses, catheters, and stents; here, they provide lubricity and cell-repellent properties [17–19]. In addition, mucin hydrogels [20] and coatings [21] have been used as versatile drug depots: The presence of different chemical groups (e.g., sialic acids, cysteines, and amines) in the mucin glycoprotein allows for specific and unspecific interactions with drug molecules [22]. Recently, mucin nanoparticles were introduced, and it was demonstrated that their structure and size can be controlled very well [23]. Interestingly, a wide range of pharmaceuticals can be loaded into those mucin-based carriers including hydrophobic and hydrophilic/charged molecules [24–27]. As the onset of the drug release process from drug carriers as well as the kinetics of this process needs to be specifically tailored for each cargo molecule (and its mode of action) and their use case, it is important to develop a nanoparticulate drug carrier system where those properties can be adjusted. Here, we describe a detailed method of how to fabricate stable mucin nanoparticles based on two different crosslinking mechanisms (Fig. 1). For those two mucin-based carrier systems, different drug release profiles (i.e., on-demand vs. sustained release) can be obtained, which is a direct consequence of the crosslinking strategy used to stabilize them.

2 Materials

Unless indicated otherwise, prepare all solutions using Ultrapure water and analytical grade reagents. Prepare and store all reagents at room temperature, unless indicated otherwise.

2.1 Mucin Purification

1. Mucus dilution buffer: Prepare 10 mM sodium phosphate-buffered saline (PBS, pH = 7.0) containing 170 mM NaCl and 0.04% (w/v) sodium azide. In detail, add ~100 mL of water to a 1-L glass beaker. Add 1.09 g of disodium hydrogen phosphate dihydrate (Na_2HPO_4) and 0.54 g of sodium hydrogen phosphate monohydrate (NaH_2PO_4) to the beaker and stir until the salts have fully been dissolved. Add 9.93 g of NaCl to

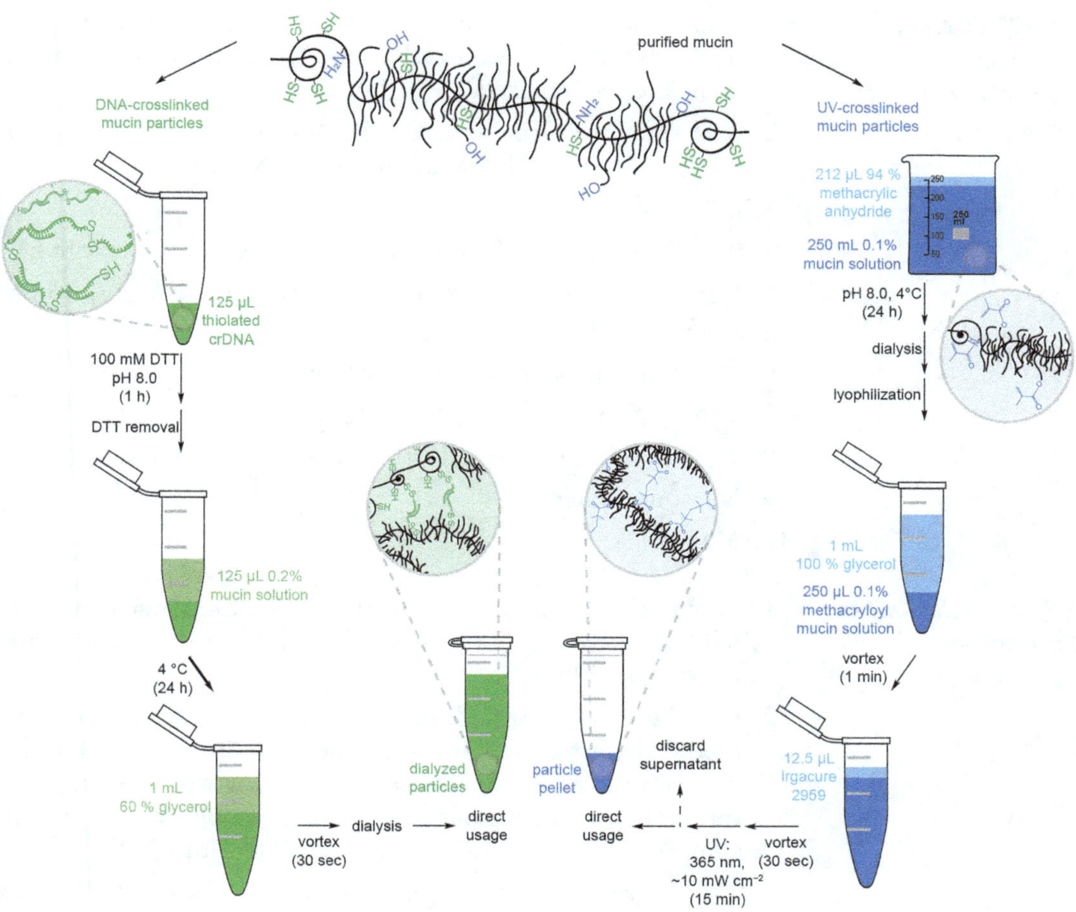

Fig. 1 Schematic overview of the different fabrication steps required to obtain DNA-stabilized (green) and covalently crosslinked (blue) mucin nanoparticles, respectively

adjust the NaCl concentration to 170 mM. Add 2 g of sodium azide (*see* **Notes 1 and 2**). Add more water until the total volume reaches 1 L. Store at room temperature or at 4 °C.

2. Stainless-steel grids with a mesh size of 1 mm, 500 µm, 200 µm, and 125 µm.
3. ÄKTA purifier system (GE Healthcare, Munich, Germany) equipped with an XK50/100 column (GE Healthcare).
4. Sepharose 6FF resin (GE Healthcare).
5. Crossflow filtration (MWCO: 100 kDa; QuixStand benchtop crossflow system, GE Healthcare) equipped with a hollow fiber module (MWCO: 100 kDa; Xampler ultrafiltration cartridge, GE Healthcare; *see* **Note 3**).
6. Conductometer.
7. Freeze dryer (*see* **Note 4**).

2.2 Methacrylic Anhydride Functionalization of Mucins

1. Mucin solution: Solubilize 10 mg of lyophilized mucin in 1 mL of Ultrapure water. Stir thoroughly until the mucins are completely solubilized.
2. 5 M sodium hydroxide (NaOH): Add 30 mL of 45% NaOH (v/v) to 70 mL of Ultrapure water (*see* **Note 5**).
3. pH meter.
4. 94% (v/v) methacrylic anhydride (MA) solution (*see* **Note 6**).
5. Centrifuge 5430 (Eppendorf SE, Hamburg, Germany); rotor: F-35-6-30 (Eppendorf SE).
6. Freeze dryer (same as in Subheading 2.1).
7. Glass beakers (Ø 105 mm, depth of 145 mm).
8. 2 mL plastic containers (Ø 10.8 mm, depth of 40 mm).
9. Ultrapure water.
10. Magnetic stirrer.

2.3 Preparing DNA-Crosslinked Nanoparticles

1. Mucin solution: Solubilize 20 mg of lyophilized mucin in 1 mL of Ultrapure water. Stir thoroughly at 4 °C until the mucins are completely solubilized.
2. Crosslinker DNA strands (crDNA): This sequence needs to be modified with a thiol group at one terminus and should not form any secondary structures at the desired working temperature; it needs to contain a partially self-complementary region comprising a minimum of 6 base pairs (*see* **Note 7**).
3. Tris(2-carboxyethyl)phosphine hydrochloride (TCEP; *see* **Note 8**).
4. RNase-free water containing 0.1 mM ethylenediaminetetraacetic acid (EDTA).
5. Glycerol.
6. Dialysis tubes (MWCO = 300 kDa).
7. 150 mM NaCl solution: Dissolve 8.77 g of NaCl in 1 L of Ultrapure water.
8. 2 mL plastic containers (Ø 10.8 mm, depth of 40 mm).
9. Vortex mixer.

2.4 Preparing UV-Crosslinked Nanoparticles

1. Methacrylated mucin (Muc-MA) solution: Solubilize 10 mg of lyophilized muc-MA in 1 mL of Ultrapure water. Stir thoroughly until the mucins are completely solubilized.
2. Glycerol.
3. Photoinitiator: 2-hydroxy-4′-(2-hydroxyethoxy)-2-methylpropiophenone. Solubilize 200 mg of this powder in 1 mL of 70% (v/v) ethanol (*see* **Note 9**).
4. UV light source (365 nm; *see* **Note 10**).

5. Centrifuge 5430 (Eppendorf SE, Hamburg, Germany); rotor: FA-45-30-11 (Eppendorf SE).
6. 2 mL plastic containers (Ø 10.8 mm, depth of 40 mm).
7. Vortex mixer.

2.5 Dynamic Light Scattering and Electrophoretic Light Scattering

1. 4-clear sided polystyrene cuvettes (10 × 10 × 45 mm, *see* **Note 11**).
2. Omega cuvettes with male Luer plugs (*see* **Note 12**).
3. Dynamic light scattering device (Litesizer500, Anton Paar; *see* **Note 13**).

2.6 Drug Release Tests

1. Tetracycline hydrochloride.
2. Beaker (Ø 42 mm, depth of 60 mm).
3. Dialysis tubes (same as in **2.3**).
4. Magnetic stirrer.
5. Displacement DNA strands (dDNA): This sequence should not form any secondary structures at the desired working temperature; it needs to contain a base sequence that is fully complementary to the crDNA strands (*see* **Note 7**).
6. 24-well plates (Ø 1.6 cm, depth of 1.8 cm).
7. Trans-well inserts (cellQART PET translucent; MWCO, 0.4 μm; Sabeu GmbH, Northeim, Germany).
8. Plate reader.

3 Methods

Unless specified otherwise, all procedures should be carried out at room temperature.

3.1 Porcine Gastric Mucin (PGM) Purification

1. Obtain fresh pig stomachs from a slaughterhouse on the day of processing.
2. Rinse fresh pig stomachs with tap water, cut the stomachs in half at the median plane, and collect the raw mucus by manual scraping the inner surface of the gastric tissue using a spoon.
3. Dilute the crude mucus five-fold in 10 mM sodium phosphate-buffered saline (PBS, pH = 7.0) containing 170 mM NaCl and 0.04% (w/v) sodium azide and stir the mixture at 4 °C overnight.
4. Filter the mucus through a cascade filtering procedure using stainless steel grids with mesh sizes of 1 mm, 500 μm, 200 μm, and 125 μm, respectively [28].

5. Isolate mucins from the other mucus constituents by size-exclusion chromatography (SEC) using an ÄKTA purifier system equipped with an XK50/100 column packed with Sepharose 6FF resin; record the absorption signal at a wavelength of 280 nm and collect the mucin proteins (= the first visible peak).
6. Collect all the mucin-containing fractions and increase the NaCl concentration to 1 M.
7. Dialyze the pooled fractions against Ultrapure water and concentrate the pooled material by crossflow filtration equipped with a hollow fiber cartridge (molecular weight cutoff: 100 kDa).
8. Lyophilize the mucin concentrate and store it at −80 °C until further use.

3.2 Functionalization of Mucins with Methacrylic Anhydride (muc-MA)

1. Solubilize purified mucins (obtained as described in Subheading 3.1) in Ultrapure water to a concentration of 10 mg/mL.
2. Place the solution on ice and titrate it with 5 M sodium hydroxide (NaOH) until a pH value of 8.0 is reached.
3. Mix 1 mL of the mucin solution with 8 μL of a 94% methacrylic anhydride (MA) solution and stir the mixture on ice for 24 h.
4. Maintain the pH of the solution at 8.0 ± 0.2 for 24 h to ensure optimal reaction conditions between the MA and the amino- and hydroxyl groups of the mucins.
5. Separate unbound MA from the methacrylated mucins (muc-MA) by centrifugation (4300 × *g* at 4 °C for 10 min) followed by size-exclusion chromatography (ÄKTA purifier system).
6. Lyophilize the obtained muc-MA and store it at −80 °C until further use.

3.3 Preparing DNA-Crosslinked Mucin Nanoparticles

1. To obtain DNA-crosslinked mucin nanoparticles, incubate 100 μM of synthetic crosslinker DNA (crDNA; this sequence needs to be modified with a thiol group at one terminus and should not form any secondary structures at the desired working temperature; it needs to contain a partially self-complementary region comprising a minimum of 6 base pairs) [26] in 10 μM tris(2-carboxyethyl)phosphine-hydrochloride (TCEP) dissolved in RNase-free water containing 0.1 mM ethylenediaminetetraacetic acid (EDTA) at RT for 2 h.
2. Solubilize purified mucins in Ultrapure water to a final concentration of 15 mg/mL.
3. Add the prepared crDNA solution to the mucin solution in a volume ratio of 1:1; with this, you obtain a final mucin concentration of 10 mg/mL. Incubate the mixture under constant shaking at room temperature overnight. To obtain drug-loaded mucin nanoparticles, add a drug (e.g., tetracycline

hydrochloride [TCL]) to this solution during this step. For the example drug (TCL) described here, the drug feed concentration is chosen to be 0.06 mg TCL/mL.

4. Place 1 mL of a 30% (v/v) glycerol solution into a tube and add 250 μL of the prepared crDNA-mucin solution (which, in our example, also contains TCL) to the wall of the tube. Vortex for 30 s. This step forces the mucins to assume a condensed (= particle) state.
5. To remove the glycerol, place the generated particle dispersion into a dialysis tube (Spectra/Por Float-A-Lyzer G2, MWCO: 300 kDa). Place the tube in 2 L of a 150 mM NaCl solution and dialyze it at 4 °C for 2 days. After the first 4 h, replace the NaCl solution with a fresh NaCl solution.
6. Collect the mucin nanoparticle dispersion from the dialysis tube. Store the dispersion at 4 °C until further use.

3.4 Preparing Covalently Crosslinked Mucin Nanoparticles

1. To prepare drug-loaded, covalently crosslinked mucin nanoparticles, solubilize muc-MA (10 mg/mL) in the appropriate amount of drug solution (in our example, we use TCL as a model drug and a feed concentration of 0.06 mg TCL/mL of Ultrapure water; *see* **Note 14**) [25].
2. Mix 1 mL of glycerol with 250 μL of the prepared muc-MA solution while stirring for 1 min. This step forces the mucins to assume a condensed (= particle) state.
3. Add 12.5 μL of the radical photoinitiator (use the prepared stock solution of 200 mg mL^{-1} prepared in 80% EtOH) to the prepared glycerol/muc-MA solution and mix thoroughly for 30 s.
4. Induce photo-crosslinking of the mucin particles by UV treatment (365 nm, ~10 mW cm^{-2}) for 15 min.
5. Remove the glycerol from the particle dispersion via centrifugation (20,817 × *g* for 10 min). Discard the supernatant and replace it with Ultrapure water. Store the particle dispersion at 4 °C until further use.

3.5 NP Characterization by Dynamic Light Scattering and Electrophoretic Light Scattering

1. To measure the hydrodynamic size of the particles, place 850 μL of a nanoparticle dispersion into a 4-clear sided polystyrene cuvette (*see* **Note 15**).
2. Place the cuvette into the measuring module of the DLS device. Set the measurement temperature to 20 °C and the detection angle to 175° (*see* **Notes 16 and 17**).
3. To measure the zeta potential of the particles, fill the omega cuvette provided by the manufacturer as follows: Insert the syringe filled with 1 mL of a sample into one of the sample ports; inject the sample; insert the cuvette stoppers (*see* **Note 12** for details). Obtained data are summarized in Fig. 2.

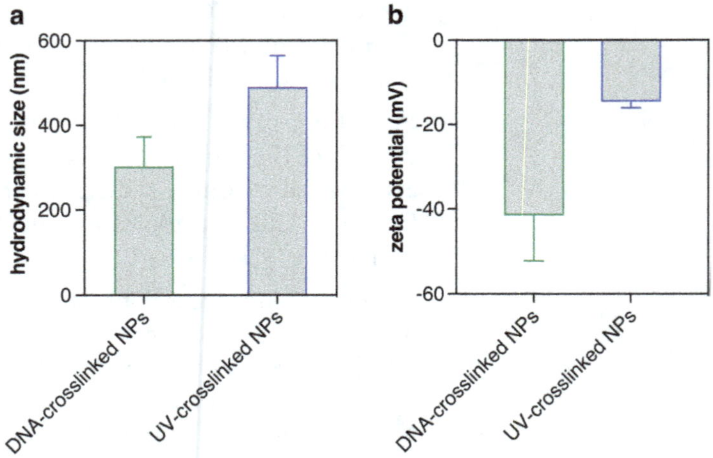

Fig. 2 Hydrodynamic size **(a)** and zeta potential **(b)** of DNA- and UV-crosslinked mucin nanoparticles. Data shown represents mean values and error bars denote the standard deviation as obtained from $n \geq 3$ independent samples

3.6 Quantifying Drug Release from the NPs

1. Prepare a standard curve by measuring the absorbance behavior of serially diluted drug solutions (for TCL, use an absorption wavelength of 356 nm; for other drugs, run a spectrum first to find the wavelength of optimal absorbance) using a spectrometric plate reader (*see* **Note 18**). In the concentration range tested (for TCL: 3.9×10^{-3} mg/mL to 62.5×10^{-3} mg/mL), a linear relation between the measured absorbance values and the drug concentration should hold (Fig. 3a and b).
2. (a) For DNA-crosslinked NPs: Place 5 mL of a suspension of drug-loaded nanoparticles (we recommend a particle concentration corresponding to a mucin concentration of about 1 mg/mL) into dialysis tubings and immerse them into 50 mL of PBS (pH = 7.4) while stirring this buffer solution at 350 rpm and 37 °C. To initiate the release of TCL from the DNA-crosslinked mucin nanoparticles, add 50 μL of a 100 mM dDNA solution to the dialysis tube (with this amount, the molar ratio of dDNA to crDNA is roughly about 1; this step will trigger NP opening and thus trigger drug release).

   (b) For UV-crosslinked NPs: Add 500 μL of PBS into each well of a 24-well plate. Add trans-well inserts into the wells (which contain a semipermeable polycarbonate membrane) and add 100 μL of a suspension of drug-loaded nanoparticles (we recommend a particle concentration corresponding to a muc-MA concentration of 1 mg/mL) on top of each membrane. With this configuration, the PBS will receive the released drug from the NPs added to the top of the membrane insert as the liberated drug molecules will diffuse across the membrane into the PBS reservoir (*see* **Note 19**).

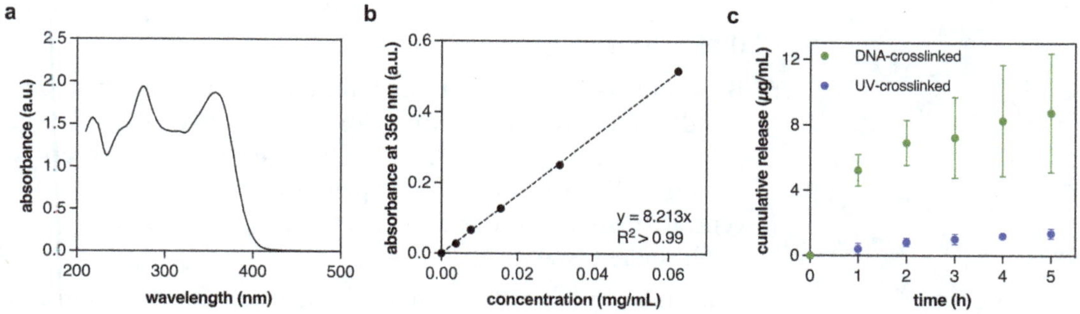

Fig. 3 Quantifying drug release from the mucin nanoparticles. (**a**) To find a suitable wavelength for detecting the drugs released from the mucin nanoparticles, an absorption spectrum of the chosen drug (in our example: tetracycline hydrochloride [TCL]) prepared as a 1 mg/mL solution in Ultrapure water is determined in a wavelength range of 210–500 nm. (**b**) Afterwards, a TCL standard curve is obtained by measuring the absorbance values of serially diluted TCL solutions (3.9×10^{-3} mg/mL to 62.5×10^{-3} mg/mL) at the chosen absorption peak (here: 356 nm). (**c**) Cumulative TCL release obtained for DNA-crosslinked mucin nanoparticles and covalently crosslinked mucin particles (**d**). Data shown in (**c, d**) represents mean values and error bars denote the standard deviation as obtained from $n \geq 3$ independent samples

3. At selected time intervals, remove 500 μL (when using dialysis) or 200 μL (when using the trans-well method) of the PBS buffer and replace it with the same amount of the fresh buffer.
4. Measure the absorbance values of the collected samples and convert the measured absorbance values into concentrations using the prepared standard curve (Fig. 3c).

4 Notes

1. Solubilized mucins should always be stored at 4 °C to slow down bacterial growth. Use such prepared mucin solutions within 48 h. If you require longer storage times (and if your experimental conditions allow for it), adding a bacterial growth inhibitor (e.g., sodium azide) to the prepared mucin solution can be an option. Otherwise, exposing the lyophilized mucins to UV light (for, e.g., at 254 nm for 30 min) prior to generating the mucin solution can be another option to reduce bacterial contaminations during the mucin particle preparation process.
2. Sodium azide is very toxic if swallowed, if it gets in contact with skin or if it gets inhaled. Wear protective gloves, eye protection, and protective clothing when handling sodium azide and its solutions.
3. The desalting process should be conducted until a value of approximately 30 μS/cm is achieved.
4. Pre-freeze the sample at −20 °C (or below) before placing it into the freeze dryer. Set the operating temperature of the

freeze dryer to −53 °C and the operating pressure to 0.072 mbar.

5. Be careful as NaOH solutions can cause severe skin burns and eye damage. Wear protective gloves, protective clothing, eye protection, and face protection. Be aware of the exothermic reaction: When mixing NaOH pellets (or when diluting concentrated NaOH solutions) in water, cooling in an ice bucket might be necessary.
6. Harmful if swallowed or if inhaled; can cause skin irritation and serious eye damage. Wear protective gloves, eye protection, and face protection.
7. The crosslinker and displacement DNA sequences (crDNA and dDNA, respectively) needed to prepare DNA stabilized mucins are published in Nowald et al. [29]. Of course, those sequences can be adjusted to meet the needs of your experiment. For the example discussed in this book chapter, the crDNA and dDNA sequences are as follows:

   crDNA: AAAAGAAGCAAAGACAACCCGGGTAA (sequence from 5′ to 3′; modification: 5′ thiol modifier C6 S-S),

   dDNA: TTACCCGGGTTGTCTTTGCTTC (sequence from 5′ to 3′; no further modification).
8. Causes severe skin burns and eye damage. Wear protective gloves, protective clothing, eye protection, and face protection.
9. Toxic to aquatic life with long-lasting effects. Avoid release to the environment and dispose to an approved waste disposal plant.
10. Wear protective clothing, eye protection, and face protection to avoid UV radiation.
11. The polystyrene cuvettes used for hydrodynamic size measurements should be changed for each measurement. Ideally, the sample volume should be around 1 mL; it must not be less than 0.85 mL and not more than 3 mL.
12. The Omega cuvettes are made from polycarbonate and thus should be cleaned only with water. Hold the cuvette upside down when injecting the first half of the sample. This will help avoiding the formation of bubbles. Once half of the cuvette has been filled, carefully turn the cuvette around and continue injecting the sample until it is full. If there are any air bubbles, tap the cuvette to dislodge them. Note that high salt concentrations will damage the electrodes of the cuvette, so avoid using high-salt conditions.
13. Before starting a measurement, wait for 10 min after switching on the device (to allow the laser to warm up).

14. Of course, similar to the DNA-stabilized particles, you can also first dissolve the mucin and then mix this mucin solution with a drug stock solution.
15. The chosen particle concentration might affect the outcome of the DLS measurements. The recommended nanoparticle dispersion concentration for mucin nanoparticles is 12.5 μg mucin/mL. Higher concentrations might induce the light scattered by each particle to be further scattered by other particles and thus lead to false results.
16. Mixing samples by vortexing or ultrasonication before measuring their hydrodynamic size is recommended. In our example, the input parameters for particle size measurements were set as follows: Measurement cell, disposable cell; measurement angle, 175°; target temperature, 20 °C; equilibration time, 2 min; analysis model, general; cumulant model, ISO 22412.
17. For measurements conducted at room temperature, the recommended equilibration time is 2 min. However, these measurements can be performed in a temperature range of 0 to 90 °C. Add an additional minute of equilibration time for every °C different from ambient temperature. For measurements above 70 °C, a special quartz cell must be used.
18. Aromatic amino acids of mucins absorb ultraviolet light (at 280 nm); thus, mucins give a peak around this wavelength when their absorbance is scanned in the range of 190–900 nm. If there is a spectral overlap between the mucin solution and the (loaded) drug molecules, use an unloaded (=drug-free) mucin nanoparticle dispersion as a reference for the absorbance measurements.
19. The trans-well insert setup requires a lower volume of buffer in the acceptor well; thus, the released drugs are diluted less when compared to the use of dialysis tubings. This allows for detecting released drugs with a higher sensitivity than when using dialysis.

Acknowledgements

This project was funded by the Federal Ministry of Education and Research (BMBF) and the Free State of Bavaria under the Excellence Strategy of the Federal Government and the Länder through the ONE MUNICH Project Munich Multiscale Biofabrication.

References

1. Davidson HJ, Kuonen VJ (2004) The tear film and ocular mucins. Vet Ophthalmol 7(2): 71–77
2. Johansson ME, Sjövall H, Hansson GC (2013) The gastrointestinal mucus system in health and disease. Nat Rev Gastroenterol Hepatol 10(6):352–361
3. Lagow E, DeSouza MM, Carson DD (1999) Mammalian reproductive tract mucins. Hum Reprod Update 5(4):280–292
4. Bansil R, Turner BS (2006) Mucin structure, aggregation, physiological functions and biomedical applications. Curr Opin Colloid Interface Sci 11(2–3):164–170
5. Linden SK, Sutton P, Karlsson NG et al (2008) Mucins in the mucosal barrier to infection. Mucosal Immunol 1(3):183–197
6. McGuckin MA, Lindén SK, Sutton P et al (2011) Mucin dynamics and enteric pathogens. Nat Rev Microbiol 9(4):265–278
7. Ma L, Gaisinskaya-Kipnis A, Kampf N et al (2015) Origins of hydration lubrication. Nat Commun 6:6060
8. Käsdorf BT, Weber F, Petrou G et al (2017) Mucin-inspired lubrication on hydrophobic surfaces. Biomacromolecules 18(8):2454–2462
9. Marczynski M, Jiang K, Blakeley M et al (2021) Structural alterations of mucins are associated with losses in functionality. Biomacromolecules 22(4):1600–1613
10. Lieleg O, Vladescu I, Ribbeck K (2010) Characterization of particle translocation through mucin hydrogels. Biophys J 98(9):1782–1789
11. Yakubov GE, McColl J, Bongaerts JH et al (2009) Viscous boundary lubrication of hydrophobic surfaces by mucin. Langmuir 25(4): 2313–2321
12. Song J, Winkeljann B, Lieleg O (2019) The lubricity of mucin solutions is robust toward changes in physiological conditions. ACS Appl Bio Mater 2(8):3448–3457
13. Lieleg O, Lieleg C, Bloom J et al (2012) Mucin biopolymers as broad-spectrum antiviral agents. Biomacromolecules 13(6):1724–1732
14. Caldara M, Friedlander RS, Kavanaugh NL et al (2012) Mucin biopolymers prevent bacterial aggregation by retaining cells in the free-swimming state. Curr Biol 22(24):2325–2330
15. Rickert CA, Lutz TM, Marczynski M et al (2020) Several sterilization strategies maintain the functionality of mucin glycoproteins. Macromol Biosci 20(7):e2000090
16. Rickert CA, Bauer MG, Hoffmeister JC et al (2022) Effects of sterilization methods on the integrity and functionality of covalent mucin coatings on medical devices. Adv Mater Interfaces 9(3):2101716
17. Rickert CA, Wittmann B, Fromme R et al (2020) Highly transparent covalent mucin coatings improve the wettability and tribology of hydrophobic contact lenses. ACS Appl Mater Interfaces 12(25):28024–28033
18. Winkeljann B, Bauer MG, Marczynski M et al (2020) Covalent mucin coatings form stable anti-biofouling layers on a broad range of medical polymer materials. Adv Mater Interfaces 7(4):1902069
19. Song J, Lutz TM, Lang N et al (2021) Bioinspired dopamine/mucin coatings provide lubricity, wear protection, and cell-repellent properties for medical applications. Adv Healthc Mater 10(4):2000831
20. Duffy CV, David L, Crouzier T (2015) Covalently-crosslinked mucin biopolymer hydrogels for sustained drug delivery. Acta Biomater 20:51–59
21. Kimna C, Winkeljann B, Song J et al (2020) Smart biopolymer-based multi-layers enable consecutive drug release events on demand. Adv Mater Interfaces 7(19):2000735
22. Marczynski M, Kimna C, Lieleg O (2021) Purified mucins in drug delivery research. Adv Drug Deliv Rev 178:113845
23. Yan H, Chircov C, Zhong X et al (2018) Reversible condensation of mucins into nanoparticles. Langmuir 34(45):13615–13625
24. Butnarasu C, Petrini P, Bracotti F et al (2022) Mucosomes: intrinsically mucoadhesive glycosylated mucin nanoparticles as multi-drug delivery platform. Adv Healthc Mater 11(15): e2200340
25. Lutz TM, Kimna C, Lieleg O (2022) A pH-stable, mucin based nanoparticle system for the co-delivery of hydrophobic and hydrophilic drugs. Int J Biol Macromol 215:102–112
26. Kimna C, Lutz TM, Yan H et al (2020) DNA strands trigger the intracellular release of drugs from mucin-based nanocarriers. ACS Nano 15(2):2350–2362
27. Fukui Y, Fukuda M, Fujimoto K (2018) Generation of mucin gel particles with self-degradable and-releasable properties. J Mater Chem B 6(5):781–788
28. Marczynski M, Rickert CA, Fuhrmann T et al (2022) An improved, filtration-based process to purify functional mucins from mucosal tissues with high yields. Sep Purif Technol 294: 121209
29. Nowald C, Käsdorf B, Lieleg O (2017) Controlled nanoparticle release from a hydrogel by DNA-mediated particle disaggregation. J Control Release 246:71–78

Chapter 34

Evaluation of Rheological Properties of Saliva by Determining the Spinnbarkeit

Taro Mukaibo and Mikio Yamada

Abstract

Saliva is crucial to maintaining oral health and facilitating chewing, swallowing, and speech functions. Decreased saliva secretion, known as hyposalivation, impairs these functions and increases the risk of dental caries and other infectious diseases in the oral cavity.

Saliva exhibits various rheological properties, with mucin being a factor in determining these properties. Alterations in these properties can also affect the sensation of dry mouth. In this article, we focus on the spinnbarkeit of saliva using the Neva Meter instrument and provide a methodology for fully understanding the appropriate conditions for its use.

Key words Spinnbarkeit, Mucin, Rheology, Xerostomia, Dry mouth, Hyposalivation

1 Introduction

Saliva exhibits various rheological properties, including viscosity, elasticity, adherence, and spinnbarkeit, and several mechanisms contribute to its protective functions, such as antimicrobial effects, bacterial aggregation, and buffering system [1]. The spinnbarkeit of saliva reflects its ability to adhere to oral surfaces and serves a protective role while facilitating lubrication [2]. Alterations in the spinnbarkeit of saliva may lead to the loss of adhesiveness or the ability to bind to surfaces, potentially contributing to oral dryness associated with Sjogren's syndrome and oral mucositis. The Neva Meter, a device designed to measure the liquid spinnability of a specimen, is suitable for quickly measuring the spinnbarkeit of saliva chairside in a clinical setting. The Neva Meter has dimensions of 120 × 120 × 240 mm, weighs 3.1 kg, and operates by automatically stretching a saliva sample placed in the testing dish of the device at a constant rate, followed by the application of an optical laser running along the measuring probe which stops at the moment the thread breaks, displaying the results in millimeters.

Akihiko Kameyama (ed.), *Mucins: Methods and Protocols*, Methods in Molecular Biology, vol. 2763,
https://doi.org/10.1007/978-1-0716-3670-1_34,

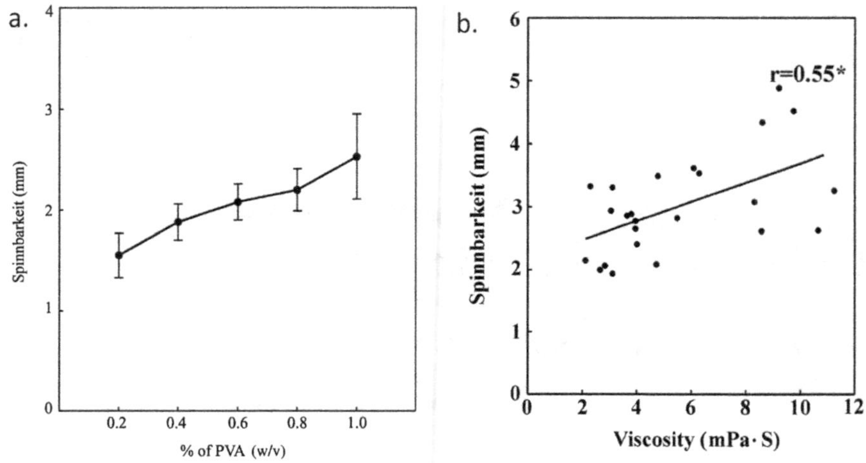

Fig. 1 (**a**) The spinnbarkeit of PVA (concentration: 0.2–1.0 w/v%) used as a standard solution was measured with a Neva Meter, yielding values ranging from 1.6 to 2.5 mm depending on concentration. (**b**) The relationship between spinnbarkeit and viscosity of resting saliva was examined for 24 individuals. Spinnbarkeit values in resting saliva ranged from 1.9 to 4.9 mm, with a significant correlation between the two variables. (Reproduced from [1] with permission from Elsevier)

The study verified the consistency of the instrument in measurement using polyvinyl alcohol (PVA) in water (Fig. 1a) and correlated the spinnbarkeit (mm) with viscosity (mPS) in saliva samples (Fig. 1b). Mucins, responsible for surface wetting and effectively maintaining the moisture retention of the oral mucosa [3], play a role in microbial defense by preventing pathogens from reaching and binding to mucosal surfaces as decoys for microbial-binding sugar chains. Human salivary glands secrete two types of mucins: high-molecular-weight MUC5B (>2–4 × 10^4 kDa) and low-molecular-weight MUC7 (130–180 kDa) [4]. However, the same parameters do not always determine viscosity and spinnbarkeit; MUC5B contributes to viscosity in human saliva and MUC7 to spinnbarkeit [5]. Bicarbonate and pH concentrations in saliva also have a prominent effect on spinnbarkeit [6]. This article describes the collection and measurement of saliva, including spinnbarkeit measurement, as one of the major rheological properties in saliva measured by the Neva Meter.

2 Materials

Various methods for saliva sample collection have been described previously [6]. In this study, we introduce two main forms of saliva collection: unstimulated whole saliva (UWS) and taste-stimulated submandibular and sublingual gland saliva; the latter is used due to the recent RNA-seq analysis revealing the predominant expression of MUC5B and MUC7 genes in the submandibular and sublingual glands [7].

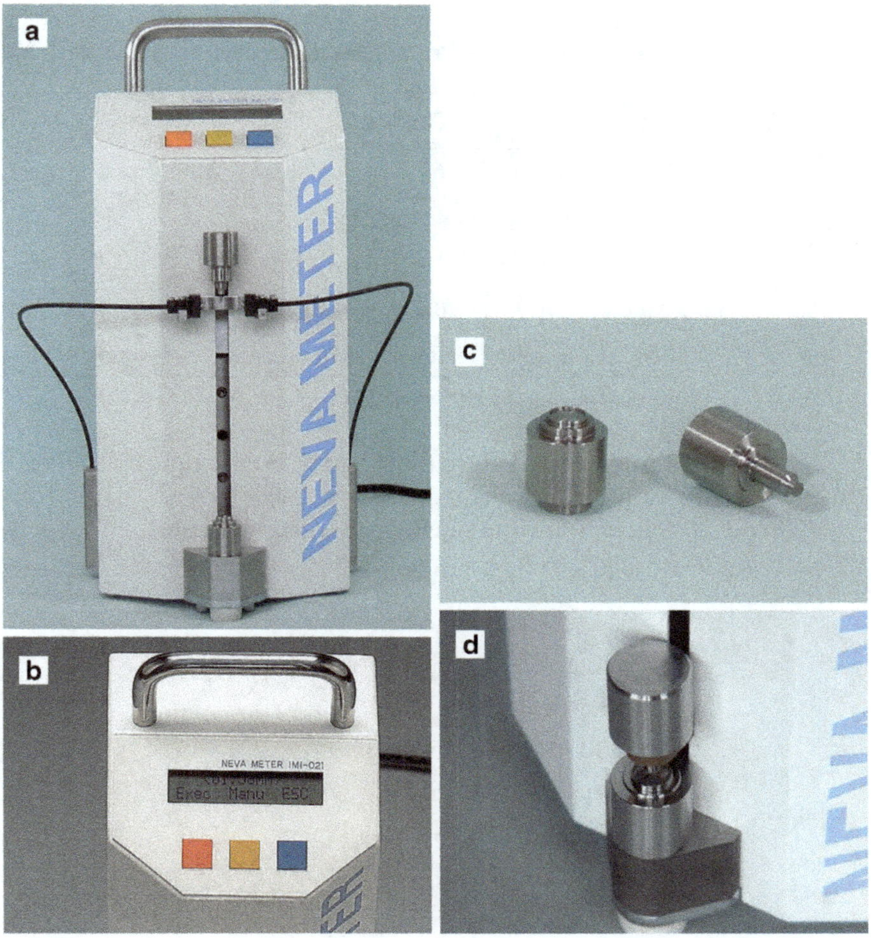

Fig. 2 The Neva Meter (IMI-0501, Ishikawa Iron Works, Japan) is an instrument that measures the length of liquid threads drawn when fluid or wet substances are stretched. (**a**) The main unit of the instrument is shown. (**b**) The liquid crystal display is accompanied by three buttons below. (**c**) The 80 μL liquid capacity testing dish is shown on the left, with the 3 mm-diameter measuring probe on the right. (**d**) The main unit, testing dish, and measuring probe are shown set up in one place

The Neva Meter (IMI-0501, Ishikawa Iron Works, Japan) consists of three components: a main unit, a testing dish, and a measuring probe (Fig. 2). The measured value is displayed on an interactive monitor. The measurement process involves four steps and requires as little as 60 μL of the sample.

Before using the device, ensure that it is placed on a flat surface and not tilted. The testing dish has a capacity of up to 80 μL, and the results have been consistent for samples ranging from 60 to 80 μL. The sample should be placed in the testing dish at room temperature (16–28 °C), and the measurement should be performed immediately after sample collection, within 10 min. If this is not possible, add a protease inhibitor cocktail to the sample to

Table 1
Factors that affect spinnbarkeit results determined by Neva Meter

Factor	Condition
Loading volume	60 μL (60–80 μL)
Temperature	16–28 °C
Relative humidity	10–90%
Wind velocity	0.3 m/s or less
Measurement time	Immediately after sample collection (within 10 min)

Reproduced from [8] with permission from The Japanese Society for Oral Health

prevent degradation immediately after sample collection. The conditions, which have been previously tested [8] and resulted in consistent measurements, are listed in Table 1.

1. Timer.
2. Wet ice in a Styrofoam cup.
3. 50 mL sterile conical polypropylene tube.
4. 2% citric acid.
5. 2–30 mL medicine cups.
6. Sterile cotton swabs.
7. Two dental cotton rolls.
8. Plastic pipette tools.
9. 15 mL sterile conical polypropylene tube.
10. cOmplete, ULTRA, Mini, EASYpack protease inhibitor tablets (Roche, Basel, Switzerland, cat# 05892970001) (*see* **Note 1**).
11. Double-distilled or Mili-Q quality water.
12. 10 mL pipette.
13. Disposable pipette tips to fit pipettes.
14. 14 mL polypropylene tube for preparing the stock solution.
15. Neva Meter (IMI-0501, Ishikawa Iron Works, Japan).

3 Methods

3.1 Unstimulated Whole Saliva Collection

1. The subject should not have eaten for 90 min before the collection procedure. If the patient has eaten or drunk anything before the collection, wait for an additional 90 min.
2. Saliva should be collected between 8:00 a.m. and 12:00 noon.
3. During the collection period, the subject should be seated with eyes open and head tilted slightly forward.

4. The subject should be instructed to avoid orofacial movements to minimize the influence on salivary flow (the subject should not swallow or speak during the collection process).
5. The subject is instructed to swallow immediately before the collection begins.
6. The subject allows the saliva to accumulate on the floor of the mouth for 60 s without swallowing.
7. The subject should empty the accumulated saliva into a 50 mL sterile conical polypropylene tube in a pre-weighed container.

The procedure should be repeated four more times for a total collection time of 5 min. Subjects should be instructed not to swallow or speak during the 5 min collection period.

3.2 Stimulated Submandibular/ Sublingual Gland Saliva Collection

1. Apply a 2% citric acid solution to the posterior lateral surface of the tongue with a cotton applicator twice consecutively on each side every 30 s to stimulate secretion. Citric acid stimulation should be continued at 30 s intervals throughout the collection process.
2. The orifices of the parotids are blocked with dental cotton rolls to prevent parotid saliva from flowing to the floor of the mouth.
3. After 2 min of stimulation, collect saliva from the left and right submandibular/sublingual glands at the orifice of the duct in a pre-weighed 15 mL sterile conical polypropylene tube on ice. After saliva collection, weigh the tube.

3.3 Protease Inhibitor Preparation and Addition in the Saliva Sample

1. To make a 2× stock solution, dissolve one cOmplete ULTRA tablet in 5 mL of distilled water.
2. Push the tablet through the foil packaging using the base of your thumb (not the fingernail) to prevent the breakage of tablets.
3. Add 50 μL of stock solution per 1 mL of saliva (*see* **Note 2**).

3.4 Setup of Neva Meter

1. Remove the testing dish and measuring probe before turning on the power.
2. Ensure that the measuring element holder and light sensor holder are in the bottom position.
3. Attach the power supply cord.
4. Turn on the power, and the LCD screen will temporarily display the following.

   【IMI-0501】【Ver. 20.01】
5. The instrument performs internal initialization. It takes about 10 s, and the following display appears on the LCD screen (*see* **Note 3**).

【Initializing】

6. The LCD screen displays the following.

   【Ret Org Pos OK?】

7. Install the measuring probe (*see* **Notes 4 and 5**).
8. Press the red button.
9. The sensor holder rises with the measuring probe and returns to its original position.
10. The optical sensor holder then rises and stops at the specified position, with the following message displayed on the LCD screen. Do not touch the holder or optical sensor at this time.

    【Select Mode Pse.】【DRY WET COA】

11. Select the measurement mode from DRY WET COA (*see* **Note 6**).
12. The main menu is displayed on the LCD screen.
13. Install the testing dish.

3.5 Sample Installation

1. Fill the measuring dish with the sample (*see* **Note 7**).
2. Place the sample-filled testing dish on the main unit.

3.6 Measurement

1. Setting of parameters: the number of measurements, measuring speed, depth of immersion of the measuring element in the liquid, waiting time at the top, and immersion time at the bottom.
2. Start the measurement by pressing the red button. Set the number of measurements.
3. Press the red button to start the measurement.
4. During the measurement, the number of times, the measurement result, average value, and operating instructions are displayed.
5. Turn off the power of the main unit (*see* **Note 8**).

4 Notes

1. A commercially available inhibitor cocktail containing PMSF, BZA, EDTA, and NEM is used to avoid the degradation of MUC7 and MUC5B in saliva samples [4].
2. The stock solution in water is stable for 1 day, stored at 2–8 °C, and for at least 4 weeks at −15 to −25 °C.
3. After turning on the instrument, allow a warm-up time of approximately 10 min.
4. The measuring dish is not installed at this time.

5. The measuring probe must be kept clean and free of dirt.
6. The wet mode (continuous measurement without cleaning the tip of the measuring probe) is suitable for saliva samples. By default, 10 consecutive measurements are taken in the wet mode. The dry mode is applicable when saliva is mucus-like and sticky and threads on the tip of the inspection probe after measurement. In this case, the tip must be cleaned dry before each measurement.
7. It is important to avoid air bubbles when installing the sample.
8. After completing the measurements, promptly remove the testing dish and measuring probe from the main unit before turning off the power.

Acknowledgments

This work was supported by a Grant-in-Aid for Scientific Research (KAKENHI 21K09983) from the Japan Society for the Promotion of Science (JSPS).

References

1. Gohara K, Ansai T, Koseki T et al (2004) A new automatic device for measuring the spinnbarkeit of saliva: The neva meter. J Dent 32:335–338
2. Yamada M, Masaki C, Mukaibo T et al (2022) Altered rheological properties of saliva with aging in mouse sublingual gland. J Dent Res 101:942–950
3. Pramanik R, Osailan SM, Challacombe SJ et al (2010) Protein and mucin retention on oral mucosal surfaces in dry mouth patients. Eur J Oral Sci 118:245–253
4. Takehara S, Yanagishita M, Podyma-Inoue KA et al (2013) Degradation of muc7 and muc5b in human saliva. PLoS One 8:e69059
5. Inoue H, Ono K, Masuda W et al (2008) Rheological properties of human saliva and salivary mucins. J Oral Biosci 50:134–141
6. Vijay A, Inui T, Dodds M et al (2015) Factors that influence the extensional rheological property of saliva. PLoS One 10:e0135792
7. Saitou M, Gaylord EA, Xu E et al (2020) Functional specialization of human salivary glands and origins of proteins intrinsic to human saliva. Cell Rep 33:108402
8. Gohara K, Koseki T, Kakinoki Y et al (2002) Examination of conditions involved in the measurement of salivary towability with the Neva Meter, a thread property measuring instrument: Usage of a newly developed salivary property measuring instrument. J Oral Health 52:582–583

Chapter 35

Mechanical Characterization of Mucus on Intestinal Tissues by Atomic Force Microscopy

Momoka Horikiri, Mugen Taniguchi, Hiroshi Y. Yoshikawa, Ryu Okumura, and Takahisa Matsuzaki

Abstract

Mucus is part of the innate immune system that defends the mucosa against microbiota and other infectious threats. The mechanical characteristics of mucus, such as viscosity, elasticity, and lubricity, are critically involved in its barrier function. However, assessing the mechanical properties of mucus remains challenging because of technical limitations. Thus, a new approach that characterizes the mechanical properties of mucus on colonic tissues needs to be developed. Here, we describe a novel strategy to characterize the ex vivo mechanical properties of mucus on colonic tissues using atomic force microscopy. This description includes the preparation of the mouse colon sample, AFM calibration, and determining the elasticity (Young's modulus, E [kPa]) of the mucus layer in the colon.

Key words Atomic force microscopy (AFM), Mucus layer, Elasticity, Young's modulus (E kPa)

1 Introduction

In the intestine, continuously flowing viscous mucus protects the epithelial surface from commensal bacteria to prevent an excessive immune response [1]. In particular, a thick mucus layer covers the epithelia in the colon because numerous bacteria exist [2]. In the mucus layer, mucin polymers create a netlike structure that prohibits the penetration of bacteria through this layer [3]. In addition to the network structure, the mechanical properties of mucus, such as viscosity, elasticity, and lubricity, are critical for its barrier function [4]. However, experimental protocols for characterizing the mechanical properties of mucus layers in the colon have not been established. In this regard, we recently developed a novel approach to evaluate the elasticity (Young's modulus, E [kPa]) of mucus in the colon using atomic force microscopy (AFM). AFM is a scanning probe microscopy that is a widely known imaging tool for obtaining nanoscale topography of biological specimens [5, 6]. An important

Akihiko Kameyama (ed.), *Mucins: Methods and Protocols*, Methods in Molecular Biology, vol. 2763,
https://doi.org/10.1007/978-1-0716-3670-1_35,

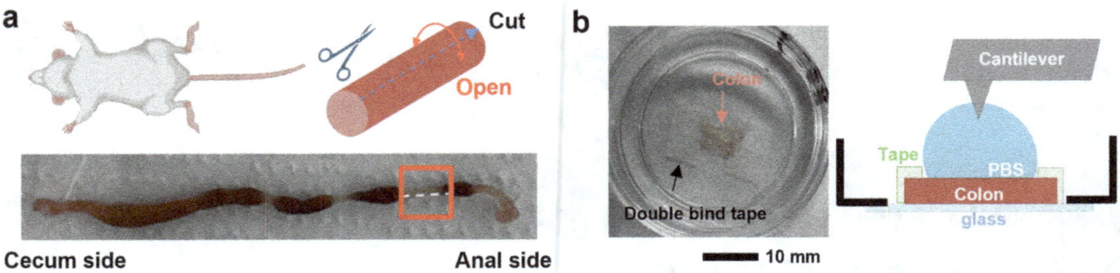

Fig. 1 Preparation of a mouse colon explant on a glass-bottom dish. (**a**) Schematic illustration of the exposing procedures for the colon surface and colon explant (bottom). A region without feces (red, rectangle) was used for the AFM measurement. BioRender was used for the mouse schematic. (**b**) Fixation to the glass-bottom dish before the AFM measurement

aspect of AFM is determining elasticity from the force curves based on elastic theory [7, 8]. A force curve can be obtained by applying forces (F) to specimens via scanning probes (i.e., cantilevers) with defined deformation (δ). In particular, in biological research, mechanical characterization of isolated adhered cells [9] and tissues [10] clearly indicates that the stiffness of cells serves as a mechanical indicator to distinguish cancer cells from healthy cells. Such a concept can be expanded to other varieties of cells [11–14], such as erythrocytes [15], liver [16], intestinal organoids [17], and the embryogenic gut [18]. However, regarding the mechanical analysis of intestinal mucus using AFM, only one previous report demonstrates ex vivo force spectrometry of ileal mucus by AFM [19]. Recently, we also assessed the mechanical properties of colonic mucus by AFM successfully, demonstrating the influence of sialylation on the mechanical property of mucus [20]. Here we described the preparation of colon biopsies and detailed mechanical analysis of mucus layers using AFM. In brief, the procedures for preparing the mouse colon explant attached to a glass-bottom dish (Fig. 1) are described. We then summarize the basic measurement protocols (Fig. 2) and analytical protocols (Fig. 3) required to obtain the spatial distribution of Young's modulus for the mucus layer.

2 Materials

2.1 Mouse Colon Explant

1. Scissors and tweezers.
2. Double-sided tape (for fixing the colon explant on the glass-bottom dish).
3. Sterile phosphate-buffered saline (PBS) (pH 7.4).
4. Glass-bottom dish (ø, 3.5 cm; observation area, ø, 27 mm) (*see* **Note 1**).

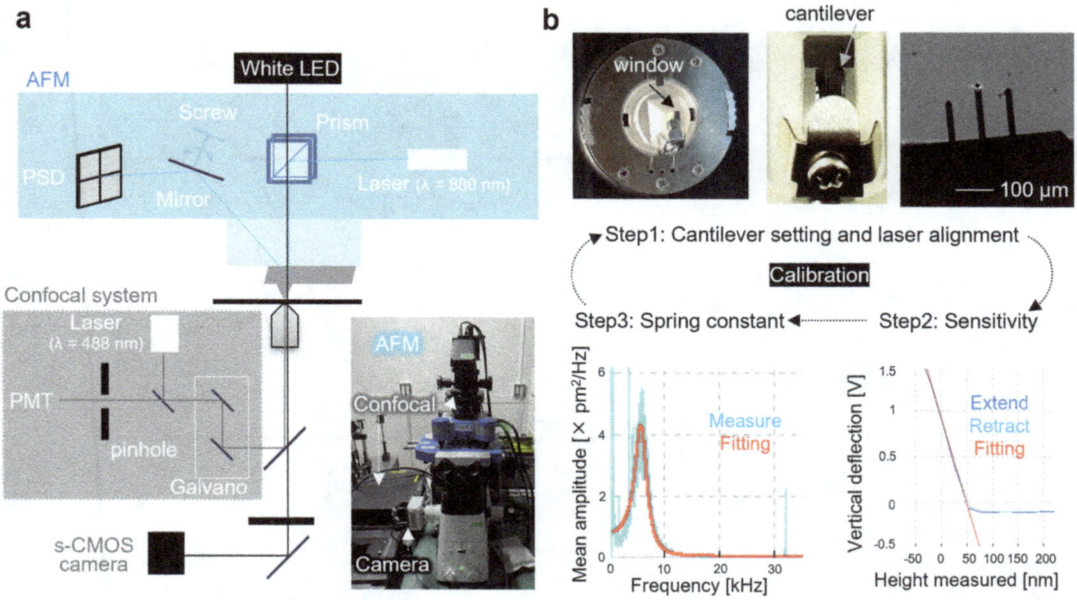

Fig. 2 Calibration of AFM. (**a**) AFM setup combining confocal microscopy. (**b**) Overview of the calibration process. Setting the cantilever to the AFM head and laser alignment (**step 1**) and calibration for sensitivity (**step 2**) and spring constant (**step 3**) for cantilevers

2.2 AFM Measurement and Analysis

Unless stated otherwise, use Ultrapure water as deionized water.

1. AFM from JPK Instruments (now Bruker, NanoWizard 3, Billerica, MA, USA) combined with an inverted confocal microscope (Nikon, Ti2-E, Tokyo, Japan). The microscope should be placed on an active anti-vibration table or optical bench (Hertz, Kanagawa, Japan). The option of introducing the manual stage of the JPK CellHesion module for extending the *z*-piezo range (100 μm in max) is strongly recommended (*see* **Note 2**). SPM JPK control software (V.4) and Nis-Elements C (Nikon) are used for the analysis.
2. For the laser alignment targeted at the top of the surface of the cantilever, a CCD camera (Imaging Source, DFK31AF03) or Orca-Spark (Hamamatsu, strongly recommended) is used.
3. For fluorescence imaging, a confocal microscope (A1HDR25 or C2Si) for thicker samples is used because epi-fluorescence systems with a high-pressure mercury lamp are insufficient for illuminating the top surface of the mucus layer.
4. PBS.
5. Nitrile glove.
6. Ethanol (99.8%).
7. Nitrogen gas.

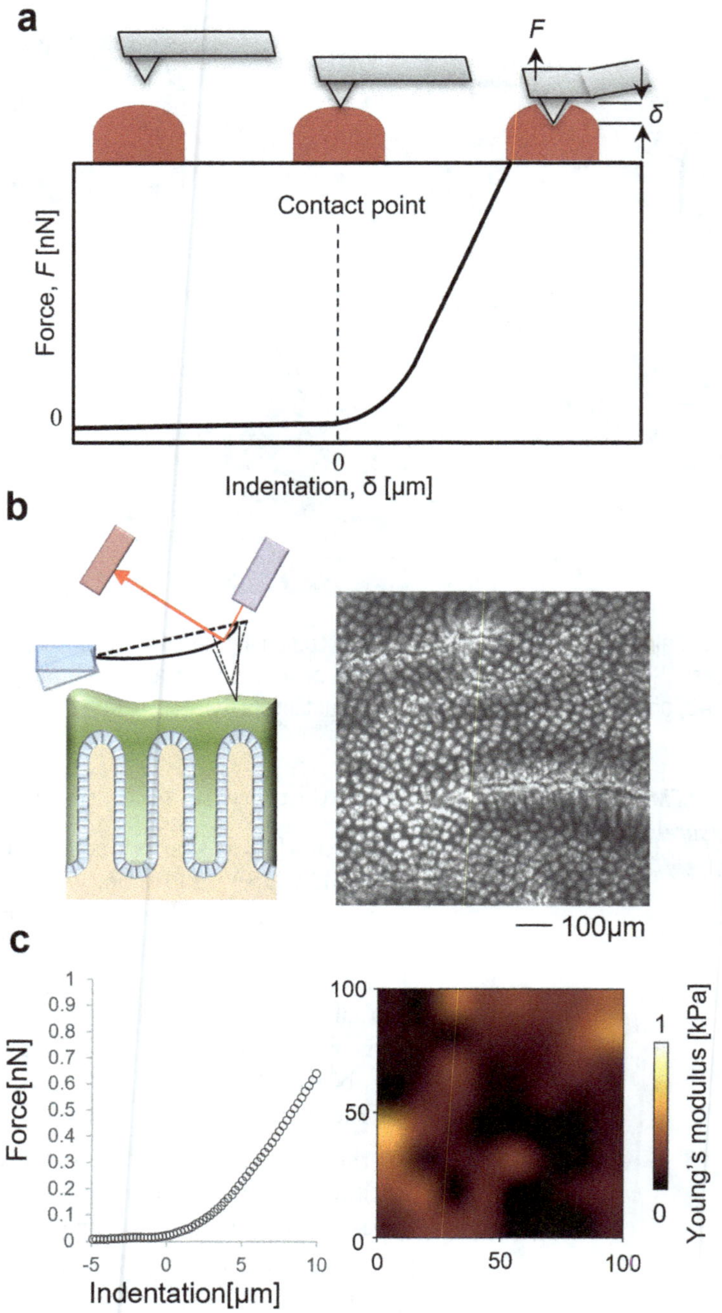

Fig. 3 AFM measurement of mucus layers in the colon. **(a)** Schematics showing how force curves are obtained for elastic materials and (**b**) mucus layer of the colon. Bright-field images of the colon surface are displayed in the right panels. (**c**) Representative force curves of the colon mucus layer and Young's modulus images of colon mucus layers

8. Lens paper.
9. JPK data processing software.
10. Igor Pro 64.

3 Methods

3.1 Preparation of the Colon Explant

The mechanical characterization of mucus in the colon requires exposure of the mucosal surface without luminal contents (Fig. 1a) and fixing on a glass-bottom dish to ensure the cantilever touches the mucus layer directly (Fig. 1b).

1. Resect the colon from mice euthanized with anesthetic agents.
2. Cut a 5 mm-long piece without feces off the resected colon (Fig. 1a).
3. Place the colon piece on the glass-bottom dish and cut the colon longitudinally without touching the mucosal surface as much as possible.
4. Stretch the sheeted colon and fix to the glass-bottom dish using double-sided tape over the colon. Ensure that the mucus layer on the epithelial is exposed upward (Fig. 1b).
5. Add 0.2 mL of PBS into the dish to prevent the mucus from drying. To perform mechanical characterization of the colon explant immediately, it is recommended to perform AFM calibration before the sacrifice. The calibrated cantilever is stably immersed in deionized water overnight.

3.2 AFM Calibration

Simultaneous imaging of biological and physical properties of the mucus layer of the colon can be obtained by JPK AFM combined with inverted Nikon confocal microscopy (Fig. 2a). Before the measurement of the biopsies, AFM needs three-step calibration procedures (Fig. 2b).

1. Turn on the CellHesion system, AFM controller, PC, JPK software, and devices related to the optical microscope (Nikon-Ti2E, A1 and CMOS) (optical setup: Fig. 2a). The main power switch of the AFM controller should typically be switched on. Ensure the manual stage with CellHesion modules have spaces for one lens (*see* **Note 3**).
2. Wear nitrile gloves to clean the glass block (cantilever holder) with deionized water and ethanol (99.8%), and then dry with a nitrogen gas flow. Removing debris is critical to avoid disturbing the laser transmittance when determining cantilever sensitivity. A combination of lens paper and lens cleaner is recommended to remove such debris (*see* **Note 4**).

3. Adjust the cantilever to the appropriate position on the transparent regions (*see* **Note 5**).
4. Attach the cantilever holder to the AFM head, so the cantilever tip points upwards. Turn the holder 90° clockwise so that tip points to the right. Lock the holder by moving the two-finger grips on the rim of the steel disk (Fig. 2b, **step 1**).
5. Place the glass-bottom dish on the manual stage and add 2 mL of deionized water (or PBS). Place the AFM head onto the manual stage. Be sure that the Z position of the piezo is at the minimum position, and the Z-stepper motor of the AFM head is sufficiently high to not attach to the surface of the glass directly.
6. Keep the cantilever at 20 °C (room temperature) for at least 30 mins before the measurement (*see* **step 7** below, which can be done in parallel).
7. Obtain the bright-field image using a CCD camera, and adjust the laser position to the end of the cantilever such that the reflected signals at PSD (photosensor diode) should be sum maximum. For a CMOS camera, the original high sensitivity detects the transmitted light path through a low-pass filter (λ = 880 nm). Thus, the exact point of the laser can be visualized easily (*see* **Note 6**).
8. The vertical and lateral deflections should be optimized to zero (*see* **Note 7**).
9. Place the cantilever near the surface of the glass substrate using the software (JPK NanoWizard Control) and the following approach conditions at the panel of feedback control: approaching speed 20 μm/s, IGain (10 Hz), PGain (0.0001), set point = 1 V.
10. Calibrate the sensitivity using the calibration manager in the NanoWizard software (*see* **Note 8**) (Fig. 2b, **step 2**).
11. Initially, maintain the distance of the piezo (confirm that the piezo has shrunk after the measurement). Then adjust the distance to 100 μm from the glass surface by tuning the stepping motor. The next step involves fluctuating the cantilever; thus, the cantilever near the surface disrupts the piezo.
12. Calibrate the spring constant using the calibration manager in the NanoWizard software (*see* **Notes 9** and **10**) (Fig. 2b, **step 3**).

3.3 AFM Measurements of Biopsies

1. Immediately after covering the surface (*see* **step 1** in Subheading 3.1 and **Note 11**), a minimum of 1–2 force curves should be obtained in one targeted area. The total size of the analysis area (x, y) should be 100 × 100 μm (which is restricted by the potential piezo movement inside the AFM) (*see* **Notes 12** and **13**).

2. After taking AFM measurements, it is better to simultaneously take the measurement area by bright-field/fluorescence images (confocal systems are preferable) (*see* **Note 14**).
3. (Optional) The laser position is sensitive to the cantilever sensitivity. Thus, if the position of laser is moved, recalibrate (repeat procedures of Subheading 3.2) and move to another sample.

3.4 AFM Data Analysis

The use of commercial JPK data processing software is recommended for determining the mechanical properties from the force (F)–indentation (δ) curves (i.e., force curves). Exporting the original text files to Igor Pro 64 for analysis is also possible. The analysis steps are summarized by describing the basic concept of force curves (Fig. 3a), schematic illustration of biopsies (Fig. 3b), and then the representative mechanical distribution of the mucus layer (Fig. 3c).

1. Confirm that the sensitivity (nm/V) and spring constant (N/m) are correctly entered. The vertical axis should then be converted from deflection voltage (V) to force (N).
2. Perform baseline correction by fitting with a liner function to obtain a plateau baseline (*see* **Note 15**).
3. The x-axis is changed from the distance of the piezo to indentation (δ m) (*see* **Note 16**).
4. Perform data fitting using the Snedden-modified Hertz model [7, 8]. Curves are well-fitted by this model, and batch processing protocols should be applied to all data (including the spatial mapping of data) (*see* **Note 17**).
5. (Optional) We recommend performing the fitting depth δ to range from 1 μm, 5 μm, and 10 μm. Different fitting depths reflect the mechanical distribution along the optical axis (*see* **Note 18**).

4 Notes

1. We recommend using a manufactured glass-bottom dish with a large observation area such as Iwaki (No1S, cat # 3970-035, Shizuoka, Japan), so as not to prevent the collision of cantilever holder with the edge of the glass-bottom dish.
2. We have developed our protocol based on this setup. We also have experience with alternative setups developed on Asylum Research Technologies (now Oxford, MFP 3-D bio, Wiesbaden, Germany) with an Olympus confocal system (now Evident,IX71, Tokyo, Japan).

3. The 4× Plan Apo Lambda objective (Nikon) is recommended for calibrating the cantilevers because of its long working distance, WD = 20 mm. For the observation of samples, an objective lens should be selected for the targets. In our protocols, we used a 10× Plan Apo Lambda objective (Nikon) for the target. After setting the objective lens to the manual stage, using a motorized lens revolver is not recommended because the automated rotation of the lens may damage the lens, stage and revolver.
4. We also recommend cleaning the holder by sonication for 15 min with a float and a 1% (w/v) sodium dodecyl sulfate solution. Be careful not to scratch the transparent part of the cantilever holder (top and bottom sides) where the laser is transmitted.
5. The cantilever should be positioned so that the end of the cantilever protrudes slightly into the transparent area (Fig. 2b). The guideline for selecting the optimal cantilever is described below. We recommend using CSC-37-B/No-Al (Mikro Masch, Tallinn, Estonia) 0.3 N/m as the gold standard cantilever for measuring isolated cells to tissues. A soft cantilever is recommended for measuring soft samples; however, samples that are too soft suppress detachment of the cantilever from the surface, and automated mapping is difficult to obtain. If several varieties of samples are measured, we recommend starting measurements with the soft samples (most difficult) and ending measurements with the stiffest samples (easiest). A rigid cantilever may penetrate the samples. Thus, depending on the thickness and softness of the samples, the set point (N) (applying force from the cantilever to the sample) should be carefully determined to avoid the cantilevers penetrating the samples.
6. The sum represents the reflected intensity of the laser light at the cantilever-air surface. Thus, dirt on the cantilever will reduce the sum (exchange the cantilever with a new one).
7. Three combinations of the screws of the AFM head change the mirror angles to reflect the laser onto the photodiode correctly. This process is difficult to achieve for beginners. Thus, we recommend contacting the instrument manufacturer for support. Alternatively, obtain training in the alignment of lasers by experts.
8. Sensitivity represents the values that convert deflection voltage (V) to deflection (m).
9. Measuring the spring constant (N/m) involves converting the deflection (m) to force (N). According to the thermal fluctuation protocols built into the JPK processing software, good

fitting of the first or second frequency peak to the equation below is required. The theoretical background is described in **Note 9**.

10. The thermal noise method is used to determine the spring constant of the cantilever. The thermal noise method involves measuring the free vibration of a cantilever and relating it to the spring constant using the law of energy equi-distribution. According to the law of energy equi-distribution, the thermal energy of a shape can be expressed as follows:

$$E = \frac{1}{2}k_B T = \frac{1}{2}k\langle q^2\rangle, \quad k = \frac{k_B T}{\langle q^2\rangle}$$

where k_B is Boltzmann's constant and T is the absolute temperature. Therefore, mean vertical deflection q with defined temperature gives the value of spring constant k.

11. A small volume of PBS does not cover the cantilever surface (< 100 μL), which makes it challenging to obtain force curves. However, a greater volume of PBS (1 mL) causes the detachment of the samples from the substrate. In previous experiments examining the mucus layer of the intestine [21], repeating the washing process detaches the excreta to leave only a strongly bonded layer of mucus. Depending on the thickness of the biospecimens, please ensure that the cantilever is distal from the surface of the biopsies. In our system, we set the cantilever 2 mm away from the glass surface.

12. Spatial mapping should be taken after optimizing the approaching and measurement conditions for a single force curve. Representative condition for the CSC37-B (0.3 mN/m) to measure the colon mucus layer as follows: approaching speed 20 μm/s, IGain (10 Hz), PGain (0.0001), set point = 10 nN.

13. The approaching condition sometimes cannot obtain the plateau background because the detachment process of the force curves fails. In this situation, please decrease the set point (mild attachment of the cantilever) or detach ~5 μm above the surface using the stepping motor of the AFM instrument. Ideally, using CellHesion allows the extension of the original *z*-piezo length (15 μm) to 100 μm, which ensures that the cantilever completely detaches (~50 μm) from the sample surface.

14. Direct overlay systems (commercial plugin for JPK data processing based on the conventional sCMOS camera and analysis) are not recommended for tissue systems. Basically, the principles of the system rely on the image obtained by the cantilever position (manual setting) and fluorescence signals acquired from the sCMOS camera. For detecting the position of the tips, a magnified objective lens with a magnification of at least

10–20× is required. For higher-precision detection, a water immersion lens (with a longer WD = 0.27 mm, Nikon CFI Plan Apo VC 60×, NA = 1.27) or Plan Apo Lambda objective (40× WD = 0.21 mm, NA = 0.95) is suitable. Focusing on the surface of the samples using shorter working lengths (i.e., high magnification with high numerical aperture) is challenging and not recommended. The most dangerous situation exists while focusing on the surface of the biopsies, where the objective lens can push samples upwards, leading to mechanical damage (e.g., break) of the cantilever.

15. Thermal equilibrium and solution evaporation can constantly change the baseline gradient. The software can offset this change, but it is necessary to confirm the equilibrium step (*see* **step 6** in Subheading 3.2), evaporation of the medium, and physical attachment of cantilever holder to dish can occur.

16. During the use of the calibration manager, force curves are obtained for the glass substrate (hard material, difficult to bend over the measurement range in this study [nN]). In this case, the indentation δ should be zero. If there is plus or minus indentation for the force curves of the glass substrate, please check the sensitivity values obtained in **step 3**.

17. The theoretical background is summarized below. The force curve is obtained by potentially bringing the cantilever needle into contact with the sample from a state where the needle does not touch the sample and then pushing the needle further into the sample (Fig. 3a). Here, the vertical axis of the force curve represents force (nN) and the horizontal axis represents indentation (m). There are several models to calculate Young's modulus of a sample, but herein, the Hertz model is used, which is a model to obtain the relationship between the force applied to the sample and its displacement. As shown in Fig. 3b, the probe tip is approximated as a cone, and it is assumed that the cantilever and the sample are in surface contact. The relationship between the force on the sample and its displacement is expressed by the following equation:

$$F = \frac{E}{1 - \nu^2} \frac{2 \tan \alpha}{\pi} \delta^2$$

where F is the force applied to the sample, α is the half-angle of the probe at the tip of the cantilever, ν is Poisson's ratio, E is Young's modulus of the sample, and δ is the indentation amount. Of these, α is a constant term because this term is probe-specific and υ is sample-specific. In other words, when the obtained force curve is fitted with the Hertz model, it becomes a quadratic function in the range of indentation >0, and the sample Young's modulus E is the value that makes this quadratic function valid.

18. Our previous collaborative studies describing the dependency of Young's modulus along the indentation depth unveiled unique mechanical distributions inside cells and organoids possibly caused by internal cytoskeletal and/or organelles networks [22]. Thus, mechanical assessment at different indentation depths should contribute to revealing the hidden mechanical fingerprints of cells and tissues.

Acknowledgments

This work was supported by grants from the Ministry of Education, Culture, Sports, Science and Technology of Japan (JP18K15187 [RO], JP21KK0195 [TM and HYY], J21H03790 [TM]) and the Japan Science and Technology Agency, FOREST Program (JPMJFR205N [TM]). TM acknowledges the Nakatani Foundation for Advancement of Measuring Technologies in Biomedical Engineering and the Uehara Memorial Foundation for research incentive grants. We thank Edanz (https://jp.edanz.com/ac) for editing a draft of this manuscript.

References

1. Carlson TL, Lock JY, Carrier RL (2018) Engineering the mucus barrier. Annu Rev Biomed Eng 20:197–220. https://doi.org/10.1146/annurev-bioeng-062117-121156
2. Johansson ME, Phillipson M, Petersson J et al (2008) The inner of the two Muc2 mucin-dependent mucus layers in colon is devoid of bacteria. Proc Natl Acad Sci U S A 105(39):15064–15069. https://doi.org/10.1073/pnas.0803124105
3. Hansson GC (2012) Role of mucus layers in gut infection and inflammation. Curr Opin Microbiol 15(1):57–62. https://doi.org/10.1016/j.mib.2011.11.002
4. Wagner CE, Wheeler KM, Ribbeck K (2018) Mucins and their role in shaping the functions of mucus barriers. Annu Rev Cell Dev Biol 34:189–215. https://doi.org/10.1146/annurev-cellbio-100617-062818
5. Kodera N, Noshiro D, Dora SK et al (2021) Structural and dynamics analysis of intrinsically disordered proteins by high-speed atomic force microscopy. Nat Nanotechnol 16(2):181–189
6. Fujioka Y, Alam JM, Noshiro D et al (2020) Phase separation organizes the site of autophagosome formation. Nature 578(7794):301–305
7. Hertz H (1882) On contact between elastic bodies. J Reine Angew Math 92:156–171
8. Sneddon IN (1965) The relation between load and penetration in the axisymmetric Boussinesq problem for a punch of arbitrary profile. Int J Eng Sci 3(1):47–57
9. Cross SE, Jin Y-S, Rao J et al (2009) Applicability of AFM in cancer detection. Nat Nanotechnol 4:72–73
10. Plodinec M, Loparic M, Monnier CA et al (2012) The nanomechanical signature of breast cancer. Nat Nanotechnol 7(11):757–765
11. Li M, Dang D, Liu L et al (2017) Atomic force microscopy in characterizing cell mechanics for biomedical applications: a review. IEEE Trans Nanobioscience 16(6):523–540. https://doi.org/10.1109/TNB.2017.2714462
12. Efremov YM, Okajima T, Raman A (2020) Measuring viscoelasticity of soft biological samples using atomic force microscopy. Soft Matter 16(1):64–81. https://doi.org/10.1039/c9sm01020c
13. Ouchi R, Togo S, Kimura M et al (2019) Modeling steatohepatitis in humans with pluripotent stem cell-derived organoids. Cell Metab 30(2):374–384.e376. https://doi.org/10.1016/j.cmet.2019.05.007
14. Deng X, Xiong F, Li X et al (2018) Application of atomic force microscopy in cancer research. J Nanobiotechnol 16(1):102. https://doi.org/10.1186/s12951-018-0428-0

15. Dulińska I, Targosz M, Strojny W et al (2006) Stiffness of normal and pathological erythrocytes studied by means of atomic force microscopy. J Biochem Biophys Methods 66(1–3): 1–11
16. Ouchi R, Togo S, Kimura M et al (2019) Modeling steatohepatitis in humans with pluripotent stem cell-derived organoids. Cell Metab 30(2):374–384.e376
17. Norman MDA, Ferreira SA, Jowett GM et al (2021) Measuring the elastic modulus of soft culture surfaces and three-dimensional hydrogels using atomic force microscopy. Nat Protoc 16(5):2418–2449
18. Chevalier NR, Gazguez E, Dufour S et al (2016) Measuring the micromechanical properties of embryonic tissues. Methods 94:120–128
19. Sotres J, Jankovskaja S, Wannerberger K et al (2017) Ex-vivo force spectroscopy of intestinal mucosa reveals the mechanical properties of mucus blankets. Sci Rep 7(1):7270. https://doi.org/10.1038/s41598-017-07552-7
20. Taniguchi M, Okumura R, Matsuzaki T et al (2023) Sialylation shapes mucus architecture inhibiting bacterial invasion in the colon. Mucosal Immunol 16:624. https://doi.org/10.1016/j.mucimm.2023.06.004
21. Sotres J, Jankovskaja S, Wannerberger K et al (2017) Ex-vivo force spectroscopy of intestinal mucosa reveals the mechanical properties of mucus blankets. Sci Rep 7(1):7270
22. Kobayashi N, Togo S, Matsuzaki T et al (2020) Stiffness distribution analysis in indentation depth direction reveals clear mechanical features of cells and organoids by using AFM. Appl Phys Express 13(9):097001

Index

Akihiko Kameyama (ed.), *Mucins: Methods and Protocols*, Methods in Molecular Biology, vol. 2763,
https://doi.org/10.1007/978-1-0716-3670-1,

Printed in the USA
CPSIA information can be obtained
at www.ICGtesting.com
CBHW071925080624
9724CB00022B/407

9 781071 636695